MW01064645

Cholinesterases and Cholinesterase Inhibitors

Edited by

Ezio Giacobini MD, *PhD*
Institutions Universitaires de Gériatrie de Genève
Thonex-Genève
Switzerland

MARTIN DUNITZ

First published in the United Kingdom in 2000 by
Martin Dunitz Ltd
The Livery House
7–9 Pratt Street
London NW1 0AE

Tel: +44-(0)20-7482-2202
Fax: +44-(0)20-7267-0159
E-mail: info@mdunitz.globalnet.co.uk
Website: http://www.dunitz.co.uk

Reprinted in 2001

A CIP catalogue record for this book is available from the British Library

ISBN 1-85317-910-8

Distributed in the United States by:
Blackwell Science Inc.
Commerce Place, 350 Main Street
Malden MA 02148, USA
Tel: 1-800-215-1000

Distributed in Canada by:
Login Brothers Book Company
324 Salteaux Crescent
Winnipeg, Manitoba R3J 3T2
Canada
Tel: 1-204-224-4068

Distributed in Brazil by:
Ernesto Reichmann Distribuidora de Livros, Ltda
Rua Coronel Marques 335, Tatuape 03440-000
Sao Paulo,
Brazil

Composition by Wearset, Boldon, Tyne and Wear
Printed and bound in Italy by Printer Trento

Cover: Ribbon diagram of the crystal structure of the acetylcholinesterase molecule. The structure of huperzine-A, a cholinesterase inhibitor of natural origin, is seen in the active site of the enzyme (carbon molecules in yellow, oxygen molecules in red and nitrogen molecules in blue). Courtesy of Joel L Sussman, Weinzmann Institute, Rehorot, Israel. Composition by Dave Banthooa, Department of Geriatrics, University of Geneva, Switzerland.

CONTENTS

Contributors vii

Preface *Ezio Giacobini* ix

1. Cholinesterase inhibitors: an introduction *Bo Holmstedt* 1

2. Structural studies on acetylcholinesterase *Israel Silman, Joel L Sussman* 9

3. Rational design of cholinesterase inhibitors *Mario Brufani, Luigi Filocamo* 27

4. Novel roles for cholinesterases in stress and inhibitor responses *Hermona Soreq, David Glick* 47

5. The genes encoding the cholinesterases: structure, evolutionary relationships and regulation of their expression *Palmer Taylor, Z David Luo, Shelley Camp* 63

6. Molecular forms and anchoring of acetylcholinesterase *Jean Massoulié* . . . 81

7. Mechanism of action of cholinesterase inhibitors *Elsa Reiner, Zoran Radić* 103

8. Neuroanatomy of cholinesterases in the normal human brain and in Alzheimer's disease *Marsel Mesulam* 121

9. Measurement of cholinesterase activity *Israel Hanin, Bertalan Dudas* 139

10. Preclinical pharmacology of cholinesterase inhibitors *Giancarlo Pepeu* 145

11. Synaptic, behavioral, and toxicological effects of cholinesterase inhibitors in animals and humans *Alexander G Karczmar* 157

12. Cholinesterase inhibitors: from the Calabar bean to Alzheimer therapy
Ezio Giacobini 181

13. Cholinesterase inhibitors do more than inhibit cholinesterase
Anne-Lie Svensson, Ezio Giacobini 227

14. The effect of cholinesterase inhibitors studied with brain imaging
Agneta Nordberg 237

15. Use of cholinesterase inhibitors in the therapy of myasthenia gravis
Morris A Fisher 249

Index 263

Contributors

Mario Brufani PhD
Dipartimento di Scienze Biochimiche 'A Rossi Fanelli', Università 'La Sapienza', Roma, Italy

Shelley Camp MD
Department of Pharmacology, University of California, San Diego, CA, USA

Bertalan Dudas MD
Department of Pharmacology and Experimental Therapeutics, Loyola University Chicago, Stritch School of Medicine, Maywood, IL, USA

Luigi Filocamo PhD
Dipartimento di Scienze Biochimiche 'A Rossi Fanelli', Università 'La Sapienza', Roma, Italy

Morris A Fisher MD
Department of Neurology, Loyola University, Stritch School of Medicine, Maywood IL, and Department of Veterans Affairs, Edward Hines Jr Hospital, Hines, IL, USA

Ezio Giacobini MD, PhD
Department de Geriatrie, Institutions Universitaires de Gériatrie de Genève, Thonex-Genève, Switzerland

David Glick PhD
Department of Biological Chemistry, The Institute of Life Sciences, The Hebrew University of Jerusalem, Israel

Israel Hanin PhD
Department of Pharmacology and Experimental Therapeutics, and Neuroscience and Aging Institute, Loyola University Chicago, Stritch School of Medicine, Maywood, IL, USA

Bo Holmstedt MD PhD
Department of Toxicology, Karolinska Institutet, Stockholm, Sweden

Alexander G Karczmar PhD
Research Services, Hines VA Hospital, Hines, IL, and Department of Pharmacology and Experimental Therapeutics, Loyola University Medical Center, Maywood, IL, USA

Z David Luo MD
Department of Pharmacology, University of California, San Diego, CA, USA

Jean Massoulié PhD
Laboratoire de Neurobiologie Cellulaire et Moleculaire, École Normale Supérieure, CNRS URA, Paris, France

Marsel Mesulam MD
Cognitive Neurology and Alzheimer's Disease Center, Northwestern University Medical School, Chicago, IL, USA

Agneta Nordberg MD PhD
Professor, Division of Molecular Neuropharmacology, Department of Clinical Neuroscience, Occupational Therapy and Elderly Care Research (NEUROTEC), Karolinska Institutet, and Geriatric Clinic, Huddinge University Hospital, Huddinge, Sweden

Giancarlo Pepeu MD
Professor of Pharmacology, Dipartimento di Farmacologia Preclinica e Clinica "Maria Aiazzi Mancini", Università degli Studi di Firenze, Firenze, Italy

Zoran Radić PhD
Department of Pharmacology, School of Medicine, University of California at San Diego, San Diego, CA, USA

Elsa Reiner PhD
Institute for Medical Research and Occupational Health, Zagreb, Croatia

Israel Silman PhD
Weizmann Institute of Science, Rehovot, Israel

Hermona Soreq PhD
The Institute of Life Sciences, The Hebrew University of Jerusalem, Israel

Joel L Sussman PhD
Weizmann Institute of Science, Rehovot, Israel

Anne-Lie Svensson PhD
Division of Molecular Neuropharmacology, Department of Clinical Neuroscience, Occupational Therapy and Elderly Care Research (NEUROTEC), Karolinska Institutet, Huddinge University Hospital, Huddinge, Sweden

Palmer Taylor PhD
Department of Pharmacology, University of California, San Diego, CA, USA

Preface

Every book should have a reason to exist and a story to tell. What about a story of cholinesterases and their inhibitors? What can be fascinating about a story of enzymes? The reason for assembling in a single volume the genetics, biochemistry, pharmacology and clinical applications of an enzyme can be many. For me they were three. First, I felt that, given the complexity of the subject, spanning from genes to clinical applications, a comprehensive review that could be used as a reference book would be useful to many specialists. Secondly, the recent introduction of cholinesterase inhibitors in the treatment of Alzheimer's disease makes this book particularly well-timed and motivated. Thirdly, I felt that forty years of research in this field gave me some kind of privilege in telling the story. This is obviously presumptuous, since in science seniority in a field alone is not reason enough to feel expert. Perhaps one could more modestly affirm that following the development of a field as large and productive as the cholinergic system for four or five decades leaves one with strong memories of fellow scientists, of impressions from hot debates, of searches for new approaches and of exciting and contradicting hypotheses.

In my personal experience I look back with a sense of gratitude to those colleagues I met during my long journey in cholinergic research, particularly those who knew (and still know) more than myself and were willing to share their knowledge and expertise. Some of them, such as the late KB Augustinsson (the master of comparative enzymology of cholinesterases) and the late GB Koelle (the father of cholinergic histochemistry), would have been a natural choice as authors of two major chapters in this book. I was fortunate that two other great masters of the field, B Holmstedt and A Karczmar, accepted to be part of this publication and to write chapters which include historical perspectives. I invited fifteen authors to help me in putting together a complete a publication as possible. All of them have contributed pivotal papers and have taken an active part in major advances made in the field. Each chapter tells a different story about cholinesterases and their inhibitors (but the real 'story' is told in the chapter by B Holmstedt). I hope the reader will agree with me that no other enzyme (and its inhibitors) have such an intriguing historical background and fascinating connections to real life as do the cholinesterases.

The story originates with the calabar bean in tribal rituals in West Africa more than a century ago, continues all the way through the Second World War, to the subway of Tokyo, and to the desert of Kuwait during the recent Gulf War. No other inhibitors have been manufactured in such secrecy and kept in such tight concealment for decades by the major military powers of the world. No other enzymatic inhibitors have redeemed themselves from such a sinister past so successfully. They have mended the original sins of lethal nerve gases in becoming the first drugs to efficaciously treat two major degenerative diseases of the nervous system, myasthenia gravis and Alzheimer's disease, which afflict millions of individuals.

What kind of title could one give to such a book? When I proposed 'From the Calabar

Bean to Alzheimer's Disease', some co-authors told me it was too obscure and not 'serious' enough for a scientific publication. Therefore, I have kept this title for my own chapter, but I am still convinced that it tells the whole story better than the present (more scientific) title.

Ezio Giacobini

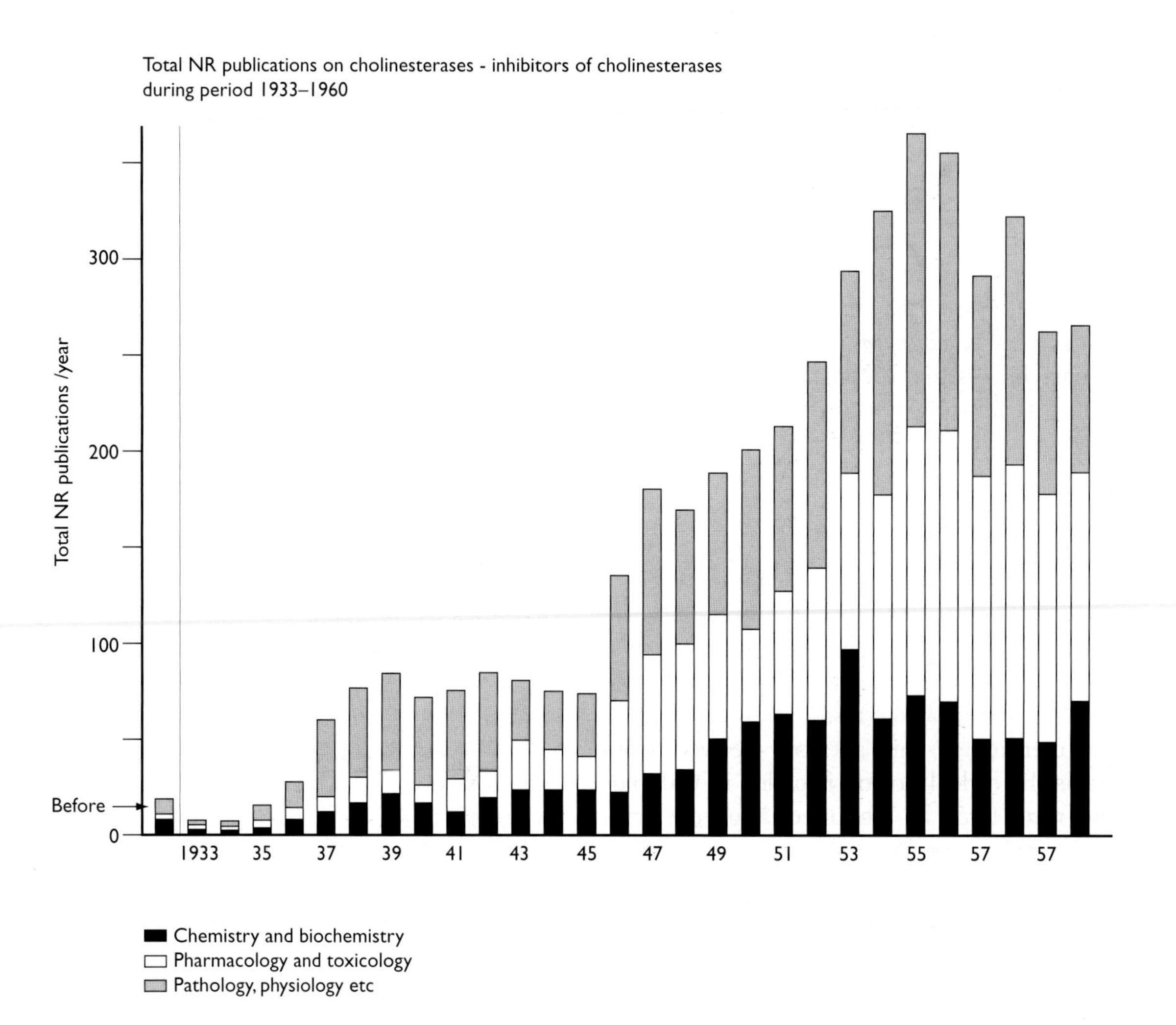

1

Cholinesterase inhibitors: an introduction

Bo Holmstedt

Historical development of carbamates[1,2]

The first anticholesterinerase agent known to man was physostigmine obtained from *Physostigma venenosum*, the ordeal bean of Calabar. Its history (details of which can be found in Holmstedt[1]) is closely linked with the Department of Pharmacology, Edinburgh. Robert Christison (1797–1882) carried out both animal experiments and heroic self-experiments with the Calabar extract (Fig. 1.1).

In 1863 Christison's successor, Thomas Fraser (1841–1920), separated from the bean an amorphous active principle with the properties of a vegetable alkaloid. He proposed the name eserine for this alkaloid derived from *ésere* as the ordeal poison is called in Calabar. Some time later he obtained a purer, crystalline form of the alkaloid. The main alkaloid present in the seeds of the Calabar bean was first isolated in a completely pure form in 1864 by Jobst and Hesse (1864), who called it physostigmine. One year later it was obtained in a crystalline form also by Vée (1865) who called it eserine. Both names are still used to designate the base.

The elucidation of the structural formula of physostigmine proved difficult. A complete description of the many blind alleys and the painstaking work leading to final synthesis has been given by Marion (1952). The chemical structure of the ring system was established by Stedman and Barger (1952). Appropriately this work was carried out in Edinburgh, where so many studies on the Calabar bean had been

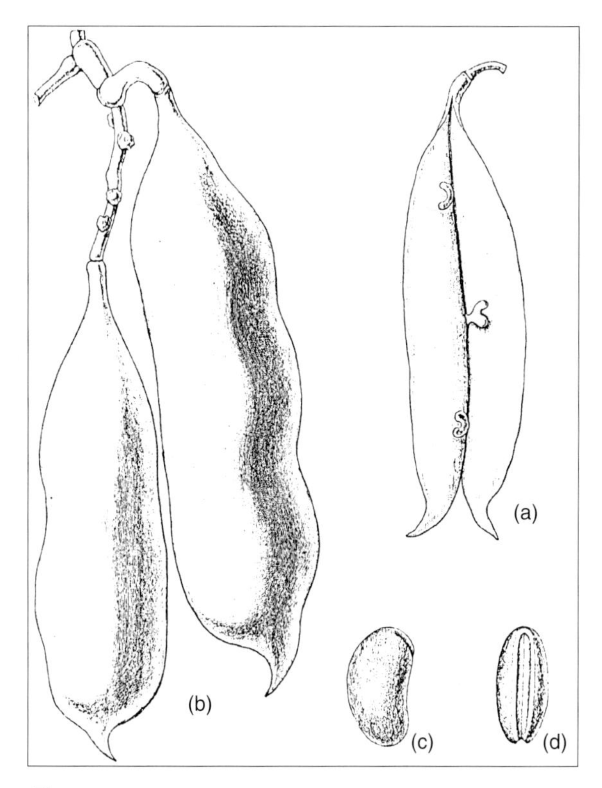

Figure 1.1
Physostigma venenosum Balf. (a) Young pod of Physostigma venenosum, with three ovules. (b) Full-grown pods of P. venenosum. (c) Seed of ordeal bean seen laterally. (d) As (c), showing the sulcate and extended hilum on the convex edge. All figures are natural size. (Reproduced from Balfour JP, Trans R Soc Edinburg 1861; 305–312.)

made. The final proof of the structure was obtained through the complete synthesis achieved by Julian and Pikl (1935). They developed a simple route of synthesis which had the advantage over those already described in that it gave rise to the same isomer as the natural base. For a complete description of these developments, see Holmstedt.[1,2]

The property of physostigmine to produce miosis and other symptoms was attributable to the urethane group, since it is absent in its hydrolysis product, eseroline. This observation has prompted the preparation of a number of substituted urethanes, the best known being neostigmine, which was synthesized by Aeschliman and Reinert (1931). Neostigmine was used in the treatment of *myasthenia gravis* at an early stage.

The use of carbamates as insecticides started only in the 1950s with their development as insect repellents. Dr H Gysin, a senior research chemist at Geigy Company, synthesized a number of cycloaliphatic carbamates as potential insect repellents and passed them on to the Biological Department for testing. However, the biologists Wiesman and Lotmar observed that one of these substances, 5,5-dimethyldihydroresorcinol dimethylcarbamate, which was later given the common name dimetan, showed only a weak repellent effect but a promising insecticidal activity. As a consequence, a considerable number of carbamates analogous to dimetan were synthesized and tested. Heterocyclic carbamates were found to be superior to the cycloaliphatic carbamates, one compound in this series even displaying systemic insecticidal activity.

The first marketed products in the carbamate field were the *N,N*-dimethylcarbamates of a variety of heterocyclic enols. The versatility of the carbamates was expanded by the introduction of the *N*-methylcarbanates, of which the 1-naphthylester, Carbaryl or Sevin, is one of the most widely used broad-spectrum insecticides. Its remarkable success is attributable to its low acute and chronic toxicity to mammals, and its environmental degradability. Another compound, Baygon, has become a standard item for household pest control and for residual spray in malaria eradication programmes.

Fraser carried out a detailed study of the action of the Calabar extract in animals (1870). After a hint from his friend Thomas Fraser, another Scottish physician, Argyll-Robertson (1837–1909), turned his attention to the Calabar bean as an important agent in the treatment of eye conditions. A host of other experiments with physostigmine were carried out during the 19th century. However, all these individual observations did not merge into a general picture of the toxicology of cholinesterase inhibitors, including the Calabar extract, until some 100 years later. Death due to this group of compounds is complex and involves central effects (convulsions and paralysis of respiration), respiratory effects (increased secretion and bronchospasm) and neuromuscular block with early involvement of respiratory muscles. The effects on the circulatory system involve bradycardia, decreased cardiac output and peripheral vascular phenomena.

A study of the part played by the cardiovascular system has confirmed what many 19th-century authors reported. When death, due to a large dose, comes quickly, the circulation is still relatively unimpaired when respiration fails. When death is delayed, it is impossible to make such a distinction, since mounting depression and final failure then involve both systems equally. Death appears to be primarily by asphyxia in some instances and primarily cardiovascular in others, and in some cases failures of both systems seems to coincide.

Fraser's original paper quoted above stated this very clearly.

As is well known, the main alkaloid of the Calabar bean was destined to play an important role in the elucidation of neurohumoral transmission — i.e. how impulses are transmitted from one nerve to another or from a nerve to an end organ by means of a hormone — in this case acetylcholine. The powerful biological activity of synthetic acetylcholine was discovered as early as 1906. Physostigmine was the first alkaloid known to act through the inhibition of an enzyme. Those interested in the detailed sequence of events in the discovery of neurochemical transmission are referred to the paper by Bacq.[3]

Development of organophosphorus anticholinesterase agents[4]

Although it was not known at the time, the first organophosphate with cholinesterase-inhibiting properties had already been synthesized in 1854 (de Clermont, 1855). This was tetraethylpyrophosphate (TEPP). de Clermont reported that the product he obtained had a 'harsh taste and a characteristic smell'.

The synthesis of TEPP and analogues was repeated half a dozen times up to 1930, with no untoward effect observed on the chemists working with the products. It must be mentioned, however, that compounds of greater volatility are more liable to cause harmful effects.

From the point of view of our present knowledge of organophosphorus anticholinesterase agents, it is easy to distinguish the important incidents in the past that bear upon the present synthetic and pharmacological work. Thus, in 1873, AW von Hofman (1818–1892) synthesized methylphosphoryl dichloride. This occurred upon the return of this famous chemist to Germany after the many years he spent in London, and was his last contribution to the field of organophosphorus chemistry, most of which had been carried out in England. Methylphosphoryl dichloride constitutes the first example of the C–P linkage. This compound, unnoticed for many years, is one of the important steps in the synthesis of modern C–P compounds, although produced in other ways. Among these are insecticides and the nerve gas Sarin.

In the latter part of the 19th century the synthesis of organophosphorus compounds was closely linked with the work of CAA Michaelis (1847–1916), Professor at the University of Rostock. All his own investigations and those of his pupils were collected in monographs devoted to various parts of the field. One of these investigations bears notably upon the present subject, namely the one describing substances containing the N–P bond (Michaelis, 1903). Also reported in this monograph is a compound with a P–CN bond, diethylamidoethoxyphosphoryl cyanide. Michaelis gave no detailed description of how he made this compound, but he got it from a product obtained in a reaction worked out by one of his pupils (Schall, 1898). This account later led to the synthesis of a number of important insecticides and the nerve gas Tabun.

At about the same time, important work in organophosphorus chemistry was also being carried out in Russia. This work was, and still is, linked to the Kazan School of Chemistry, which dates back to 1806 (AE Arbusow, 1940). Of special importance is the publication produced in 1906 by Professor AE Arbusow about the isomerization reaction that still bears his name (AE Arbusow, 1906). Thus constitutes one of the most commonly used ways of forming the stable C–P bond.

The first item of importance in this regard was the obtaining of TEPP in a pure distilled form by Nylén in Uppsala (Nylén, 1930). The synthesis of TEPP in two ways, together with that of numerous other organophosphates, was described in Nylén's thesis, which constitutes one of the fundamental works for present-day studies. Nylén had no knowledge of the toxicity of his compounds when told of this by the present author 18 years afterwards. TEPP was again synthesized the year after by the Arbusows, who also produced tetraethylmonothionopyrophosphate (Arbusow and Arbusow, 1931).

The following year witnessed the most important discovery in the field from a toxicological point of view. Willy Lange, at that time Privatdozent in chemistry at the university of Berlin, and his graduate student Gerda von Krueger, synthesized some compounds containing the P–F linkage. Lange was originally an inorganic chemist. At this time fluorine had become generally available for chemical synthesis and Lange introduced it for the first time into organophosphorus compounds. During the synthesis of dimethyl- and diethylphosphorofluoridate, the workers noticed the toxic effects of the vapour on themselves. Their short account of this included at the end of a purely chemical paper is astonishing in its pharmacological correctness (Lange and Krueger, 1932). Lange had an interest in synthetic insecticides, as shown by a passage in a letter (Lange, 1952).

Later on, the IG Farbenindustrie developed an interest in synthetic insecticides, then a completely undeveloped field. The scientific leader of the company, Otto Bayer, appointed the chemist Gerhard Schrader for this work in 1934. It was not until 1936, however, that he turned his interest to the phosphorus compounds. At the turn of the year (1936–1937), Schrader also noticed miosis and discomfort during his synthetic work. At that time he was repeating the synthesis of Michaelis' pupil Schall, previously mentioned. This occurrence led directly to the long sequence of synthetic organophosphates with insecticidal and toxic properties. It also led to the synthesis in Germany of chemical warfare agents of until then unknown toxicity. In March 1937, Schrader patented the general formula for contact insecticides of this type (Schrader, 1952). The subsequent development of the organophosphate insecticides is treated in Schrader's monograph (Schrader, 1952). In the ensuing years 1938–1941, he developed the fluorine containing compounds, including diisopropylphosphorofluoridate (DFP) (Wirth, 1949) and the pyrophosphorus derivatives including TEPP. The systemically insecticidal OMPA (octamethyl pyrophosphortetramide) and related compounds were patented, together with H Kükenthal. In 1940 came Bladan, the active principle of which is probably TEPP. Schrader then turned to the thio- and thionophosphorus compounds. Paraoxon (E600) and its sulphur analog Parathion (E605) were ready in 1944. The insecticidal value of the latter compound was elucidated at the end of the Second World War. By that time Schrader is said to have synthesized around 2000 organophosphorus compounds.

From 1935 onwards, the German Government required that information about new toxic products of any importance should be submitted for investigation. By 1944, around 200 compounds were considered significantly toxic by the Ministry of Defence to be classified as secret. This sometimes could be of great discomfort to commercial firms. Among these compounds was, for example, TEPP.

Also among these secret compounds were ethyl-*N*,*N*-dimethylphosphoramidocyanidate, prepared by Schrader in 1937 and bearing the codename Tabun, and isopropylmethylphosphonofluoridate, Sarin, made in the same year. Soman (pinacolylmethylphosphonofluoridate) was only at the laboratory stage at the end of the Second World War. Numerous chemists studied the military properties of these compounds which were considered sufficiently interesting for immediate manufacture.

The investigations of the pharmacological and toxicological properties of the organophosphorus compounds in Germany were carried out by both the industrial firms working with insecticides and laboratories belonging to the armed forces. The first pharmacological experiments with these agents were carried out independently at Farbenfabriken Bayer in Elberfeld by Eberhard Gross, and at the Militärärztliche Akademie in Berlin where the department was headed by Wolfgang Wirth, who had as his co-worker J Sextel. This was started in 1937. Later on, after the beginning of the War, the group at the Militärärztliche Akademie was joined by other prominent German pharmacologists, by then drafted. Among these were L Lendle, W Koll, H Gremels and O Grindt.

Gremels is said to have recognized the anticholinesterase properties of Tabun at an early stage in 1940. Among other things he observed the potentiating effect of a Tabun injection on the drop in the cat's blood pressure evoked by constant doses of acetylcholine (Lendle, 1959). According to Schrader (1952), the enzyme-inhibiting properties of TEPP were recognized by Eberhard Gross in 1939. It is not stated whether this was done by biochemical or pharmacological investigations. In any event, the parasympathomimetic effects of the nerve gases were clearly recognized by the German workers, and atropine was established as an antidote (Collomp, 1949). The symptomatology was also well recognized both from experiments in animals of various species and by the exposure of volunteers to low concentrations of the compounds.

Long before the interrogation of German scientists after the end of the War and the publication of reports about the work carried out in Germany during the War, similar activities were carried out in England.

Kilby's survey of the open literature had led him to the passage in Lange and Krueger's paper about the toxic effects. This initiated the first synthetic and toxicological work with these compounds in England. Exposure of volunteers, including the investigators themselves, was carried out with the compounds, but was reported only after the War (Kilby and Kilby, 1947).

Later on, many homologs and related compounds were synthesized by a team under H McCombie and BC Saunders. Among these compounds was DFP. The toxicological effects of the new compounds were studied by a team under Lord Adrian, which included W Feldberg and BA Kilby. The long-lasting miosis led these investigators to think that the organosphosphates were enzyme inhibitors like eserin. This was borne out by experiments on isolated organs, including the frog rectus muscle, in 1942. The same year, a biochemical team consisting of Dixon, Mackworth and Webb compared DFP to eserine in Warburg experiments, and found it to be a strong inhibitor of horse serum cholinesterase. All the chemical, pharmacological, and biochemical experiments were published only after the War.

When at the end of the War the existence of the German nerve gases was disclosed to the allies it was still kept a secret for military reasons. The greater toxicity of Tabun and Sarin outweighs that of a compound like DFP.

The formulas of Tabun and Sarin, however, were published in 1948 by Bonnaud and by Valade and Sallé, who also reported some preliminary pharmacological experiments in the same year (Valade and Sallé, 1948). The first full account of the synthesis and pharmacology of Tabun was given in 1951 (Holmstedt, 1951).

During the 1950s, an enormous development took place in this field. This is well known even to those remotely connected with work in toxicology, pharmacology or field protection. Important events were the advent of malathion, an insecticide with very low toxicity to man (Cassaday, 1950), the introduction of Systox (Schrader and Lorenz, 1951) and the finding that a new type of sulphur containing organophosphorate was of extreme potency (Ghosh and Newman, 1955). To this may be added the synthesis of a series of pharmacologically interesting compounds related to choline esters and thiocholine esters.

Atropine as an antidote to anticholinesterases and the development of the receptor theory[1]

The antagonism between physostigmine and atropine, which Fraser found to exist with respect to their action on the pupil and the heart rate, also made it possible that a physiological antagonism might be found for other effects. This led Fraser to perform a careful study in which he showed that the lethal effect of physostigmine could be prevented by atropine (Fraser, 1870). This was outlined in a lecture which dealt with drug antagonism (Fraser, 1872), and which has been carefully commented upon by Gaddum.

Fraser pointed out that some antidotes destroy the poison in the stomach, but others act after absorption. He emphasizes the fallacies which complicate the interpretation of clinical observations on the effects of antidotes in those cases where it is not possible to know whether the patient would have died if no antidote had been given. He then describes his own work on the antagonism between physostigmine and atropine when both are injected subcutaneously in rabbits. This antagonism was itself a new concept at that time, but Fraser was interested in its quantitative aspect. After a given dose of physostigmine, there is a range of doses of atropine that will save life. If too little is given, the rabbit dies of physostigmine poisoning, and if too much is given it dies of the effects of atropine. As the dose of physostigmine is increased this range gets smaller and, if the dose of physostigmine is more than 3.5 times the lethal dose, life cannot be saved by any quantity of atropine.

Fraser's experiments showed that small doses of atropine saved life by opposing the lethal action of physostigmine, but he also obtained evidence that small doses of physostigmine caused death by increasing the lethal action of atropine. As pointed out by Gaddum (1962), unthinking persons may perhaps find this result surprising, but it was just the kind of thing that Fraser expected, and his experiments were designed to detect it. He pointed out that, as both atropine and physostigmine possess a number of separate actions, it was not unreasonable to anticipate that several of them are not mutually antagonistic, and in some cases the two drugs might be expected to work together.

Fraser's carefully carried out experiments on the antidote effects of atropine have been followed by experiments too numerous to be described here. The first human use of atropine in anticholinesterase poisoning, however, precedes that of the animal experiments.

Kleinwächter (1864), at the ophthalmic clinic in Prague, had the courage to use Calabar extract systemically as an antidote in atropine poisoning (his is the earliest published reference to this fact). He was called to a prison to see three prisoners who had drunk a solution of atropine under the impression that it was some kind of liquor. He gave Calabar extract to the one who seemed to be the most severely poisoned and kept the others as controls. In this way he obtained convincing evidence that Calabar extract and atropine were antagonistic. Interestingly enough, physostigmine has been suggested as an antidote to modern chemical warfare agents with an atropine-like action (International Defence Review/Interavia/II/1969, pp. 170–174).

The physostigmine–atropine antagonism not only paved the way for studies on the autonomic nervous system, but also for the receptor theory of drug action. Although it was suggested by several investigators in the 19th century, particularly Fraser, that the action of drugs might be due to a reaction between the drug and some constituent of the cell, the concept of drug receptors was first clearly developed at the beginning of the 20th century by JN Langley (1852–1925) in England and by Paul Ehrlich (1854–1915) in Germany. Beginning with different problems, each elaborated a theory which he was able to use to explain certain pharmacological phenomena. The development of the receptor theory has been described in detail by Parascandola *et al.*[5] The concept was first clearly stated by JN Langley in 1905 and by Paul Ehrlich in 1907. Langley's concept developed out of his investigations on the actions of nicotine and adrenaline on the body, which were occasioned by his interest in the sympathetic nervous system. Ehrlich's theory developed out of his studies on drug resistance, which were a result of his interest in the chemotherapy of trypanosomes, and ultimately out of his side-chain concept of cellular function. While their ideas were developed separately and from different sources, their thoughts did interact. Langley recognized the similarity of his concept of receptive side-chains of protoplasm (an idea which he developed further in his later work) to Ehrlich's side-chain theory of immunity. Ehrlich admitted to being influenced by Langley's work in his decision to apply the side-chain or receptor concept to drugs.

Historical development of reactivators

The basis for antidotes other than atropine and similar anticholinergic drugs is the theoretical possibility that a drug which reactivates the enzyme (if it does so soon enough) will be useful in therapy, since once the enzyme has been reactivated it will perform its normal function of destroying the accumulated acetylcholine and the symptoms of poisoning should disappear. However, there is an alternative approach. A drug may be found which when administered prophylactically will circulate in the blood and react with and destroy the entering cholinesterase inhibitor before it gets a chance to reach the vital sites in the body. On the one hand, one could look for substances which would produce a regeneration of activity in cholinesterase inactivated by reaction with inhibitors and, on the other hand, for substances which would react rapidly with inhibitors under physiological conditions, be relatively harmless if taken internally and be effective in reasonable doses.

The detailed sequence of events in the development of these antidotes is difficult to trace. It is also said that the first breakthrough in the search for such compounds came in 1951 when Jandorf found that hydroxylamine reacted rapidly and smoothly with organophosphorus inhibitors in solution under physiological conditions and detoxified the compounds.[6] In the same year, Wilson[7] reported both the spontaneous and the hydroxylamine-enhanced reactivation of

TEPP-inhibited enzyme. A follow-up of this line of research led to the use in 1953 by Wagner-Jauregg and his group of the compounds of the class known as hydroxamic acids which appeared to be promising.[8,9] Hydroxamic acids are relatively non-toxic and their reaction mechanism with cholinesterases has been investigated extensively.

In 1953, it was reported that DFP-inactivated cholinesterase could be regenerated successfully by treatment with dilute solutions of certain hydroxamic acids in vitro.[10] This was the first example of how inactivation of cholinesterase by as firmly bound a compound as DFP could be overcome in vitro.

In the development of reactivators, as in phosphoryl compounds, military security regulations make it difficult to follow the historical sequence of events. An important step was the introduction by Davies in 1955 of the oximes, particularly pyridine-2-aldoxime methiodide (PAM), as a reactivator.[11,12] The even more active bispyridinium compounds were introduced in 1958.[13–15]

References

1. Holmstedt B. In: Swain T, ed. *The Ordeal Bean of Old Calabar: The Pageant of Physostigma venenosum in Medicine*. Cambridge, MA: Harvard University Press, 1972: 303–360.
2. Holmstedt B. Historical developments of carbamates. The Third Symposium on Prophylaxis and Treatment of Chemical Poisoning, 22–24 April 1985, Stockholm, Sweden. *Fundam Appl Toxicol* 1985; **5(Suppl 2):** S1–S9.
3. Bacq ZM. Chemical transmission of nerve impulses. In: Parnham MJ, Bruinvels J, eds. *Discoveries in Pharmacology*, Vol 1: *Psycho- and Neuropharmacology*. Amsterdam: Elsevier, 1983: 49–103.
4. Holmstedt B. Structure–activity relationships of organophosphorus anticholinesterase agents. In: Eichler O, Farah A, eds. *Handbuch Der Experimentellen Pharmakologie Ergänzungswerk*. Berlin: Springer-Verlag, 1963: 427–485.
5. Parascandola J, Jasensky R. Origins of the receptor theory of drug action. *Bull Hist Med* 1974; **48:** 199–220.
6. Summerson WH. Progress in the biochemical treatment of nerve gas poisoning. *Armed Forces Chem J* 1955; **Jan–Feb:** 24–26.
7. Wilson IB. Acetylcholinesterase XI. Reversibility of tetraethyl pyrophosphate inhibition. *J Biol Chem* 1951; **190:** 111–117.
8. Stolberg M, Tweit R, Steinberg G, Wagner-Jauregg T. The preparation of O-phenycarbamyl benzohydroxamate through the Lossen rearrangement. *J Am Chem Soc* 1955; **77:** 765–767.
9. Wagner-Jauregg T. Experimentelle Chemotherapie von durch phosphorhaltige Anti-Esterase-herworgerufenen Vergiftungen. *Arzneim-Forsch* 1956; **6:** 194–196.
10. Wilson IB, Meislich EK. Reactivation of acetylcholinesterase inhibited by alkylphosphate. *J Am Chem Soc* 1953; **75:** 4628–4629.
11. Childs AF, Davies DR, Green AL, Rutland JP. The reactivation by oximes and hydroxamic acids of cholinesterase inhibited by organophosphorus compounds. *Br J Pharmacol* 1955; **10:** 462–465.
12. Wilson IB, Ginsburg A. A powerful reactivator of alkylphosphate-inhibited acetylcholinesterase. *Biochim Biophys Acta* 1955; **18:** 168–170.
13. Hobbiger F, O'Sullivan DG, Sadler PW. New potent reactivators of acetylcholinesterase inhibited by tetraethyl pyrophosphate. *Nature (London)* 1958; **182:** 1498–1499.
14. O'Leary JF, Wills JH, Carlstrom LA. Relative efficacy of various prophylactic adjuncts to atropine in sarin poisoning. *Fed Proc* 1958; **17:** 401.
15. Poziomek EJ, Hackley BE Jr, Steinberg GM. Pyridinium aldoximes. *J Org Chem* 1958; **23:** 714–717.

2

Structural studies on acetylcholinesterase

Israel Silman, Joel L Sussman

Introduction

Solution of the three-dimensional (3D) structure of *Torpedo californica* acetylcholinesterase (*Tc*AChE) in 1991,[1] opened up new horizons in research on an enzyme which was already the subject of intensive investigation.[2] The unanticipated structure of this extremely rapid enzyme, in which the active site was found to buried at the bottom of a deep and narrow gorge, lined by aromatic residues, led to a revision of the views then held concerning substrate traffic, recognition and hydrolysis. This led to a series of theoretical and experimental studies, which took advantage of recent advances in theoretical techniques for the treatment of proteins, such as molecular dynamics and electrostatics, and of site-directed mutagenesis, utilizing suitable expression systems.

The realization that AChE, together with several other enzymes whose 3D structures were solved at about the same time,[3] was a member of a new family of proteins sharing a common fold, the α/β-hydrolase fold, likewise opened up a fertile field of research in which state-of-the-art techniques of sequence alignment and homology modelling were utilized.[4] The burgeoning number of members of this new family resulted in the foundation of the ESTER database at Montpellier (http://www.montpellier.inra.fr:70/cholinesterase),[5] to collate and check the large amounts of data accumulating, and to make them available and accessible to a large body of users. The fact that the α/β-hydrolase fold family included several members that lacked one or more of the residues in the catalytic triad characteristic of the enzymes in the family, and which had earlier been shown to be adhesion proteins, suggested alternative functions for AChE. These, in turn, necessitated fresh approaches towards the analysis of the structure[5] of the members of this family.[6]

Finally, at a more practical level, the 3D structure itself, followed not long thereafter by the 3D structures of complexes with a number of ligands,[7] provided the basis for a rational structure-based approach to the design of drugs, such as anti-Alzheimer medications,[8] to the treatment of organophosphate (OP) intoxication,[9] and to the development of new classes of insecticides,[10] especially taken in conjunction with the recently solved 3D structures of human AChE (hAChE),[11] and of the *Drosophila melanogaster* enzyme (*Dm*AChE).[12]

In this chapter, after giving a brief background, we first analyse the overall 3D structure of AChE in the context of the α/β-hydrolase fold family. We then go on to discuss the quaternary structure of the oligomeric forms of the enzyme, dimers and tetramers. This is followed by an outline of current views concerning the mode of action of the enzyme. Special attention is paid to traffic of substrates and products through the active-site gorge, and to such controversial issues as the role of electrostatics in catalysts and the possible existence of a 'back door' through which substrate and/or products might pass. We then analyse the role of electrostatics in putative alternative biological roles for AChE and for other members of the α/β-hydrolase fold family. Finally, we discuss structure–function relationships in the interaction of reversible ligands and covalent agents with AChE.

Three-dimensional structure: fold and function

Crystal structure

The 3D structure of *Tc*AChE was solved in 1991,[1] subsequent to solubilization, purification and crystallization of the glycophosphatidylinositol (GPI)-anchored dimer that is present in large amounts in *Torpedo* electric organ.[13] Solution of the 3D structure indeed proved that the cholinesterases (ChEs) contain a catalytic triad, albeit with a glutamate in place of the aspartate found in the serine proteases. However, the 3D structure displays a number of unexpected features (Fig. 2.1). The active site is deeply buried, being located almost 20 Å from the surface of the catalytic subunit, at the bottom of a long and narrow cavity. This cavity was named the active-site gorge or, since over 60% of its surface is lined by the rings of conserved aromatic residues, the aromatic gorge.[1,14] Despite the prediction that the 'anionic' site would contain several negative charges,[15] in fact only one negative charge is close to the catalytic site, i.e. that of Glu199, which is adjacent to the active-site serine (Ser200). Based both upon docking of acetylcholine (ACh) within the active site[1] and upon affinity labelling,[16] the quaternary group of ACh appears to be interacting, via a cation–π-electron interaction,[17,18] with the indole ring of one of the conserved aromatic residues (Trp84). Another conserved aromatic residue (Phe330) is also involved in the interaction.[7]

α/β-Hydrolase fold

At about the same time that the 3D structure of *Tc*AChE was solved, several other hydrolase structures were elucidated which were shown to share a common fold with AChE. This fold was named the α/β-hydrolase fold,[3] since all its members are α/β-proteins which contain a very similar structure, in which a central β-sheet is surrounded by loops and helices. Although the catalytic triad is very similar to that found in the serine proteases, the fold is completely novel. While AChE shares sequence homology with some members of the family, in other cases no significant homology is detectable.[3,4] The α/β-hydrolase fold family contains enzymes displaying a broad range of specificities. These include neutral lipases and peptidases as well as proteins which, based on their sequence alignment, display substantial sequence homology with the ChEs, but lack one or more of the residues of the catalytic triad.[19] These proteins share the common feature of having been identified as adhesion proteins, and are discussed below in connection with possible non-catalytic functions of the ChEs.

Functional binding sites

The unexpected structure of the active-site itself, and of the gorge leading to it, have been explored experimentally by two approaches: crystallographic studies of complexes of *Tc*AChE with a repertoire of ligands;[7,20–23] and site-directed mutagenesis (for literature see Doctor *et al.*).[24]

Crystallographic studies, using suitable qua-

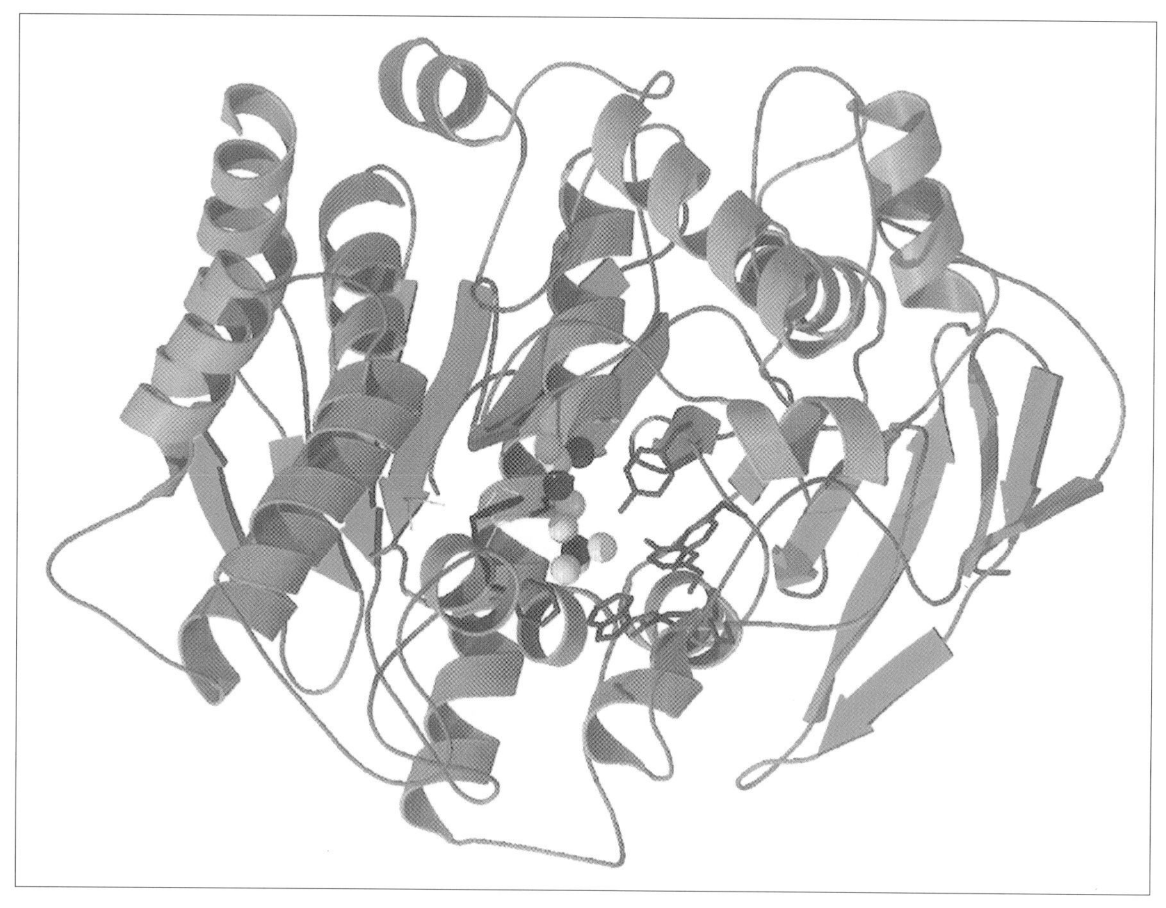

Figure 2.1
Ribbon diagram of the 3D structure of TcAChE. Green arrows represent β-strands, and brown coils represent α-helices. The side-chains of the catalytic triad and of key aromatic residues in the active-site gorge are indicated as purple stick figures. ACh, manually docked in the active site, is represented as a space-filling model, with carbon atoms shown in yellow, oxygen atoms in red and the nitrogen atoms in blue. The quaternary group of the ACh faces the indole of Trp84.

ternary ligands,[7,21] confirmed the prediction mentioned above, made on the basis of computerized docking of ACh, by clearly showing that such ligands interact with Trp84 and, to a lesser degree, with Phe330. Elucidation of the structure of a complex with a powerful transition-state analogue[21] also bore out the prediction of a three-pronged 'oxyanion hole'.[1] Modelling studies[25] strongly suggested that the acyl pocket for the acetyl group of ACh is provided by two more conserved aromatic residues, Phe288 and Phe290, as was later confirmed by inspection of the complex with the transition-state analogue.[21] Finally, inspection of the structure of a complex with the elongated bisquaternary ligand, decamethonium, which was shown to bind along the active-site gorge, permitted identification of the 'peripheral' anionic site at the entrance to the active-site gorge. The 'peripheral' site involves three more conserved aromatic residues: Tyr70, Trp279 and Tyr121.[7]

Solution of the structure of a complex of *Tc*AChE with the snake venom polypeptide fasciculin, which was known to serve as a peripheral site inhibitor,[26] showed that it exerts its inhibitory action by binding to the surface of the enzyme at the top of the gorge, thereby blocking almost completely entrance of substrate.[20] Similar data were presented by Bourne *et al.*[27] for a complex of fasciculin with mouse AChE, the structure of which, moreover, closely resembles that of *Tc*AChE, as was predicted from the high degree of sequence identity (ca. 50%) and similarity (ca. 70%).[1,25] Fasciculin, which is purified from green mamba venom, is a member of the large family of three-fingered toxins of which the nicotinic antagonist, α-bungarotoxin, and the homologous cobra α-neurotoxin are the most prominent members.[28] Solution of the structure of the complex of fasciculin with AChE is the first example of the structure of a complex of a three-fingered toxin with its target, and demonstrates the large contact area and multiple interactions which account for their high affinity and specificity.

Correlation with site-directed mutagenesis studies

The extensive site-directed studies, carried out primarily in the laboratories of Shafferman and Taylor (see, for example, Ordentlich *et al*,[29] Shafferman *et al.*[30] and Taylor and Radic[31]) have supplemented the structural data. A complete list of mutations is available from the ESTHER server (see above). We mention here a few of the key issues addressed. Site-directed mutagenesis of Phe288 and Phe290, the two aromatic residues shaping the acyl pocket, was shown to broaden specificity, thus enabling AChE to hydrolyse butyrylcholine, normally only hydrolysed by butyrylcholinesterase (BuChE), which on the basis of sequence alignment and molecular modelling, possesses a larger acyl pocket in which these two aromatic residues are replaced by aliphatic residues.[25,29,32] Mutation of Trp84 to an aliphatic residue clearly demonstrated the key role played by this residue in recognition of the quaternary group of ACh.[29] Mutation of the aromatic groups in the peripheral site drastically reduced affinity for peripheral-site ligands, such as propidium.[25,33] Radic and co-workers[26] elegantly demonstrated the contribution of these residues to the binding of fasciculin. Thus the triple mutant in which the residues in mouse AChE equivalent to Tyr70, Trp84 and Tyr121 were all eliminated bound fasciculin 10^8 times more weakly than the wild type (WT) enzyme, with an affinity equal to that of BuChE, which lacks all three of these aromatics and, consequently, also the peripheral site.[25]

Quaternary structure

AChE displays structural polymorphism, being expressed as a repertoire of molecular forms, the pattern of which differs from tissue to tissue, even in the same animal.[34,35] Most vertebrates contain a single gene coding for AChE, and alternative splicing gives rise to two principal catalytic subunits: H and T. H subunits form GPI-anchored dimers,[36] whereas T subunits occur as monomers, dimers and tetramers, the latter being dimers of disulphide-linked dimers. T subunits also associate with structural subunits (P and Q), to form membrane-anchored tetramers and asymmetric forms, in which 1–3 tetramers are attached to a collagen-like tail.[34,35] X-ray crystallography can provide direct information about the motifs and forces involved in the formation of the oligomeric species.

The Torpedo californica AChE dimer

*Tc*AChE is a dimer of H subunits in which the two disulphide bonds are covalently linked by a disulphide bridge near the COOH terminus. Although the disulphide bridge was not resolved in the original 3D structure,[1] a four-helix bundle was clearly also involved in the intersubunit interaction. Constructs in which the cysteine residue responsible for interchain disulphide bond formation had been removed by site-directed mutagenesis were employed to generate H subunit monomers of mouse and human AChE, which were then crystallized as stoichiometric complexes with fasciculin.[11,27] Even though both the mouse and the human AChE are monomeric in solution, inspection of their 3D structures revealed that, in the crystal, they pack as dimers which are isomorphous with the disulphide linked dimer of *Tc*AChE in the crystals of the fasciculin–*Tc*AChE complex.[20]

The Electrophorus electricus AChE tetramer

The only tetrameric form of AChE for which any crystallographic data are available is the tetramer obtained by tryptic digestion of the asymmetric species purified from electric organ tissue of the electric eel, *Electrophorus electricus* (*Ee*AChE).[37] This tetramer, being derived from an asymmetric form, is composed of T subunits. A data set to 4.4 Å was collected for orthorhombic crystals of the eel enzyme, but work on this structure was discontinued when the structure of *Tc*AChE was published. Since, in higher vertebrates, the predominant molecular forms of AChE are based on T subunits, it is important to understand the molecular basis for their association with each other and with the P and Q structural subunits. Accordingly, Raves *et al.*[38] attempted to solve the structure by means of molecular replacement on the basis of the *Tc*AChE structure. The solution so obtained revealed a biological dimer, with a four-helix bundle at the interface, just like that in the *Torpedo* dimer. A second two-fold axis, perpendicular to the first, reveals the arrangement of the dimers in a tetramer, with a 20° twist from planarity (Fig. 2.2(a)). There is a large gap between the two dimers, about 30 Å wide, where the four carboxy termini (C-termini) point towards each other. This space is most likely occupied by the extra 4 × 40 residues of the T subunit, which are lacking in the H subunit. The structure thus proposed is in good agreement with a model recently constructed for the tetramer, in which it was proposed that the four subunits form a pseudo-square arrangement, with highly conserved sequences at the C termini generating amphipathic helices which, in turn, form two additional four-helix bundles (Fig. 2.2(b)).[39] The crystallographic structure, as well as the structural model, are in good agreement with the model recently proposed by Massoulié and co-workers[40] for assembly of the tetramer around the proline-rich attachment domain (PRAD), through the amphipathic sequences at the C termini of the T subunits.

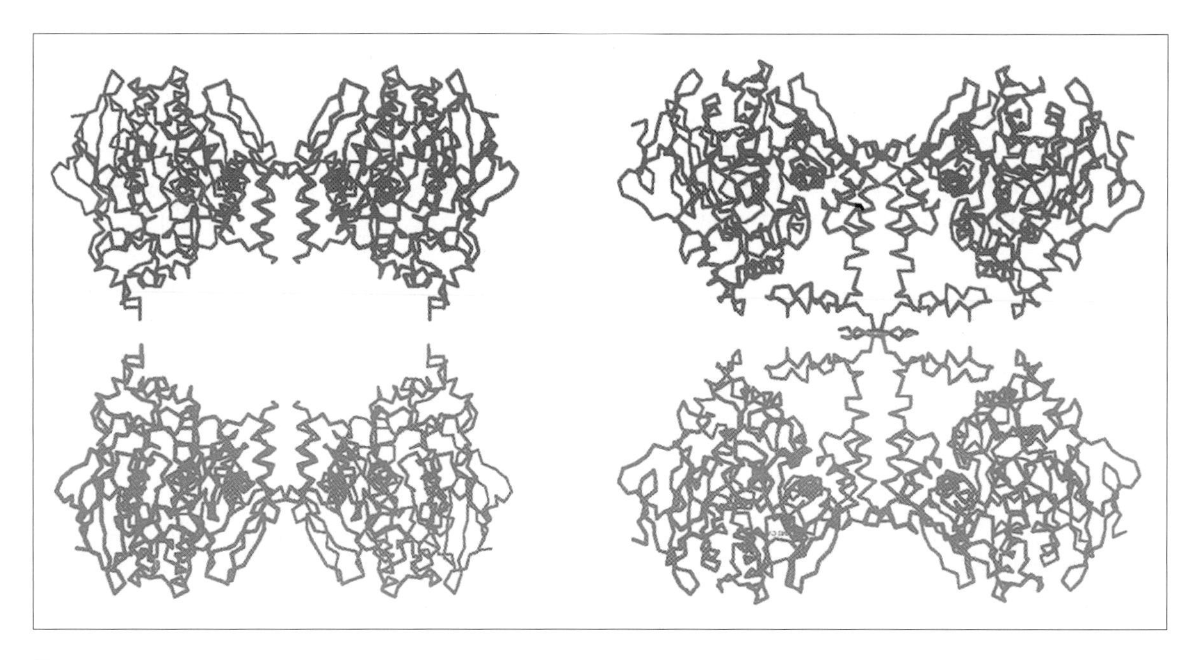

Figure 2.2
Quaternary structure of the EeAChE tetramer: (a) crystal structure; (b) modelled structure.

Electrostatic characteristics: role in catalysis and traffic through the active-site gorge

Role in catalysis

An important development was the discovery, by calculations utilizing the 3D structure of *Tc*AChE, that the ChEs possess a highly asymmetric charge distribution, resulting in a large dipole moment, oriented approximately along the axis of the active-site gorge.[41–43] Electro-optical measurements on the monomeric AChE of *Bungarus fasciatus* provided direct experimental evidence for this dipole moment.[44] Its direction is such that the positively charged ACh molecule would be drawn down the gorge, towards the active site. This suggested that the electric field contributed to the high catalytic activity of AChE. Such a contention was, however, challenged by site-directed mutagenesis. Seven negative residues, surrounding the entrance to the active-site gorge, were eliminated, without producing a major change in catalytic activity.[30] This resulted in a heated controversy, which has not yet been fully resolved.[45,46] A recent theoretical study by Botti *et al.*[47] proposes an integral model for traffic of ligands, substrates and products through the active-site gorge. These authors invoke two modules in their theoretical treatment:

(a) a surface trap for cationic species at the entrance to the gorge, which operates via local, short-range interactions, and is independent of ionic strength; and
(b) an ionic-strength-dependent steering mechanism generated by long-range electrostatic interactions arising from the overall charge distribution.

Their calculations show that diffusion of charged ligands towards the active site, relative to neutral isosteric analogues, is enhanced about 10-fold by the surface trap. Electrostatic steering contributes only a 1.5- to two-fold enhancement at physiological ionic strength.

The 'back door' controversy

The electrostatic characteristics of AChE also raised questions regarding the movement of charged products out of the active site. Molecular dynamics suggest that one or more routes of entrance to (or exit from) the gorge exist,[48,49] which may provide alternate routes for entrance of substrate and/or possible escape routes for products. This possibility is reinforced by kinetic studies showing residual catalytic activity in the presence of fasciculin, which, as mentioned above, appears to completely block the entrance to the gorge.[20,27] Site-directed mutagenesis experiments, designed to test the existence of such putative 'back doors' or 'side doors', have not provided support for their existence.[50,51] However, a recent crystallographic study, on a conjugate of *Tc*AChE with the long-chain physostigmine analogue MF268 provides evidence in favour of a back door.[52] It is hoped that time-resolved crystallography,[53] using suitable 'caged' compounds,[54] will provide a direct experimental approach to this issue.

Electrostatic characteristics in relation to non-cholinergic functions

Both the temporal and spatial appearance of the ChEs has raised the possibility that they may play biological roles other than termination of synaptic transmission by hydrolysis of ACh.[55–57] In recent years it has been noticed that a number of adhesion proteins display significant sequence homology with the ChEs. These include three such proteins from *Drosophila* (neurotactin,[58] glutactin[59] and gliotactin[60]) and the mammalian protein neuroligin.[61] This prompted the suggestion that AChE itself, and perhaps BuChE, may also serve as adhesion proteins.[19] This would be plausible in view of the tightly regulated temporal appearance and disappearance of both enzymes during early stages of embryogenesis in the chick.[57] Solution of the 3D structure of *Tc*AChE, and the concomitant recognition of the α/β-hydrolase fold family of proteins,[3] revealed that these adhesion proteins were all members of this family, even though they are most likely devoid of hydrolase activity due to the absence of one or more of the residues of the catalytic triad.[4] The strong homology with the ChEs was in the extracellular domain of all four of these fusion proteins. In an elegant study performed by Darboux *et al.*,[62] a chimera was constructed in which the extracellular domain of neurotactin was replaced by the catalytic subunit of *Tc*AChe. This chimeric protein could substitute for neurotactin itself in causing aggregation of cultured *Drosophila* cells, providing clear evidence that AChE can indeed play a role as an adhesion protein. Recently, Botti *et al.*[6] have shown that an electrostatic motif, at the mouth of the active-site gorge of AChE, is also present in the non-catalytic adhesion proteins. They have suggested that this motif contributes to long-range electrostatic interactions associated with their adhesive function, and have suggested that they constitute a novel class of adhesion proteins, which they have named the electrotactins.

Three-dimensional structures of complexes with nerve agents and drugs

The drugs and toxins which display anticholinesterase action can be broadly divided into two classes: those which bind non-covalently and reversibly; and those which form a covalent bond, the action of which is irreversible, or which act as slow substrates,[63] which include the OPs and carbamates. These reagents, which act by blocking the active-site serine, comprise a rather homogeneous class of inhibitors, since basically they must mimic efficiently the first stage of substrate hydrolysis.[64] Thus they must display the correct steric and mechanistic characteristics for carbamylating or phosphorylating the active site, with possible enhancement of their affinity by recognition of suitable groups within the substrate-binding site, such as those which recognize the acyl group and quaternary nitrogen atom of ACh. In contrast, the ligands which inhibit ChEs reversibly are a heterogeneous class of molecules, some of which bear no obvious resemblance to ACh. Consequently, if one is considering ligands of potential use as drugs or insecticides, structure-based design may be essential, since quantitative structure–activity relationship (QSAR) studies or computerized docking of the ligand with the enzyme may not provide the correct orientation. This was, indeed, the case for fasciculin, as mentioned above, and for huperzine A and E2020, as discussed below.

In the following, we briefly review the recently solved structures of conjugates with *Tc*AChe of OP nerve agents,[65] and of the two anti-Alzheimer drugs Exelon (ENA-713)[66] and the physostigmine analogue MF268.[52] We then analyse the subtle interactions which characterize the interaction of the anti-Alzheimer drugs huperzine A and E2020 with the enzyme.

Conjugates with organophosphate nerve agents

Organophosphate (OP) nerve agents form stable covalent conjugates with AChE, which may then undergo an internal dealkylation reaction called 'ageing'.[67] The stability of such 'aged' conjugates renders them notoriously resistant to the oxime therapy used to treat OP intoxication.[68] To understand the basis of the powerful inhibitory action of these reagents and of the resistance of the 'aged' conjugates, we solved the crystal structures of 'aged' conjugates obtained by reaction with *Tc*AChe of soman, sarin and diisopropylphosphorofluoridate (DFP).[65] In all three conjugates, the OP oxygen atoms were within hydrogen-bonding distance of four potential donors from catalytic subsites of the enzyme, suggesting that electrostatic forces significantly stabilize the 'aged' enzyme. The active sites of the 'aged' sarin–*Tc*AChe and soman–*Tc*AChe conjugates were essentially identical, and provided structural models for the negatively charged tetrahedral intermediate that occurs during deacylation with the natural substrate, ACh.[69] Phosphorylation with DFP caused an unexpected movement in the main chain of a loop that includes residues Phe288 and Phe290 of the acyl pocket (Fig. 2.3). This is the first major conformational change reported in the active site of any AChE–ligand complex, and offers a structural explanation for the substrate selectivity of AChE.

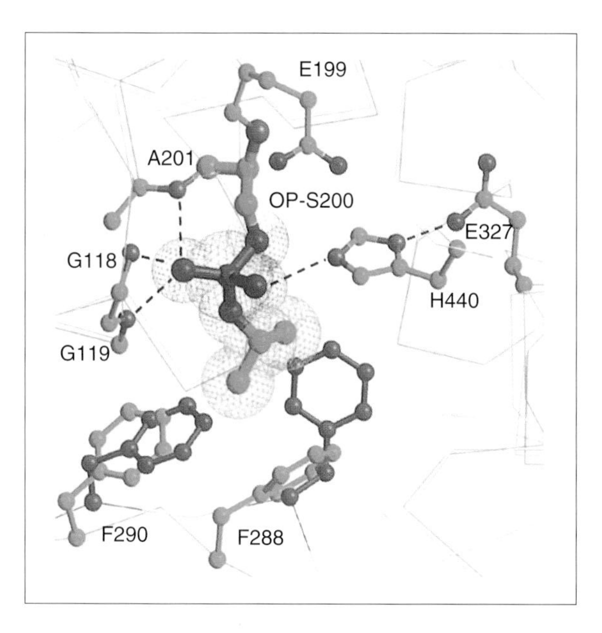

Figure 2.3
Active site region of the 3D structure of monoisopropylphosphoryl–TcAChE, the 'aged' conjugate obtained by reaction of DFP with TcAChe. Note the salt bridge with His440, and the significant movements in the positions of residues Phe288 and Phe290 in the OP conjugate (green) relative to their positions in the native enzyme (purple). (Adapted from Millard and co-workers.[65])

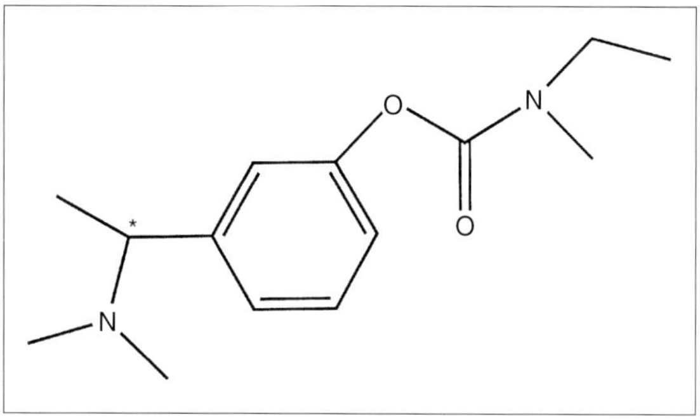

Figure 2.4
Chemical formula of ENA-713.

Conjugate with ENA713 (Exelon)

The anti-Alzheimer drug, (+)(*S*)-*N*-ethyl-3-[(1-dimethylamino)ethyl]-*N*-methyl-phenylcarbamate (ENA713; Exelon; Fig. 2.4), belongs to a series of miotine derivatives all displaying inhibitory action towards AChE both in vitro and in vivo.[70] Compared to other clinically useful carbamates, it has a longer duration of action in vivo, and preferentially inhibits AChE of the hippocampus and cortex.[71] Kinetic studies revealed slow inhibition and a significant irreversible component in its interaction with *Tc*AChe,[66] whereas the analogue in which the bulky leaving group, 3-[(1-dimethylamino)ethyl]phenol (NAP), had been replaced by chloride inhibited more rapidly and yielded a conjugate that reactivated fully and rapidly upon dilution. The crystal structure revealed that the active-site serine (Ser200) was ethylmethylcarbamylated, and that the leaving group, NAP, remained bound non-covalently in the active site. It was concluded that ENA713 can inhibit both by decarbamylation and by its action as a pro-drug to deliver the leaving group, NAP, which is itself a good reversible inhibitor of AChE.

Conjugate with the physostigmine analogue MF268

8-(*cis*-2,6-Dimethylmorpholino)octylcarbamoyleseroline (MF268) is a long-chain analogue of physostigmine. Whereas AChE inhibited by physostigmine is reactivated quite rapidly, long-chain physostigmine analogues form much more stable conjugates, and

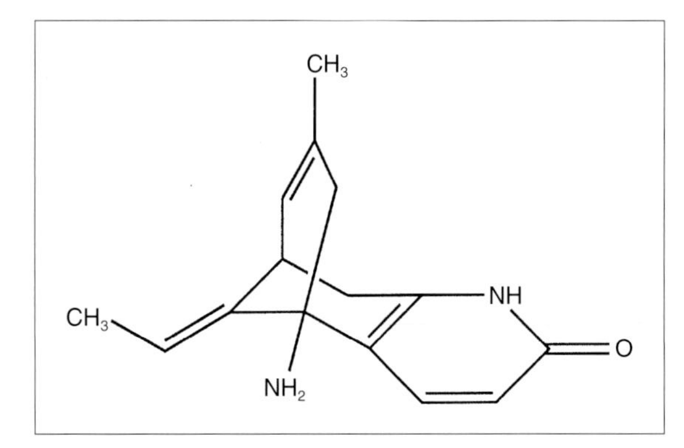

Figure 2.5
Chemical formula of HupA.

MF268 behaves in vitro as an almost irreversible inhibitor.[72] Bartolucci *et al.*[52] used X-ray crystallography to investigate this phenomenon by solving the crystal structure of the conjugate of MF268 with *Tc*AChe. In the crystal structure, the dimethylmorpholinooctylcarbamic moiety of MF268 is covalently bound to the active-site serine (Ser200). The alkyl group of the inhibitor fills the upper part of the gorge, thus blocking the entrance to the active site. This prevents the leaving group (eseroline) from exiting by this route. Surprisingly, however, the bulky eseroline moiety is not seen in the crystal structure, implying the existence of an alternative route, i.e. a 'back door', for its clearance.

Complex with huperzine A

(−)-Huperzine A (HupA; Fig. 2.5) is a nootropic alkaloid which has been used in China for centuries as a folk medicine.[73] It is a potent reversible inhibitor of AChE, and is undergoing clinical trials in China for treatment of Alzheimer's disease.[74] The structure of HupA reveals no obvious similarity to that of ACh. In fact, a number of studies, utilizing either computerized docking techniques and/or site-directed mutagenesis,[75–77] predicted various possible orientations for HupA within the active site of AChE. It seemed, therefore, desirable to solve the crystal structure of the HupA–*Tc*AChe complex, thus establishing its correct orientation, and providing the basis for future structure-based drug design.

The crystal structure of the HupA–*Tc*AChe complex (Fig. 2.6) showed an unexpected orientation for the inhibitor, with surprisingly few strong direct interactions with the protein to explain its high affinity.[22] Even though HupA has three potential hydrogen-bond donor and acceptor sites, only one strong hydrogen bond is seen, with Tyr130. The high affinity may be ascribed to the cumulative effect of a large number of hydrophobic contacts and of hydrogen bonds with water molecules within the gorge which are, themselves, hydrogen bonded to other water molecules or to backbone or side-chain atoms of the protein. Modelling Phe330 as tyrosine, the corresponding residue in hAChE, permits formation of a 3.3 Å hydrogen atom between the hydroxyl oxygen atom and the primary amino group of HupA. This additional hydrogen bond, in addition to π–cation interactions with both Trp84 and Phe(Tyr)330, may explain why HupA binds to hAChE five- to ten-fold more strongly than to *Tc*AChe, and only weakly to BuChE, which lacks an aromatic residue at this position.[75] A consequence of the large number of contacts mentioned above is that there does not appear to be much room for additional substituents without causing clashes. Nevertheless, addition of a methyl group near the amide group of HupA leads to an eight-fold increase in affinity, probably due to additional hydrophobic contacts with W84.[78]

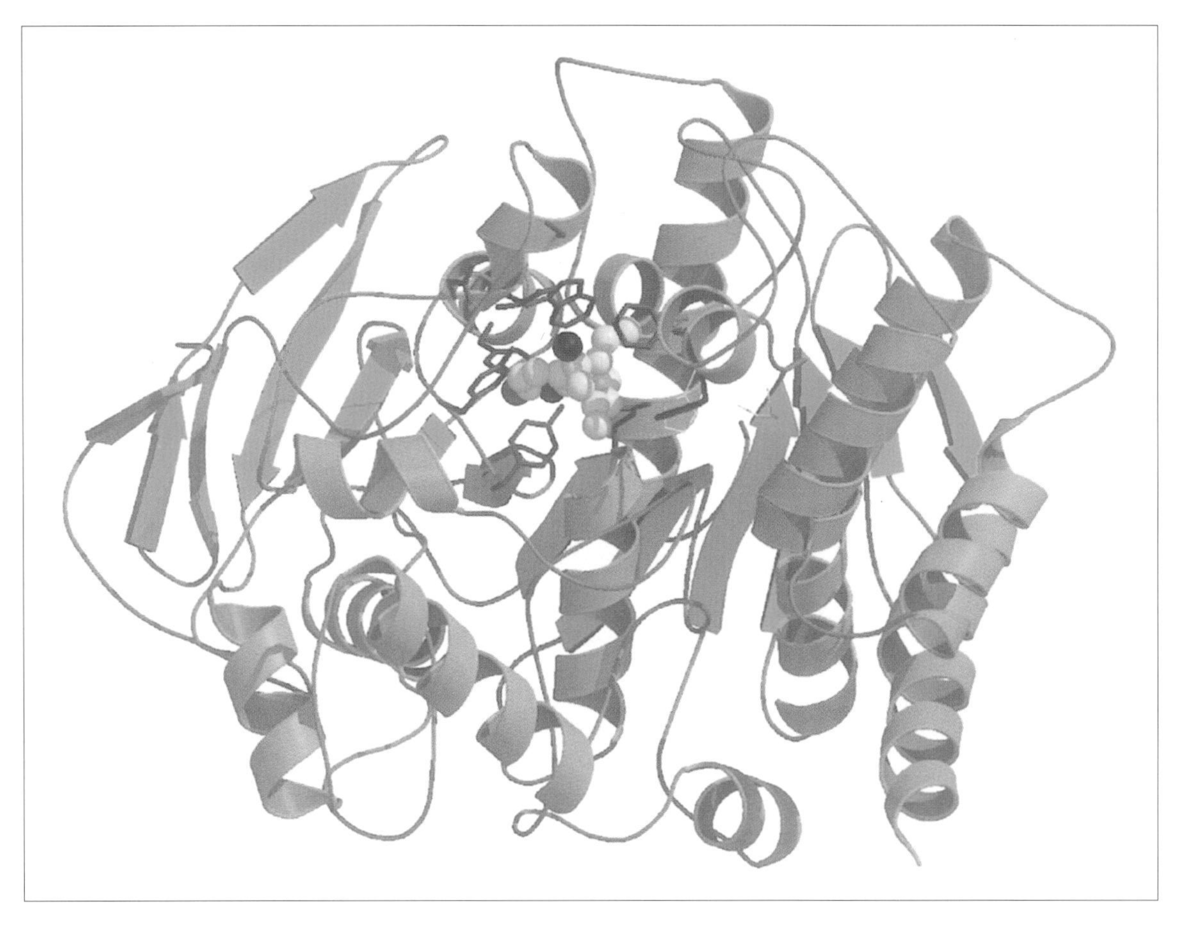

Figure 2.6
Ribbon diagram of the crystal structure of the HupA–TcAChE complex. The representation is as in Fig. 2.1, but the space-filling model of ACh is replaced by that of HupA.

Complex with E2020 (Aricept)

E2020, (*R*,*S*)-1-benzyl-4-[5,6-dimethoxy-1-indanon)-2-yl]methylpiperidine (Aricept; Fig. 2.7), is a member of a large family of *N*-benzylpiperidine based AChE inhibitors, developed, synthesized and evaluated by the Eisai Company in Japan.[79] It was approved for treatment of Alzheimer's disease by the US Food and Drug Administration in 1996,[80] and shows high selectivity for AChE relative to BuChE.[81] The earlier modelling studies attributed this differential specificity to differences in geometry within the active sites of AChE and BuChE.[82] However, more recent studies, carried out subsequent to determination of the 3D structure of *Tc*AChe, suggested that E2020 and other Eisai compounds orient along the

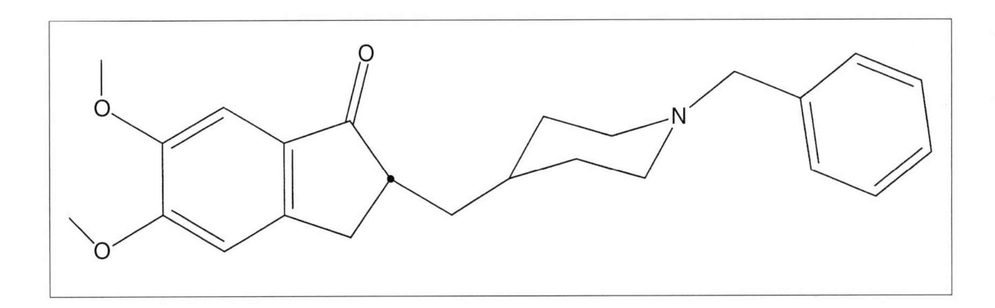

Figure 2.7
Chemical formula of E2020.

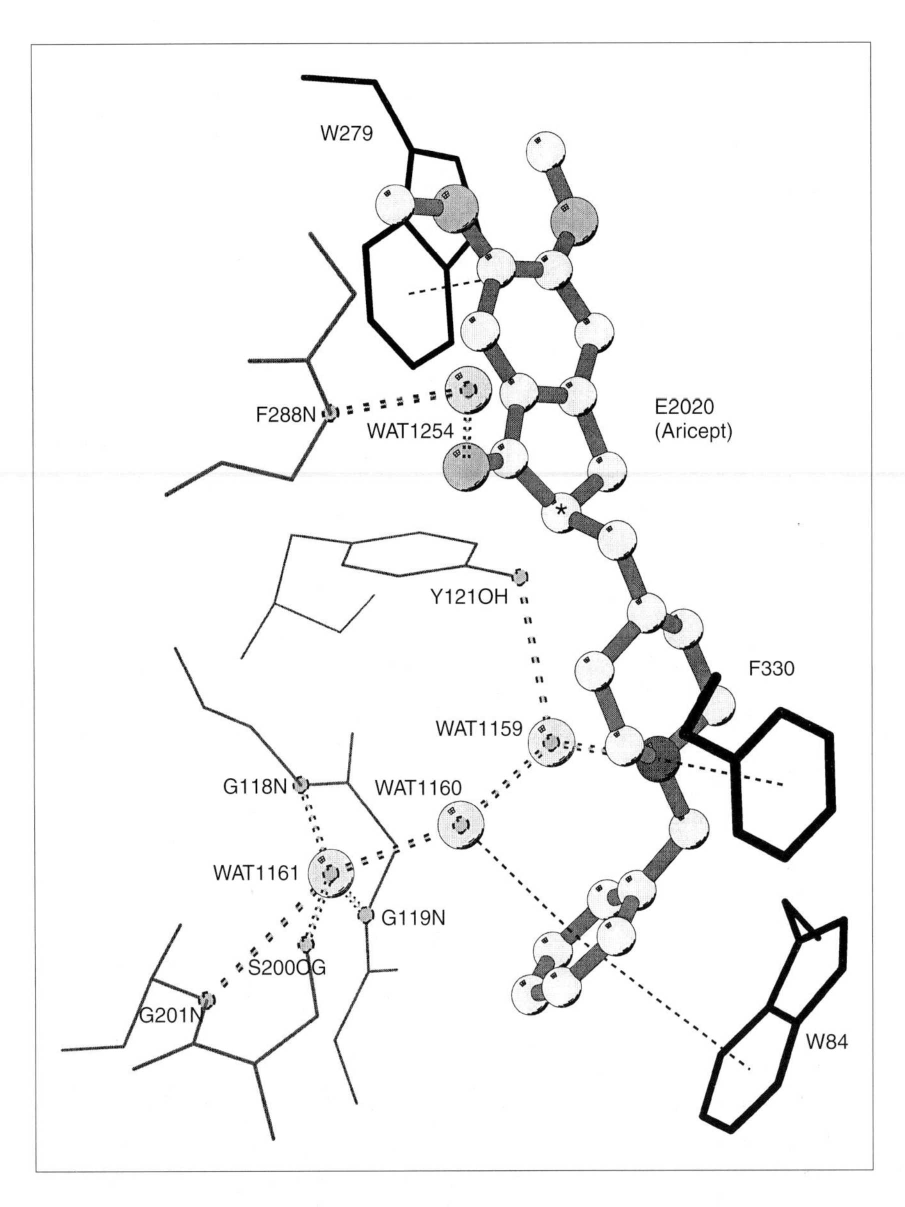

Figure 2.8
*Binding modes of E2020 to TcAChE. E2020 is displayed as a ball-and-stick model (*chiral centre). Water molecules are represented as light grey balls; 'standard' hydrogen-bonds as heavy dashed lines; aromatic hydrogen-bonds, π–cation and stacking interactions as light dashed lines. Note the multiple water-mediated contacts of E2020 atoms with the protein.*

active-site gorge, and that the differential specificity can be attributed to structural differences between AChE and BuChE at the top of the gorge, at the 'peripheral' anionic site.[79,83] The 3D structure of the E2020–*Tc*AChe complex fully confirms these latter assignments (Fig. 2.8).[23,84] It can be seen that E2020 interacts with both the 'anionic' subsite of the active site, at the bottom of the gorge, and with the 'peripheral' anionic site, near its top, via aromatic stacking interactions with conserved aromatic residues. The piperidine nitrogen atom is also involved in a cation–π-electron interaction with the phenyl ring of Phe330. E2020 does not, however, interact directly with either the catalytic triad or the 'oxyanion' hole, but only indirectly, via solvent molecules. These loci, and a finger-shaped void towards the acyl pocket, provide spaces into which substituents could fit, thus yielding analogues of E2020 with increased affinity and selectivity.

Acknowledgements

This work was supported by the US Army Medical Research Acquisition Activity under Contract No. DAMD17-97-2-7022, the European Union IVth Framework in Biotechnology, the Kimmelman Center for Biomolecular Structure and Assembly, Israel, the Nella and Leon Benoziyo Center for Neurosciences, and the generous support of Tania Friedman. I.S. is Bernstein–Mason Professor of Neurochemistry.

References

1. Sussman JL, Harel M, Frolow F *et al.* Atomic structure of acetylcholinesterase from *Torpedo californica*: a prototypic acetylcholine-binding protein. *Science* 1991; **253:** 872–879.
2. Silman I, Sussman JL. Structural and functional studies on acetylcholinesterase: a perspective. In: Doctor BP, Taylor P, Quinn DM, Rotundo RL, Gentry MK, eds. *Structure and Function of Cholinesterases and Related Proteins*. New York: Plenum, 1998: 25–34.
3. Ollis DL, Cheah E, Cygler M *et al.* The α/β-hydrolase fold. *Protein Eng* 1992; **5:** 197–211.
4. Cygler M, Schrag JD, Sussman JL *et al.* Relationship between sequence conservation and three-dimensional structure in a large family of esterases, lipases, and related proteins. *Protein Sci* 1993; **2:** 366–382.
5. Cousin X, Hotelier T, Lievin P, Toutant J-P, Chatonnet A. A CHOLINESTERASE GENES SERVER (ESTHER): a data base of cholinesterase-related sequences for multiple alignments, phylogenetic relationships, mutations and structural data retrieval. *Nucleic Acids Res* 1996; **24:** 132–136.
6. Botti SA, Felder CE, Sussman JL, Silman I. Electrotactins: a class of adhesion proteins with conserved electrostatic and structural motifs. *Protein Eng* 1998; **11:** 415–420.
7. Harel M, Schalk I, Ehret-Sabatier L *et al.* Quaternary ligand binding to aromatic residues in the active-site gorge of acetylcholinesterase. *Proc Natl Acad Sci USA* 1993; **90:** 9031–9035.
8. Fisher A, Hanin I, Yoshida M. *Progress in Alzheimer's Disease and Parkinson's Diseases*. New York: Plenum Press, 1998.
9. Millard CB, Broomfield CA. Anticholinesterases: medical applications of neurochemical principles. *J Neurochem* 1995; **64:** 1909–1918.
10. Casida JE, Quistad GB. Golden age of insecticide research: past, present, or future? *Annu Rev Entomol* 1998; **43:** 1–16.
11. Kryger G, Giles K, Harel M *et al.* 3D structure at 2.7 Å resolution of native and E202Q mutant human acetylcholinesterase complexed with fasciculin-II. In: Doctor BP, Taylor P, Quinn DM, Rotundo RL, Gentry MK, eds. *Structure and Function of Cholinesterases and Related Proteins*. New York: Plenum, 1998: 323–326.
12. Harel M, Kryger G, Sussman JL *et al.* 3D structures of *Drosophila* acetylcholinesterase: implications for insecticide design. *Abstracts IUCr XVIII World Crystallography Congress, Glasgow, 1999*. 09.04.023.
13. Sussman JL, Harel M, Frolow F *et al.* Purifica-

tion and crystallization of a dimeric form of acetylcholinesterase from *Torpedo californica* subsequent to solubilization with phosphatidylinositol-specific phospholipase C. *J Mol Biol* 1988; **203:** 821–823.
14. Axelsen PH, Harel M, Silman I, Sussman JL. Structure and dynamics of the active site gorge of acetylcholinesterase: synergistic use of molecular dynamics simulation and X-ray crystallography. *Protein Sci* 1994; **3:** 188–197.
15. Nolte H-J, Rosenberry TL, Neumann E. Effective charge on acetylcholinesterase active sites determined from the ionic strength dependence of association rate constants with cationic ligands. *Biochemistry* 1980; **19:** 3705–3711.
16. Weise C, Kreienkamp H-J, Raba R, Pedak A, Aaviksaar A, Hucho F. Anionic subsites of the acetylcholinesterase from *Torpedo californica*: affinity labelling with the cationic reagent *N,N*-dimethyl-2-phenylaziridinium. *EMBO J* 1990; **9:** 3885–3888.
17. Dougherty DA, Stauffer DA. Acetylcholine binding by a synthetic receptor: implications for biological recognition. *Science* 1990; **250:** 1558–1560.
18. Dougherty D. Cation–π interactions in chemistry and biology: a new view of benzene, phe, tyr, and trp. *Science* 1996; **271:** 163–168.
19. Krejci E, Duval N, Chatonnet A, Vincens P, Massoulié J. Cholinesterase-like domains in enzymes and structural proteins: functional and evolutionary relationships and identification of a catalytically essential aspartic acid. *Proc Natl Acad Sci USA* 1991; **88:** 6647–6651.
20. Harel M, Kleywegt GJ, Ravelli RBG, Silman I, Sussman JL. Crystal structure of an acetylcholinesterase–fasciculin complex: interaction of a three-fingered toxin from snake venom with its target. *Structure* 1995; **3:** 1355–1366.
21. Harel M, Quinn DM, Nair HK, Silman I, Sussman JL. The X-ray structure of a transition state analog complex reveals the molecular origins of the catalytic power and substrate specificity of acetylcholinesterase. *J Am Chem Soc* 1996; **118:** 2340–2346.
22. Raves ML, Harel M, Pang Y-P, Silman I, Kozikowski AP, Sussman JL. 3D structure of acetylcholinesterase complexed with the nootropic alkaloid (−)-huperzine A. *Nature Struct Biol* 1997; **4:** 57–63.
23. Kryger G, Silman I, Sussman JL. Structure of acetylcholinesterase complexed with E2020 (Aricept®): implications for the design of new anti-Alzheimer drugs. *Structure* 1999; **7:** 297–307.
24. Doctor BP, Taylor P, Quinn DM, Rotundo RL, Gentry MK, eds. *Structure and Function of Cholinesterases and Related Proteins.* New York: Plenum, 1998.
25. Harel M, Sussman JL, Krejci E *et al.* Conversion of acetylcholinesterase to butyrylcholinesterase: modeling and mutagenesis. *Proc Natl Acad Sci USA* 1992; **89:** 10 827–10 831.
26. Radic Z, Durán R, Vellom DC, Li Y, Cerveñansky C, Taylor P. Site of fasciculin interaction with acetylcholinesterase. *J Biol Chem* 1994; **269:** 11 233–11 239.
27. Bourne Y, Taylor P, Marchot P. Acetylcholinesterase inhibition by fasciculin: crystal structure of the complex. *Cell* 1995; **83:** 503–512.
28. Endo T, Tamiya N. Structure–function relationships of postsynaptic neurotoxins from snake venoms. In: Harvey AL, ed. *Snake Toxins.* New York: Pergamon, 1991: 165–222.
29. Ordentlich A, Barak D, Kronman C *et al.* Dissection of the human acetylcholinesterase active center — determinants of substrate specificity — identification of residues constituting the anionic site, the hydrophobic site, and the acyl pocket. *J Biol Chem* 1993; **268:** 17 083–17 095.
30. Shafferman A, Ordentlich A, Barak D *et al.* Electrostatic attraction by surface charge does not contribute to the catalytic efficiency of acetylcholinesterase. *EMBO J* 1994; **13:** 3448–3455.
31. Taylor P, Radic Z. The cholinesterases: from genes to proteins. *Annu Rev Pharmacol Toxicol* 1994; **34:** 281–320.
32. Vellom DC, Radic Z, Li Y, Pickering NA, Camp S, Taylor P. Amino acid residues controlling acetylcholinesterase and butyrylcholinesterase specificity. *Biochemistry* 1993; **32:** 12–17.
33. Radic Z, Pickering NA, Vellom DC, Camp S, Taylor P. Three distinct domains in the cholinesterase molecule confer selectivity for acetyl- and butyrylcholinesterase inhibitors.

Biochemistry 1993; **32:** 12 074–12 084.
34. Massoulié J, Pezzementi L, Bon S, Krejci E, Vallette F-M. Molecular and cellular biology of cholinesterases. *Prog Neurobiol* 1993; **14:** 31–91.
35. Massoulié J, Anselmet A, Bon S *et al.* Acetylcholinesterase: C-terminal domains, molecular forms and functional localization. *J Physiol (Paris)* 1998; **92:** 183–190.
36. Silman I, Futerman AH. Modes of attachment of acetylcholinesterase to the surface membrane. *Eur J Biochem* 1987; **170:** 11–22.
37. Schrag J, Schmid MF, Morgan DG, Phillips GN Jr, Chiu W, Tang L. Crystallization and preliminary X-ray diffraction analysis of 11(*S*)-acetylcholinesterase. *J Biol Chem* 1988; **263:** 9795–9800.
38. Raves M, Giles K, Schrag JD *et al.* Quaternary structure of tetrameric acetylcholinesterase. In: Doctor BP, Taylor P, Quinn DM, Rotundo RL, Gentry MK, eds. *Structure and Function of Cholinesterases and Related Proteins.* New York: Plenum, 1998: 351–356.
39. Giles K. Interactions underlying subunit association in cholinesterases. *Protein Eng* 1997; **10:** 677–685.
40. Simon S, Krejci E, Massoulié J. A four-to-one association between peptide motifs: four C-terminal domains from cholinesterase assemble with one proline-rich attachment domain (PRAD) in the secretory pathway. *EMBO J* 1998; **17:** 6178–6187.
41. Ripoll DR, Faerman CH, Axelsen P, Silman I, Sussman JL. An electrostatic mechanism for substrate guidance down the aromatic gorge of acetylcholinesterase. *Proc Natl Acad Sci USA* 1993; **90:** 5128–5132.
42. Tan RC, Truong TN, McCammon JA, Sussman JL. Acetylcholinesterase: electrostatic steering increases the rate of ligand binding. *Biochemistry* 1993; **32:** 401–403.
43. Felder CE, Silman I, Lifson S, Botti SA, Sussman JL. External and internal electrostatic potentials of cholinesterase models. *J Mol Graph Model* 1997; **15:** 318–327.
44. Pörschke D, Créminon C, Cousin X, Bon C, Sussman JL, Silman I. Electrooptical measurements demonstrate a large permanent dipole moment associated with acetylcholinesterase. *Biophys J* 1996; **70:** 1603–1608.
45. Shafferman A, Ordentlich A, Barak D, Kronman C, Ariel N, Velan B. Contribution of the active center functional architecture to AChE reactivity toward substrates and inhibitors. In: Doctor BP, Taylor P, Quinn DM, Rotundo RL, Gentry MK, eds. *Structure and Function of Cholinesterases and Related Proteins.* New York: Plenum, 1998: 203–209.
46. McCammon JA, Wlodak S, Clark T, Kirchhoff P, Scott LR, Tara S. Computer simulation studies of acetylcholinesterase dynamics and activity. In: Doctor BP, Taylor P, Quinn DM, Rotundo RL, Gentry MK, eds. *Structure and Function of Cholinesterases and Related Proteins.* New York: Plenum, 1998: 327–329.
47. Botti SA, Felder C, Lifson S, Silman I, Sussman JL. A modular treatment of molecular traffic through the active site of cholinesterases. *Biophys J* 1999; **77:** in press.
48. Gilson MK, Straatsma TP, McCammon JA *et al.* Open 'back door' in a molecular dynamics simulation of acetylcholinesterase. *Science* 1994; **263:** 1276–1278.
49. Wlodek ST, Clark TW, Scott LR, McCammon JA. Molecular dynamics of acetylcholinesterase dimer complexed with tacrine. *J Am Chem Soc* 1997; **119:** 9513–9522.
50. Kronman C, Ordentlich A, Barak D, Velan B, Shafferman A. The 'back door' hypothesis for product clearance in acetylcholinesterase challenged by site-directed mutagenesis. *J Biol Chem* 1994; **269:** 27 819–27 822.
51. Faerman C, Ripoll D, Bon S *et al.* Site-directed mutants designed to test back-door hypotheses of acetylcholinesterase function. *FEBS Lett* 1996; **386:** 65–71.
52. Bartolucci C, Perola E, Cellai L, Brufani M, Lamba D. 'Back door' opening implied by the crystal structure of a carbamoylated acetylcholinesterase. *Biochemistry* 1999; **38:** 5714–5719.
53. Ravelli RBG, Raves ML, Ren Z *et al.* Static Laue diffraction studies on acetylcholinesterase. *Acta Crystallogr, Part D* 1998; **54:** 1359–1366.
54. Peng L, Silman I, Sussman JL, Goeldner M. Biochemical evaluation of photolabile precursors of choline and carbamoylcholine for potential time-resolved crystallographic studies on cholinesterases. *Biochemistry* 1996; **35:** 10 854–10 861.

55. Layer PG. Comparative localization of acetylcholinesterase and pseudocholinesterase during morphogenesis of the chicken brain. *Proc Natl Acad Sci USA* 1983; **80:** 6413–6417.
56. Greenfield S. Acetylcholinesterase may have novel functions in the brain. *TINS* 1984; **7:** 364–368.
57. Layer PG, Rommel S, Bulthoff H, Hengstenberg R. Independent spatial waves of biochemical differentiation along the surface of chicken brain as revealed by the sequential expression of acetylcholinesterase. *Cell Tissue Res* 1988; **251:** 587–595.
58. Barthalay Y, Hipeau-Jacquotte R, de la Escalera S, Jimenez F, Piovant M. *Drosophila* neurotactin mediates heterophilic cell adhesion. *EMBO J* 1990; **9:** 3603–3609.
59. Olson PF, Fessler LI, Nelson RE, Sterne RE, Campbell AG, Fessler JH. Glutactin, a novel *Drosophila* basement membrane-related glycoprotein with sequence similarity to serine esterases. *EMBO J* 1990; **9:** 1219–1227.
60. Auld VJ, Fetter RD, Broadie K, Goodman CS. Gliotactin, a novel transmembrane protein on peripheral glia, is required to form the blood–nerve barrier in *Drosophila*. *Cell* 1995; **81:** 757–767.
61. Ichtchenko K, Hata Y, Nguyen T *et al.* Neuroligin 1: a splice site-specific ligand for β-neurexins. *Cell* 1995; **81:** 435–443.
62. Darboux I, Barthalay Y, Piovant M, Hipeau-Jacquotte R. The structure–function relationships in *Drosophila* neurotactin show that cholinesterasic domains may have adhesive properties. *EMBO J* 1996; **15:** 4835–4843.
63. Aldridge WN, Reiner E. *Enzyme Inhibitors as Substrates: Interactions of Esterases with Esters of Organophosphorus and Carbamic Acids*. Amsterdam: North-Holland, 1972.
64. Silman I, Millard CB, Ordentlich A *et al.* A preliminary comparison of structural models for catalytic intermediates of acetylcholinesterase. *Chem Biol Interactions* 1999; **119–120:** 43–52.
65. Millard C, Kryger G, Ordentlich A *et al.* Crystal structure of 'aged' phosphylated *Torpedo* acetylcholinesterase: nerve agent reaction products at the atomic level. *Biochemistry* 1999; **38:** 7032–7039.
66. Bar-On P, Harel M, Millard CB, Enz A, Sussman JL, Silman I. Kinetic and structural studies on the interaction of the anti-Alzheimer drug, ENA-713, with *Torpedo californica* acetylcholinesterase. *J Physiol (Paris)* 1998; **92:** 406–407.
67. Berends F, Posthumus CH, Sluys IVD, Deierkauf FA. The chemical basis of the 'ageing process' of DFP-inhibited pseudocholinesterase. *Biochim Biophys Acta* 1959; **34:** 576–578.
68. Loomis TA, Salafsky B. Antidotal action of pyridinium oximes in anticholinesterase poisoning, comparative effects of soman, sarin and neostigmine on neuromuscular function. *Toxicol Appl Pharmacol* 1963; **5:** 685–690.
69. Ashani Y, Green BS. Are the organophosphorous inhibitors of acetylcholinesterase transition-state analogs? In: Green BS, Ashani Y, Chipman D, eds. *Chemical Approaches to Understanding Enzyme Catalysis: Biomimetic Chemistry and Transition State Analogs*. Amsterdam: Elsevier, 1981: 169–188.
70. Weinstock M, Razin M, Chorev M, Tashma Z. Pharmacological activity of novel anticholinesterase of potential use in the treatment of Alzheimer's disease. In: Fisher A, Hanin I, Lachman P, eds. *Advances in Behavioral Biology*. New York: Plenum, 1986: 539–549.
71. Enz A, Amstutz R, Boddeke H, Gmelin G, Malanowski J. Brain selective inhibition of acetylcholinesterase: a novel approach to therapy for Alzheimer's disease. *Prog Brain Res* 1993; **98:** 431–438.
72. Perola E, Cellai L, Lamba D, Filocamo L, Brufani M. Long chain analogs of physostigmine as potential drugs for Alzheimer's disease: new insights into the mechanism of action in the inhibition of acetylcholinesterase. *Biochim Biophys Acta* 1997; **1343:** 41–50.
73. Liu J-S, Zhu Y-L, Yu C-M *et al.* The structures of huperzine A and B, two new alkaloids exhibiting marked anticholinesterase activity. *Can J Chem* 1986; **64:** 837–839.
74. Zhang RW, Tang XC, Han YY *et al.* Drug evaluation of huperzine A in the treatment of senile memory disorders. *Acta Pharmacol Sinica* 1991; **12:** 250–252.
75. Ashani Y, Grunwald J, Kronman C, Velan B, Shafferman A. Role of tyrosine 337 in the binding of huperzine A to the active site of

human acetylcholinesterase. *Mol Pharmacol* 1994; **45:** 555–560.
76. Pang Y-P, Kozikowski A. Prediction of the binding sites of huperzine A in acetylcholinesterase by docking studies. *J Comput Aided Mol Des* 1994; **8:** 669–681.
77. Saxena A, Qian N, Kovach IM *et al.* Identification of amino acid residues involved in the binding of huperzine A to cholinesterases. *Protein Sci* 1994; **3:** 1770–1778.
78. Kozikowski AP, Campiani G, Sun L-Q *et al.* Identification of a more potent analogue of the naturally occurring alkaloid huperzine A; predictive molecular modeling of its interaction with acetylcholinesterase. *J Am Chem Soc* 1996; **118:** 11 357–11 362.
79. Kawakami Y, Inoue A, Kawai T, Wakita M, Sugimoto H, Hopfinger AJ. The rationale for E2020 as a potent acetylcholinesterase inhibitor. *Bioorg Med Chem* 1996; **4:** 1429–1446.
80. Nightingale SL. Donepezil approved for treatment of Alzheimer's disease. *JAMA* 1997; **277:** 10.
81. Sugimoto H, Iimura Y, Yamanishi Y, Yamatsu K. Synthesis and structure–activity relationships of acetylcholinesterase inhibitors: 1-benzyl-4-[(5,6-dimethoxy-1-oxoindan-2-yl)methyl] piperidine hydrochloride and related compounds. *J Med Chem* 1995; **38:** 4821–4829.
82. Cardozo MG, Kawai T, Imura Y, Sugimoto H, Yamanishi Y, Hopfinger AJ. Conformational analyses and molecular-shape comparisons of a series of indanonebenzylpiperidine inhibitors of acetylcholinesterase. *J Med Chem* 1992; **35:** 590–601.
83. Pang Y-P, Kozikowski A. Prediction of the binding-site of 1-benzyl-4-[(5,6-dimethoxy-1-indanon-2-yl)methyl] piperidine in acetylcholinesterase by docking studics with the SYSDOC program. *J Comput Aided Mol Des* 1994; **8:** 683–693.
84. Kryger G, Silman I, Sussman JL. Three-dimensional structure of a complex of E2020 with acetylcholinesterase from *Torpedo californica. J Physiol (Paris)* 1998; **92:** 191–194.

3

Rational design of cholinesterase inhibitors

Mario Brufani, Luigi Filocamo

Physostigmine and its traditional synthetic analogues

Introduction

Physostigmine[1] is the natural product that has inspired the synthesis of many cholinesterase (ChE) inhibitors for more than 60 years. (−)-Physostigmine (**1**), or (3a*S*)-*cis*-1,2,3,3a,8,8a-hexahydro-1,3a,8-trimethylpyrrolo[2,3b] indol-5-ol methylcabamate (ester), also called eserine or physostol, is an alkaloid isolated from Calabar beans, the dried ripe seed of the vine *Physostigma venenosum*, a perennial plant of tropical West Africa. The same plant produces other alkaloids that are chemically correlated to physostigmine; among these, the alkaloid physovenine (**2**) has been studied extensively. The chemical and pharmacological properties of these alkaloids have recently been reviewed.[1,2]

Some pharmacological properties and the toxicity of the Calabar beans were known long before physostigmine isolation. They were used by native tribes in West Africa as a poison in trials for witchcraft, and were introduced into Europe in 1840 by Daniel, a British medicinal officer. The first pharmacological studies were performed in 1855 using an extract of the beans. The pure alkaloid was isolated in 1864 by Jobst and Hesse, and its chemical structure was determined by Stedman and Barger[3] in 1925. The structure was later confirmed by Julian and Pikl,[4] who carried out the first total synthesis of the alkaloid, and by Petcher and Pauling,[5] who solved its crystal structure.

Physostigmine has two chiral centres and its absolute configuration was determined by chemical degradation[6] and confirmed by X-ray analysis.[7] As an unnatural distomer, (+)-physostigmine has a pharmacological profile that is quite different from the natural eutomer, (−)-physostigmine. Several chiral total syntheses of the alkaloid have been realized[8–11] and some of them are suitable for industrial production.

The first clinical uses, the traditional pharmacology and the anticholinesterase (anti-ChE) activity of physostigmine and its synthetic analogues have been well described by Taylor.[12] Taylor classifies physostigmine and its analogues which carbamoylate acetylcholinesterase (AChE) as reversible inhibitors and organophosphorus anti-ChE agents as irreversible inhibitors. The classification reflects only a quantitative difference in the times of deacylation of AChE when it is methylcarbamoylated by physostigmine or phosphorylated by organophosphorus, in about 15–30 minutes for the former and several days for the latter. The in vitro reactivation times of AChE, inactivated by analogues of physostigmine with long alkyl chains on the carbamic group, such as heptylphysostigmine (17), were carefully analysed.[13] Half-lives of the order of days (11 days for heptylphosostigmine) were found. Therefore, quantitative differences exist between the AChE reactivation times of both carbamoylating agents and organophosphorus inhibitors, and they can be better classified as AChE pseudo-irreversible inhibitors.

Physostigmine was traditionally used as a drug for the treatment of glaucoma,[14] myasthenia gravis[15] and, more recently, for protection against organophosphate poisoning.[16] These and other possible uses, such as the treatment of atony of the smooth muscle of the intestinal tract and urinary bladder, all based on AChE inhibition,[17] are strongly conditioned by the high toxicity of physostigmine.

Synthetic AChE inhibitors correlated to physostigmine

The toxicity of physostigmine, its synthetic analogues and the organophosphates may be caused by both peripheral as well as central ChE inhibition. There have been few studies that have attempted to quantify the relative contribution of peripheral and central mechanisms to the lethality of physostigmine. In a relatively recent study, using centrally and peripherally acting antimuscarinic drugs to reduce the lethality of physostigmine, Janowsky *et al.*[18] found that, at least at high doses of physostigmine, the central mechanism appears to be of great importance in determining the toxicity of the alkaloid. Therefore, the first synthetic analogues of physostigmine, selected for their reduced toxicity, are compounds with an increased ratio of peripheral to central activity (relative to physostigmine). For basic substances, such as physostigmine, penetration of the blood–brain barrier (BBB) depends on lipophilicity and, above all, basicity, because the protonated charged form of the alkaloid crosses the BBB very slowly.

As a further confirmation of these statements, the synthetic analogues which have been introduced in clinical use differ from physostigmine, above all in their pharmacokinetic properties: they are either less lipophilic or more basic than physostigmine or are directly quaternary ammonium compounds. The first of them, miotine (**4**), maintains the *N*-methylcarbamoyl moiety and the basic function of physostigmine (**3**). The compound has the same inhibitory power of physostigmine, and was tested, and then discontinued, as a miotic agent.[19] Another compound, neostigmine,[20,21] (**5**) has a quaternary ammonium structure, is stable in solution and probably does not cross the BBB. It is used in the symptomatic treatment of myasthenia gravis, post-operative intestinal atony and urinary bladder atony. Pyridostigmine[22] (**6**) has the ammonium function incorporated into the pyridinium ring and has properties and uses similar to those of neostigmine. Distigmine[23] (**7**) and demecarium[24] (**8**), which can be considered dimers of pyridostigmine and neostigmine respectively, also have similar properties. The compound ambenonium[25] (**9**) has quite a different structure, but has no more similarity to physostigmine.

For many physostigmine synthetic analogues (miotine, neostigmine, pyridostigmine, distigmine and decametonium) the presence of a carbamoylating function in the molecule was considered necessary, based on the hypothesis that a pseudo-irreversible carbamoylation of the enzyme is the rational basis for AChE. The similarity between acetylcholine (ACh, **10**) and miotine, for example, led to the hypothesis that neostigmine is hydrolysed like ACh. However, the intermediate *N*-methylcarbamoyl ester bond is more stable than the corresponding acetyl ester bond formed in the reaction of ACh with AChE and is hydrolysed more slowly, resulting in a pseudo-irreversible inhibition of the enzyme (Scheme 3.1).

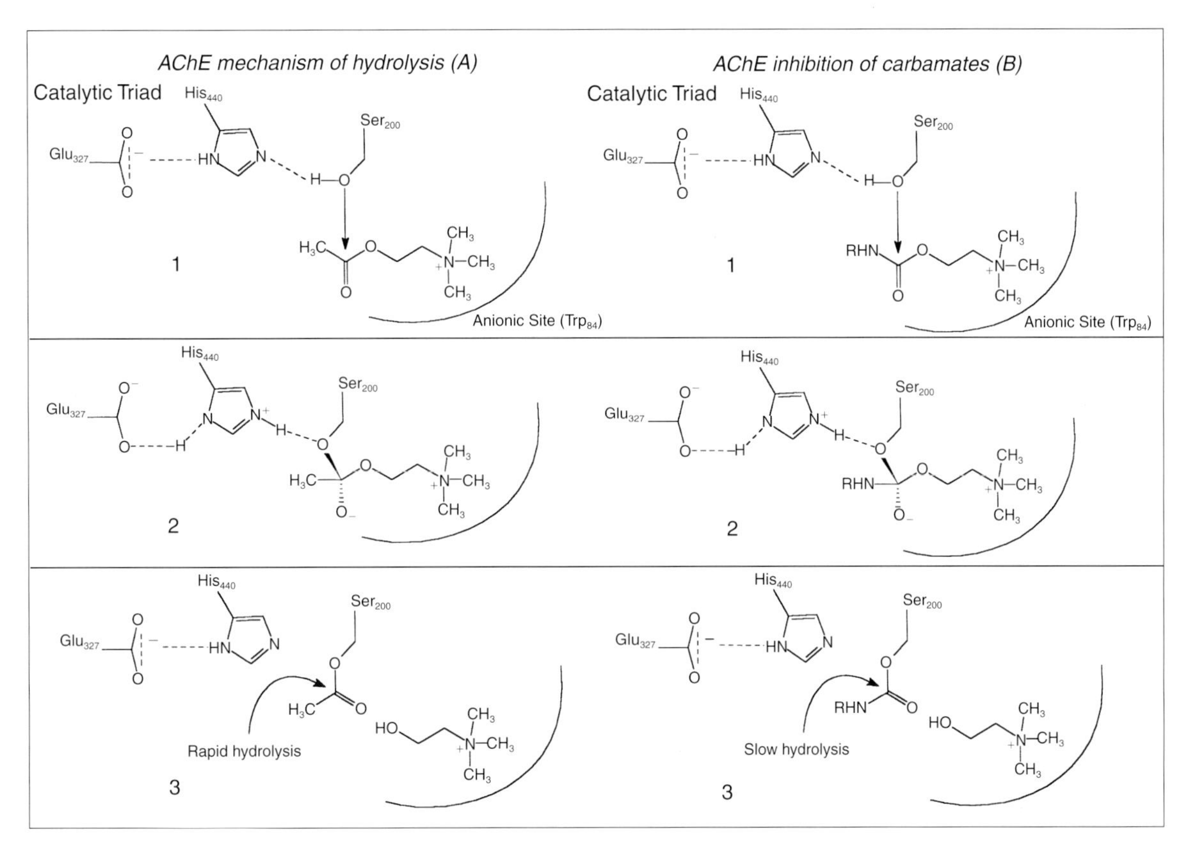

Scheme 3.1
The mechanism of AChE and the inhibition by carbamates.

Both ambenonium and, above all, edrophonium (**11**), a well-known AChE inhibitor which is used in anaesthesiology for the reversal of neuromuscular blockage,[26] as a diagnostic marker in myasthenia gravis[27] and in the treatment of oesophageal chest pain,[28] lack the carbamoylating function. Even though miotine, edrophonium and ambenonium are more or less directly derived from physostigmine, they belong to three different classes of AChE inhibitors. Miotine, like physostigmine, is a carbamoylating inhibitor, edrophonium is a reversible inhibitor which interacts non-covalently with the active site of the enzyme, and ambenonium is a reversible inhibitor that probably interacts, as do bis-quaternary ammonium compounds, with both the active site and the peripheral anionic site of AChE. The X-ray structure of AChE from *Torpedo californica* (*Tc*AChE), determined in 1991 by Sussman *et al.*[29] and described in Chapter 2, showed that AChE, like other serine hydrolases, contains a catalytic triad in the active site,[30] formed by Ser_{200}-His_{440}-Glu_{327}. Ser_{200}, activated by His_{440} and Glu_{327}, makes a nucleophilic attack on the carbonyl group of ACh (see Scheme 3.1, **1A**) and the quaternary transition state (see Scheme 3.1, **2A**) decomposes, resulting in the acetylenzyme (see Scheme 3.1, **3A**). Hydrolysis of the acetylester reactivates the enzyme.

When a carbamic ester is the substrate, the *N*-alkylcarbamoylenzyme (see Scheme 3.1, **3B**) hydrolyses slowly, resulting in a pseudo-irreversible inhibition. Edrophonium, as the X-ray structure clearly shows,[31] forms a complex with AChE in which the hydroxy group forms hydrogen bonds with His_{440} and Ser_{200} and the quaternary ammonium group interacts with the central anionic site of the enzyme.

Bis-quaternary ammonium compounds

Ambenonium is not a carbamoylating agent, because it has neither carbamic groups nor the hydroxy group of edrophonium, which are necessary for its complexation with the active site of AChE. It probably binds AChE as bis-quaternary ammonium compounds do by interacting with both the central and 'peripheral' anionic sites of the enzyme.

A large series of bis-quaternary ammonium salts has been prepared and evaluated for anti-AChE activity.[32–34] Many of them have the general formula **12**, and their AChE inhibitory power strongly depends on the length of the aliphatic chain which binds the two ammonium groups. The best activity is apparently obtained when $n = 6$.[34] The structure of the complex of decamethonium (**12**, $n = 10$) with *Tc*AChE has been solved[31] and it clearly shows that the two quaternary ammonium groups interact with the central and peripheral anionic sites. The ten carbon atom chain of decamethonium spans exactly the distance between the two anionic sites of the enzyme. Other derivatives have more complex structures and their pharmacological activity also depends on the structure and lipophilicity of the substituents on the quaternary nitrogen atoms. However the structure–activity relationships of these compounds are complicated by their direct action on ACh receptors, which is mainly responsible for the depolarizing neuromuscular blocking activity of these compounds.[35]

Simple quaternary ammonium compounds, such as tetramethylammonium salt, probably also inhibit AChE by interacting with both anionic sites of the enzyme; these compounds, however, have a short duration of action and minimal therapeutic utility.[36]

The structure–activity relationship among the synthetic analogues of physostigmine and related compounds

The X-ray structure of *Tc*AChE published in 1991[29] and the mutagenesis experiments performed directly on the human enzyme[37–40] confirmed many of the previously advanced hypotheses concerning the mechanism of ACh hydrolysis by AChE. The main difference was that the central anionic site proved not to be a carboxylic group, as previously postulated, but an 'aromatic gorge' formed by the rings of 14 conserved aromatic amino acids. One of these, Trp_{84} is considered essential for the interaction with the trimethylammonium group of ACh. A second anionic site, the so-called 'peripheral anionic site', is located at the entrance of the aromatic gorge where it allosterically modulates the catalytic activity of AChE by interacting with ACh. This peripheral anionic site is wider than the central one and contains at least five different residues (one Trp, three Tyr and one Glu) that could interact with either the ammonium group of ACh or its analogues. In the central anionic site, the presence of both a tryptophan residue, which can interact with the ammonium group of AChE inhibitors through a cation–π–electron contact,[47–50] and many aromatic residues, which can interact through π–π contacts broadens the structure–activity relationship of the inhibitors.

The distances of the amino or ammonium groups of miotine, neostigmine and pyridostigmine from the methyl or dimethylcarbamic groups are quite different, notwithstanding that all three compounds are potent AChE inhibitors. Physovenine, which lacks the amino group in position 1, is an AChE inhibitor, as is the 8-carbaphysostigmine analogue **13**, which has been prepared recently[51] and lacks the amino group in position 8. Therefore, the age-old question of whether it is the amino group in position 3 or the amino group in position 8 of physostigmine that simulates the trimethylammonium group of ACh was misdirected, since both amino groups can interact with the aromatic gorge of AChE. The peripheral anionic site is wider and the enzyme can therefore accommodate a large series of bifunctional compounds with two amino or ammonium groups, as bis-quaternary ammonium salts of different lengths.

The substitution of the physostigmine *N*-methylcarbamic group with *N*,*N*-dimethylalkylcarbamic groups does not prevent the interaction of inhibitors with the catalytic triad of AChE, because neostigmine, pyridostigmine and distigmine are all good AChE inhibitors. The nature of the alkyl carbamic groups dramatically influences the kinetics of AChE reactivation, a topic that is addressed below.

Interaction of the hydroxy group of edrophonium with the enzymatic catalytic triad is well documented by the X-ray structure of its complex with AChE,[31] while the action of the catalytic triad the acetic group of ACh and on the carbamic group of the inhibitors is supported by the design and synthesis of some transition-state analogue inhibitors of AChE.[52–56] The X-ray structure of one of these (*m*-(*N*,*N*,*N*-trimethylammonio)-2,2,2-trifluoroacetophenone (TMFA), **14**) complexed with *Tc*AChE, clearly shows that the inhibitor is covalently bonded to Ser_{200} of the enzyme.[57]

The pharmacological data reported for traditional synthetic analogues of physostigmine do not clearly distinguish between the influence of their structural features on enzymatic and pharmacokinetic activities, but the reduced toxicity of many ammonium salt inhibitors is probably due to their poor BBB penetration.

Anti-AChE organophosphorous insecticides

Among the most powerful anti-ChEs are phosphorus-containing compounds that are mainly derivatives of either orthophosphoric acid or phosphonic acid (Fig. 3.1). A wide variety of ester substituents is possible, including, but not limited to an alkyl, alkoxy, aryloxy, amido or mercaptan group bound directly to phosphorus. There is a leaving group that can be a halide, cyanide, thiocyanate, phenoxy, thiophenoxy or carboxylate group. Compounds in this category are potent and useful insecticides.

The structure–activity relationships of organophosphorous insecticides have been reviewed extensively.[58–61] The reaction between AChE and most organophosphorous inhibitors occurs only at the esteric site of the enzyme. The reaction is a transesterification, comparable to that involving the carbamate esters and ACh itself. The reaction is enhanced by the geometry of tetrahedral phosphates,

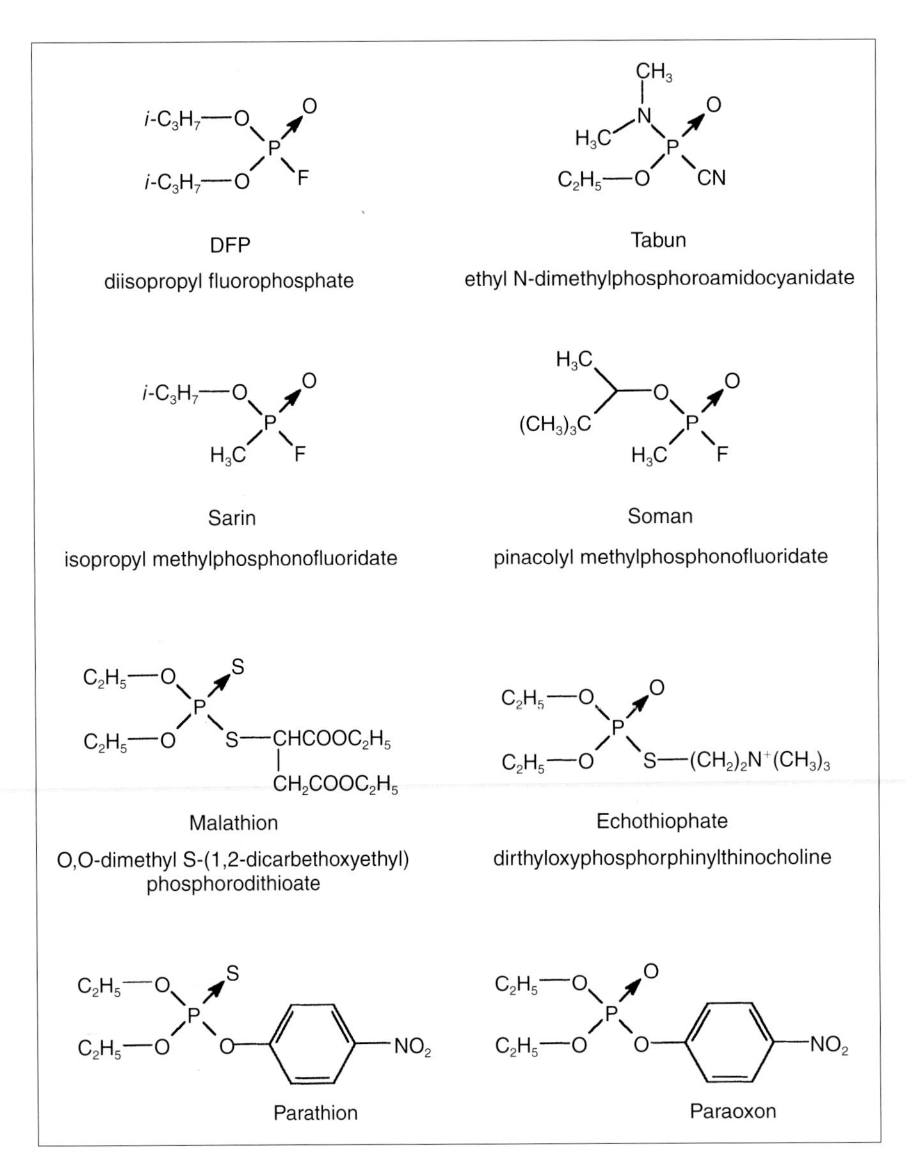

Figure 3.1 *Some phosphorus-containing AChE inhibitors.*

which resemble the transition state of acetyl ester hydrolysis, and the resultant phosphorylated or phosphonylated (phosphylated) enzyme is extremely stable.

Nucleophilic compounds such as fluoride[62–64] salts and some oximes[65–68] can reactivate organophosphate-inhibitred ChEs. Sometimes, phosphylated ChEs become progressively refractory to reactivation. In fact, organophosphoryl adducts can undergo dealkylation of an alkoxy group, which converts phosphylated ChEs into non-reactivable

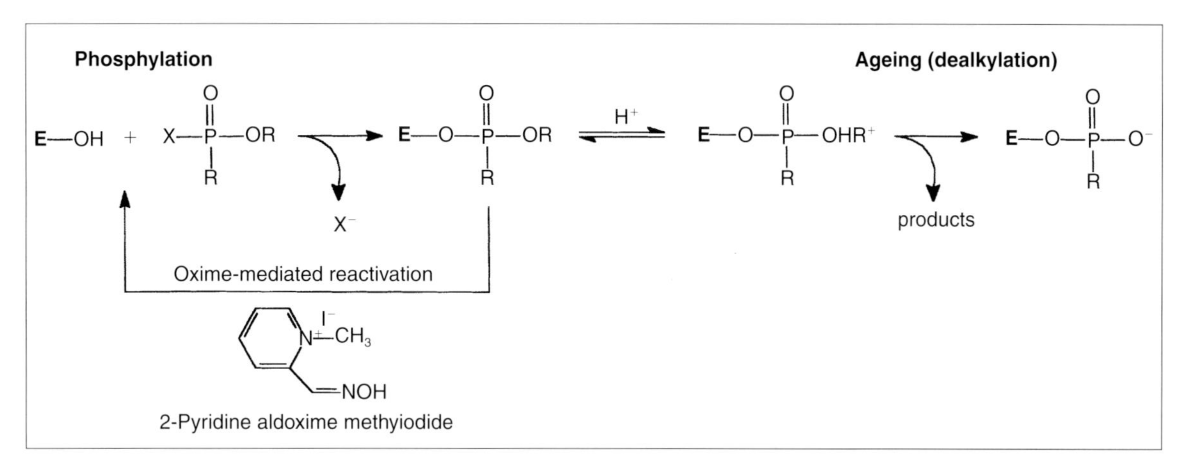

Figure 3.2
Steps in the interaction of an esterase with an organophosphorous inhibitor. The mechanism of ChE phosphylation and dealkylation (ageing) of the phosphylated enzyme.

species, referred to as 'aged'[69,70] (Fig. 3.2). This process is most pronounced for branched alkyl groups.[71] The role of the structure of the AChE active centre in facilitating reactions with organophosphate inhibitors was examined by a combination of site-directed mutagenesis and kinetic studies.[72,73] These studies identified some residues that interact directly with the ligands, suggesting that the functional architecture of the AChE active centre has a major role in the stabilization of the enzyme–phosphate complexes. The same active centre residues might be involved in accelerating the ageing process, and could explain the more efficient ageing of the phosphonyl conjugates of ChEs compared with the corresponding conjugates of other serine hydrolases.[74–76] Another important component is an arrangement of hydrogen-bond donors that can stabilize the tetrahedral transition enzyme–substrate complex through accommodation of the negatively charged carbonyl oxygen.[77] In the X-ray structure of the transition state analogue TMFA complexed with *Tc*AChE,[57] it is easy to note a three-pronged oxyanion hole formed by peptidic NH groups of Gly_{118}-Gly_{119}-Ala_{201}. The decrease in inhibitory activity toward the human enzyme mutated with alanine instead of the first glycine was very substantial for both phosphates and phosphonates (2000- to 6700-fold), irrespective of the size of the alkoxy substituents on the phosphorus atom. On the other hand, when the human AChE (hAChE) enzyme was mutated with alanine instead of the second glycine, the relative decline in reactivity to the phosphonates (500- to 460-fold) differed from the reactivity to the phosphates (12- to 95-fold). Although the formation of complexes with substrates does not seem to involve significant interaction with the oxyanion hole, interactions with this motif are a major stabilizing element in the accommodation of covalent inhibitors like organophosphates or carbamates.[78]

Diisopropylphosphorofluoridate (DFP) is probably the most thoroughly studied compound in this general class. Its high lipid solubility, low molecular weight and volatility facilitate inhalation, transdermal absorption and penetration into the cerebral nervous system. The 'nerve gases' (tabun, sarin and soman) are among the most potent synthetic toxic agents known; they are lethal in submilligram doses.[79]

Parathion (etilon, folidol and niran) is a widely used insecticide. It has a low volatility and is stable in water. Parathion itself does not inhibit AChE in vitro; paraoxon is the active metabolite.[80] Other insecticides possessing the phosphorothiolate structure are commonly employed in home, garden and agricultural uses. These insecticides include dimpylate (diazinon), fenthion and chlorpyrifos.

Malathion (chemathion and mala-spray) also requires the replacement of a sulphur atom in vivo. Phosphorotriester hydrolases and carboxylesterases can detoxify this insecticide. Phosphorotriester hydrolases (arylesterases and DFPases) catalyse the phosphorus–anhydride bond cleavage of the 'leaving group', a major route of detoxification, while carboxylesterases hydrolyse the carboethoxy group of malathion. Overall, the hydrolysis of insecticides increases the polarity of the metabolites and decreases their biological activity.[81,82] This reaction is much more rapid in mammals than in insects, giving rise to an additional degree of selective toxicity.[83]

Among the quaternary ammonium organophosphorous compounds, only echothiophate is clinically useful.[84] It is non-volatile and does not readily penetrate the skin.[85]

Semisynthetic and synthetic analogues of physostigmine and organophosphorous agents for the treatment of Alzheimer's disease

Synthesis and activity

About 20 years ago, when the cholinergic hypothesis of Alzheimer's disease (AD) was being advanced,[86,87] the pharmacokinetic requirements of the traditional AChE inhibitors were reversed. Their central action was considered necessary for AD treatment and their peripheral action was presumed to be responsible for their toxic or, at least negative, side-effects. Therefore the known synthetic analogues of physostigmine, above all the quaternary ammonium compounds, were unsuitable for AD treatment because of their poor BBB penetration. This led to the selection of physostigmine and tacrine (1,2,3,4-tetrahydro-5-aminoacridine, **15**). The latter had been previously known as a respiratory drug,[88] but its AChE inhibitory action was demonstrated in 1961.[89]

Physostigmine is a weak base with a pK_a of 7.9,[90] and therefore about a quarter of the alkaloid is in its unprotonated form at physiological pH (7.2). In addition, as the substance is relatively lipophilic, it can readily penetrate the BBB. However, absorption after oral administration, distribution and elimination of physostigmine and, therefore, the duration of ChE inhibition vary among different animal species and individuals. The ex vivo studies pursued in rats showed that physostigmine, after oral administration, produced a strong, but short-lived, inhibition of brain AChE.[91] The increases in rat brain ACh levels, determined through microdialysis techniques, were also high, but ephemeral. Despite these unfavourable pharmacokinetic and pharmacological properties, physostigmine was submitted to clinical trials for the palliative treatment of AD. The clinical trials gave unsatisfactory results.[92] In humans the drug showed a plasma half-life of 15–31 min and a plasma ChE inhibition half-life of 20–30 min, with high individual variability.[93–95] The cognitive improvements observed in the studies were inconsistent. Positive results were observed in some trials, but not in others.[96] The adverse effects observed included vomiting, diarrhoea, nausea, sweating and other cholinergic effects.

In order to increase the plasma half-life of physostigmine, a controlled-release oral tablet and systems of continuous infusion through an infusion pump[97] were submitted to clinical studies.[98] However, no preparation containing physostigmine has thus far been approved for the treatment of AD.

The limiting and adverse effects observed in the clinical studies of physostigmine clearly indicated the improvements that were necessary for the development of a synthetic or semi-synthetic analogue that could be used in the treatment of AD. The new inhibitors needed to have, in relation to physostigmine, an increased plasma half-life, an increased central/peripheral action ratio, a reduced peripheral cholinergic effect and a better and less variable oral adsorption.

The problem of AChE versus butirylcholinesterase (BuChE) inhibition selectivity was not taken into account during the preparation of the first semisynthetic physostigmine analogues (**16**), because BuChE inhibition was not considered a cause of toxicity at the time. An initial series of derivatives in which the methylcarbamic group of physostigmine was substituted by alkylcarbamil groups of increasing length was prepared. Among these, heptylphysostigmine (eptastigmine, Mediolanum Farmaceutici, **17**) was chosen as a compromise between AChE inhibitory power and the physicochemical properties hypothesized to be necessary for BBB penetration.[99–105] In fact, its inhibitory power decreases when n increases, and is low when $n > 6$ or the alkyl group is a bulky isopropyl or tert-butyl group.

Heptylphysostigmine is a potent inhibitor in vitro of both AChE and BuChE, with a fourfold preference for BuChE.[106,107] It produced, as hypothesized, a sustained brain AChE, an elevation in the extracellular level of rat brain ACh, good oral absorption, reduced toxicity, and markedly reduced peripheral cholinergic effects, all with respect to its precursor physostigmine. Moreover, it produced positive effects in animal models of memory and cognition.

Eptastigmine was subjected to clinical trials in the USA, Italy and some other European countries. Despite positive results,[108] the drug was withdrawn from these clinical studies because two patients in Phase III trials developed aplastic anaemia.[109] The causes of eptastigmine haematic toxicity are unknown, but are probably related to the presence of the heptyl chain, since physostigmine is free from this effect.

In order to reduce the lipophilicity of the eptastigmine alkyl chain and modify its pharmacokinetic and pharmacodynamic behaviours, several physostigmine derivatives were synthesized that carried amino groups of increasing lengths at the end of an alkyl chain (**18**).[110–112] Higher in vitro inhibitory activity was observed when n varied from 8 to 12 and when R contained a strongly basic group. However, brain AChE inhibition is incompatible with a strongly basic group. In fact, the highest inhibition of the brain enzyme was obtained with the compound MF268 (**19**), in which $n = 8$ and R is the weakly basic *cis*-2,6-dimethylmorpholino group.[19] Oral administration of the compound resulted in a substituted inhibition of rat-brain AChE and a significant increase in extracellular ACh cortex.[111] A kinetic study of the reactivation of electric-eel AChE inhibited by MF268 showed that this compound is a long-lasting inhibitor of the enzyme with a half-life of reactivation of several days, a property which should allow for a reduction in dosing frequency with respect to eptastigmine.[112].

Brossi and colleagues at the National Institutes of Health studied another semisynthetic analogue of physostigmine, the phenyl derivative phenserine (**20**).[113–116] Phenserine has an inhibitory power in vitro that is 50 times higher on AChE than on BuChE and can therefore be considered AChE selective. It readily enters the brain and has a half-life for rat brain AChE inhibition of about 8 h, after intravenous administration. Despite the long duration of its enzyme inhibition, phenserine is rapidly eliminated from the body with a plasma half-life of about 10 min. Its disappearance from the brain is also rapid, with a half-life similar to that of plasma. Phenserine gave positive results in the attenuation of a scopolamine-induced learning improvement (T-maze) in young and elderly rats, which showed an age-related decline in working memory.[117]

Many other derivatives and analogues of physostigmine have been prepared and tested as AChE inhibitors; the 8-carbaphysostigmine analogues prepared in the Pfizer laboratories are more potent AChE inhibitors in vitro and less toxic than physostigmine.[51] However none of these compounds has been submitted to clinical trials. A synthetic AChE inhibitor ENA713 (Rivastigmina, Exelon, **21**) has been prepared from miotine in the Sandoz laboratories.[118–121] Rivastigmina inhibits purified rat brain AChE in vitro with an apparent K_i value of 1.5 μM. This inhibition is about 1000 times weaker than that of physostigmine and 100 times weaker than that of eptastigmine. In spite of this, rivastigmina inhibits rat-brain AChE in vivo, with twice the potency of eptastigmina. Moreover, it has some degree of selectivity for the brain G_1 isoenzyme of AChE with respect to the G_4 isoenzyme. The reasons for the high activity of ENA713 in the brain are not clear. Novartis has completed the preclinical and clinical development of ENA713 and the drug has been launched in some countries with the new non-proprietary name rivastigmina.[122] Among the organophosphorus insecticides, metrifonate or (2,2,2-trichloro-1-hydroxyethyl)-phosphonic acid dimethyl ester (trichlorform, **22**) was chosen for clinical development. Metrifonate is a prodrug, which is non-enzymatically transformed into its active metabolite, 2,2-dichlorvinyldimethylphosphate (dichlorvos, DDVP, **23**).[123] DDVP is a potent irreversible AChE inhibitor.

Metrifonate was previously used as an insecticide[124] and an anthelmintic.[125] However, its relative toxicity, when compared to direct-acting organophosphates, suggested that it could be developed as a drug for the treatment of AD. Metrifonate, when administered orally to patients in a once-daily dose, readily enters the brain and inhibits AChE activity in a dose-dependent fashion.[126–128] In human blood in vitro, the half-life of metrifonate is about 60 min.[128–130] Hence, metrifonate and DDVP do not accumulate after in vivo administration and their long AChE inhibition depends on the slow enzyme reactivation, as it does for the carbamoylating inhibitors. Metrifonate, in spite of the positive results observed, has temporarily been withdrawn from clinical use because of reversible peripheral cholinergic side-effects.

Structure–activity relationships

The design of eptastigmine and its analogues was based above all on pharmacokinetic con-

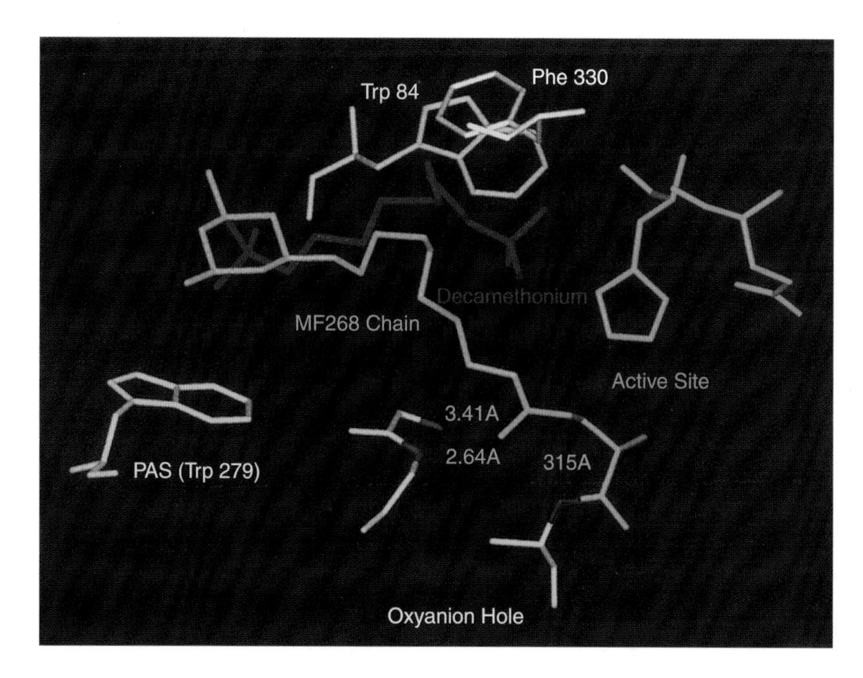

Figure 3.3
A perspective stereoview of the catalytic site and the inhibitor molecules in the superimposed refined structures of the TcAChE complexes with decamethonium (red) (4) and MF268 (green), respectively. The main and peripheral anionic sites (Trp_{84} and Trp_{279} respectively, yellow), and the oxyanion hole (peptidic NH groups of Gly_{118}-Gly_{119}-Ala_{201}, blue) are shown.

siderations. The aim of the chemical modifications was better BBB penetration and a reduction in the rate of metabolism with respect to that of physostigmine. Eptastigmine crosses the BBB very well and gives an AChE inhibition in whole rat brain that is either equal to or higher than that in red blood cells;[102] its metabolism is still rapid and, in humans, the drug reaches its C_{max} about 1 h after oral administration.[131] On the contrary, AChE inhibition is long-lasting. It is still observable after 12 h in vivo,[132] and it has a half-life of reactivation at 25°C of several days in vitro.[112] The reason for the slow reactivation of AChE after inhibition with long-chain analogues of physostigmine has recently been discovered in our laboratory by solving the X-ray structure of *Tc*AChE inactivated by MF268 (**19**).[133] This structure clearly shows that the morpholinooctylcarbamoyl moiety of MF268 is covalently bound to the Ser_{200} of AChE, filling the upper part of the catalytic pocket of the enzyme and hindering access to the active site. This strongly limits the diffusion of water molecules, thus explaining the high stability of the complex to hydrolysis. The structure also shows that the dimethylmorpholino ring of MF268 is opposed to the indole system of Trp_{279}, the main component of the peripheral anionic site of AChE, as is one of the quaternary groups of decamethonium in the corresponding complex with AChE.[31] The morpholino–Trp_{279} interaction also explains the strong dependence of the AChE inhibitory power of the aminoalkyl analogue of eptastigmine on the length of the alkyl chain and the basicity of the amino groups (Fig. 3.3).

Although the crystal structure of AChE inactivated by phenserine has not been yet determined, the AChE selectivity of this inhibitor is probably due to the interaction of its phenyl group with an aromatic residue of the AChE aromatic gorge.

The high brain AChE inhibition of ENA713 (**21**) is more difficult to explain. ENA713 is a weak in vitro inhibitor; nevertheless, it inhibits brain AChE with two times the power of eptastigmine.[121] It is probably partially transformed into a monoalkylcarbamoyl metabolite in vivo. It is also known that miotine is a stronger AChE inhibitor than physostigmine.

Reversible inhibitors

Tacrine (**15**) was the first reversible AChE inhibitor to be studied and approved for AD treatment. The drug, however, was withdrawn from the market on account of its hepatic toxicity. Several tacrine analogues were brought to preclinical or clinical evaluation, among them amiridine (**24**), 7-methoxytacrine (**25**), SM10888 (**26**)[134] and velnacrine (**27**).[135] None of them has yet reached the pharmaceutical market.

The X-ray structure shows that tacrine forms a reversible complex with *Tc*AChE,[30] in which it interacts with the Trp_{84} residue of the main anionic site, the His_{440} of the catalytic triad and the Phe_{330} of the aromatic gorge.

E2020, donezepil hydrochloride, or (±)-2,3-dihydro-5,6-dimethoxy-2-[1-(phenylmethyl)-4-piperidinyl]methyl-1*H*-inden-1-one hydrochloride (Aricept, **28**), was the first of a large series of *N*-benzylpiperidine derivatives prepared and tested as AChE reversible inhibitors.[136] E2020 is a chiral substance, but since its two enantiomers exhibit identical pharmacological profiles, including AChE inhibitory power, the drug has been developed as a racemic mixture.

Donezepil shows greater selectivity for AChE than for BuChE, and its mode of interaction with AChE differs from that of other known inhibitors, explaining its specificity. Docking studies were performed in order to predict the binding sites between donezepil hydrochloride and AChE. They showed that the *N*-benzyl substituent of the drug forms a π-stacking interaction with the indole side-chain of Trp_{84} (*Tc*AChE) and that the piperidine ring is located on the narrowest part of the active-site cavity which is formed by four amino acid residues (Tyr_{70}, Asp_{72}, Tyr_{121} and Tyr_{334}).[137] Furthermore, donezepil exhibits another interaction that involves one of its methoxy groups and the carbonyl group of the indanone ring. Another docking study suggested that the Trp_{279} residue of the AChE peripheral anionic site is involved in the interaction with the inhibitor.[138] The recently published X-ray structure of the complex between donezepil and *Tc*AChE[139] confirmed this last model, which explains the selectivity of benzylpiperidine inhibitors for AChE. In fact, the peripheral anionic site is absent in BuChE.

A large series of donezepil analogues have been synthesized and tested in vitro as AChE inhibitors, ex vivo for erythrocytes and brain AChE inhibition of treated animals, and in vivo for cognition and, in some cases, brain levels of AChE. Many compounds were prepared in the Tsukuba Research Laboratories of Eisai and the results of structure–activity relationship studies pursued on them were used in computer-assisted molecular design studies to develop guidelines for target synthesis and, retrospectively, to explain SAR behaviour.[140–142]

A series of analogues in which the indenone ring of donezepil was substituted with a benzisoxazole moiety was synthesized in the Pfizer laboratories.[143,144] Some of these derivatives gave good in vitro and in vivo results, but not one of them has been developed. From the synthetic work conducted in the Takada laboratories,[145,146] compound Tak147 (**29**) was found to be a good and selective AChE inhibitor in vitro and a specific activator of the central cholinergic system in vivo. However, this compound has not yet been developed.

Galantamine (**30**) and huperzine A (**31**) are two reversible AChE inhibitors of natural origin; the mode of interaction with the enzyme has been determined by resolving the structure of their complexes with *Tc*AChE for huperzine A only.[147] The protonated amino group of huperzine A interacts with the central anionic site of AChE as hypothesized, but the inhibitor does not have any contact with the peripheral anionic site, as predicted by docking studies.[148]

Features of an ideal inhibitor

Donezepil and rivastigmine, the last generation ChE inhibitors to be studied extensively in large clinical trials, are described as AChE selective and brain AChE selective, respectively. There is no doubt that brain selectivity is the major prerequisite for a ChE inhibitor to be used in AD treatment. The peripheral cholinergic effects, above all cardiovascular effects, that limit the tolerated doses would be reduced, if not eliminated by brain selectivity. Less evident are the advantages of AChE/BuChE selectivity. Not one of the peripheral side-effects observed in the treatment of AD patients with ChE inhibitors can be attributed with certainty to BuChE inhibition. Moreover, BuChE is present in the brain; where it could also play a role in the degeneration of the cholinergic system observed in AD. Some complications could derive from the individual variability of BuChE activity, above all in patients with BuChE genetic defects. The unsettled BuChE activity could affect the fraction of an unspecific ChE inhibitor available for AChE inhibition, and therefore the level of AChE inhibition.

It is even more difficult to compare between an irreversible and a reversible inhibitor. An irreversible, long-lasting inhibitor could guarantee a constant AChE inhibition and a reduced number of daily administrations. However, the possibility that AChE, BuChE or other proteins carbamoylated by lipophilic physostigmine analogues could act as antigens in some patients cannot be excluded. The molecular mechanism of the haematic toxicity of eptastigmine is still unknown.

Therefore, we propose that an ideal ChE inhibitor for the treatment of AD must be reversible, long-lasting, to some degree selective for AChE, centrally acting and somewhat selective for the tetramic G_1 form of AChE. This is the most abundant molecular form of AChE found in the brain.

Some research groups have tried to synthesize symbiotic molecules with AChE inhibition and other, cholinergic and non-cholinergic, activities. However, none of these substances has undergone an extensive clinical evaluation. The modest benefits observed in the treatment of AD patients with the known AChE inhibitors have discouraged the development of many new cholinergic agents.

References

1. Takano S, Ogasawa K. Alkaloids of the Calabar bean. *Alkaloids* 1989; **36:** 225–251.
2. Brossi A. Alfred Burger Award Address. Bioactive alkaloids. 4. Results of recent investigations with colchicine and physostigmine. *J Med Chem* 1990; **33(9):** 2311–2319.
3. Stedman E, Barger J. Physostigmine (Eserine): Part III. *J Chem Soc* 1925; **57:** 247–258.
4. Julian PL, Pikl J. Studies in the indole series. V. The complete synthesis of physostigmine (eserine). *J Am Chem Soc* 1935; **57:** 755–757.
5. Petcher TJ, Pauling P. Cholinesterase

inhibitors: structure of eserine. *Nature* 1973; **241(5387):** 277.
6. Longmore RB, Robinson B. The absolute configurations of the alkaloids of Physostigma venenosum seeds. *J Pharm Pharmacol* 1969; **21(Suppl):** 118S–125S.
7. Brossi A. Further explorations of unnatural alkaloids. *J Nat Prod* 1985; **48(6):** 878–893.
8. Takano S, Moriya M, Iwabuchi Y, Ogasawara K. A chiral route to both enantiomers of physostigmine and the first synthesis of (−)-norphysostigmine. *Chem Lett* 1990; 109–112.
9. Schönenberger B, Brossi A. Fragmentation of optically active (1-phenylethyl)- and (1-naphtylethyl)ureas in refluxing alcohols: easy preparation of optically active imines of high optical purity. *Helv Chim Acta* 1986; **69:** 1486–1497.
10. Node M, Hao X-J, Fuji K. A chiral total synthesis of (−)-physostigmine. *Chem Lett* 1991; 57–60.
11. Takano S, Moriya M, Ogasawa K. Enantiocontrolled total synthesis of (−)-physovenine and (−)-physostigmine. *J Org Chem* 1991; **56:** 5982–5984.
12. Taylor P. Anticholinesterase agents. In: Gilmann AG, Rall TN, Niess AS, Taylor P, eds. *The Pharmacological Basis of Therapeutics*, 8th edn. New York: Pergamon, 1990: 131–149.
13. Perola E, Cellai L, Lamba D, Filocamo L, Brufani M. Long chain analogs of physostigmine as potential drugs for Alzheimer's disease: new insights into the mechanism of action in the inhibition of acetylcholinesterase. *Biochim Biophys Acta* 1997; **1343(1):** 41–50.
14. Axelsson U. Glaucoma, miotic therapy and cataract. IV. Chronic simple glaucoma and cataract formation. *Acta Ophthalmol (Copenh)* 1969; **47(1):** 55–79.
15. Walker MB. Treatment of myastenia gravis with physostigmine. *Lancet* 1934; **i:** 1200–1201.
16. Deshpande SS, Viana GB, Kaffman FC, Rickett DL, Albuquerque EX. Effectiveness of physostigmine as a pretreatment drug for protection of rats from organophosphate poisoning. *Fundam Appl Toxicol* 1986; **6(3):** 566–577.
17. Wilson IB, Hatch MA, Ginsburg S. Carbamylation of acetylcholinesterase. *J Biol Chem* 1960; **235:** 2312–2315.
18. Janowsky DS, Risch SC, Berkowitz A, Turken A, Drennan M. Central virus peripheral antagonism of cholinesterase inhibitors induced lethality. In: Hann I, ed. *Dynamics of Cholinergic Function*. New York: Plenum, 1985: 791–798.
19. Bülbring E, Chou TC. Relative activity of prostigmine homologs and other substances as antagonists to tubocurarine. *Br J Pharmacol* 1947; **2:** 8–22.
20. Calvey TN, Wareing M, Williams NE, Chan K. Pharmacokinetics and pharmacological effects of neostigmine in man. *Br J Clin Pharmacol* 1979; **7(2):** 149–155.
21. Randall LO, Lehman G. Pharmacological properties of some neostigmine analogs. *J Pharmacol Exp Ther* 1950; **99:** 16–32.
22. Schwab RS. Management of myasthenia gravis. *N Engl J Med* 1963; **268:** 717–719.
23. Hetting G, Lillie CH, Harbich I. Ausscheidung der radioactivität nach intravenöser and peroraler Verabreichung des mit [^{3}H]markierten Hexamariums, einen lange wirkwnden Cholinesterase-Hemmkörpus, beider ratten. *Arzneimittelforschung* 1968; **18:** 479–481.
24. Joshi LD, Parmar SS. Structural consideration in the inhibition of rat brain acetylcholinesterase. *J Pharm Pharmacol* 1964; **16:** 763–765.
25. Phillips AP. Dicarboxylic acids bis-β-tertiaryaminoalkyl amides and their quaternary ammonium salts as curare substitutes. *J Am Chem Soc* 1951; **73:** 5822–5824.
26. Fisher DM, Cronnelly R, Sharma M, Miller RD. Clinical pharmacology of edrophonium in infants and children. *Anesthesiology* 1984; **61(4):** 428–433.
27. Nicholson GA, McLeod JG, Griffiths LR. Comparison of diagnostic tests in myasthenia gravis. *Clin Exp Neurol* 1983; **19:** 45–49.
28. Richter JE, Hackshaw BT, Wu WC, Castell DO. Edrophonium: a useful provocative test for esophageal chest pain. *Ann Intern Med* 1985; **103(1):** 14–21.
29. Sussman JL, Harel M, Frolow F *et al.* Atomic

structure of acetylcholinesterase from *Torpedo californica*: a prototypic acetylcholine-binding protein. *Science* 1991; **253(5022):** 872–879.
30. Silman I, Harel M, Eichler J, Sussman JL, Anselmet A, Massoulié J. Structure–activity relationship in the binding of reversible inhibitors in the active-site gorge of acetylcholinesterase. In: Becker R, Giacobini E, eds. *Alzheimer Disease: Therapeutic Strategies.* Boston: Birkäuser, 1994: 88–92.
31. Harel M, Schalk I, Ehret-Sabatier L *et al.* Quaternary ligand binding to aromatic residues in the active-site gorge of acetylcholinesterase. *Proc Natl Acad Sci USA* 1993; **90(19):** 9031–9035.
32. Holmstedt BR. Pharmacology of organophosphorus cholinesterase inhibitors. *Pharmacol Rev* 1959; **11:** 567–688.
33. Fulton MP, Mogey GA. Selective inhibitors of true cholinesterase. *Br J Pharmacol* 1954; **9:** 138–144.
34. Cavallito CJ, Sandy P. Acetylcholinesterase activities of some bis(quaternary ammonium salts). *Biochem Pharmacol* 1959; **2:** 233–242.
35. Taylor P. Agents acting at the neuromuscular junction and autonomic ganglia. In: Gilmann AG, Rall TN, Niess AS, Taylor P, eds. *The Pharmacological Basis of Therapeutics*, 8th edn. New York: Pergamon Press, 1990: 166–186.
36. Cannon JG. Cholinergics. In: Wolff ME, ed. *Burger's Medicinal Chemistry and Drug Discovery*, Vol 7, 5th edn. New York: Wiley, 1996: 3–58.
37. Shafferman A, Velan B, Ordentlich A *et al.* Substrate inhibition of acetylcholinesterase: residues affecting signal transduction from the surface to the catalytic center. *EMBO J* 1992; **11(10):** 3561–3568.
38. Shafferman A, Ordentlich A, Barak D, Stein D, Ariel N, Velan B. Aging of phosphylated human acetylcholinesterase: catalytic processes mediated by aromatic and polar residues of the active center. *Biochem J* 1996; **318(3):** 833–840.
39. Ordentlich A, Barak D, Kronman C *et al.* Dissection of the human acetylcholinesterase active center determinants of substrate specificity. Identification of residues constituting the anionic site, the hydrophobic site, and the acyl pocket. *J Biol Chem* 1993; **268(23):** 17 083–17 095.
40. Schalk I, Ehret-Sabatier L, Bouet F, Goeldner M, Hirth C. Trp279 is involved in the binding of quaternary ammonium at the peripheral site of *Torpedo marmorata* acetylcholinesterase. *Eur J Biochem* 1994; **219(1–2):** 155–159.
41. Shafferman A, Kronman C, Flashner Y *et al.* Mutagenesis of human acetylcholinesterase. Identification of residues involved in catalytic activity and in polypeptide folding. *J Biol Chem* 1992; **267(25):** 17 640–17 648.
42. Radic Z, Gibney G, Kawamoto S, MacPhee-Quigley K, Bongiorno C, Taylor P. Expression of recombinant acetylcholinesterase in a baculovirus system: kinetic properties of glutamate 199 mutants. *Biochemistry* 1992; **31(40):** 9760–9767.
43. Selvood T, Feaster SR, States MJ, Pryor AN, Quinn DM. Parallel mechanism in acetylcholinesterase-catalyzed hydrolysis of choline esters. *J Am Chem Soc* 1993; **115:** 10 477–10 482.
44. Quinn DM. Acetylcholinesterase: enzyme structure, reaction dynamics, and virtual transition states. *Chem Rev* 1987; **87:** 955–979.
45. Massoulié J, Pezzementi L, Bon S, Krejci E, Vallette FM. Molecular and cellular biology of cholinesterases. *Prog Neurobiol* 1993; **41(1):** 31–91.
46. Taylor P, Radic Z. The cholinesterases: from genes to proteins. *Annu Rev Pharmacol Toxicol* 1994; **34:** 281–320.
47. Ma JC, Dougherty DA. The cation–π interaction. *Chem Rev* 1997; **97:** 1303–1324.
48. Kim KS, Lee JY, Lee SJ, Ha T-K, Kim DH. On binding forces between aromatic ring and quaternary ammonium compound. *J Am Chem Soc* 1994; **116:** 7399–7400.
49. Dougherty DA, Stauffer DA. Acetylcholine binding by a synthetic receptor: implications for biological recognition. *Science* 1990; **250(4987):** 1558–1560.
50. Kearney PC, Mizoue LS, Kumpf RA, Forman JF, McCurdy A, Dougherty DA. Molecular recognition in aqueous media. New binding studies provide further insights into the

cation–π interaction related phenomena. *J Am Chem Soc* 1993; **115:** 9907–9919.

51. Chen YL, Nielsen J, Hedberg K *et al.* Syntheses, resolution, and structure–activity relationships of potent acetylcholinesterase inhibitors: 8-carbaphysostigmine analogues. *J Med Chem* 1992; **35(8):** 1429–1434.
52. Allen KN, Abeles RH. Inhibition kinetics of acetylcholinesterase with fluoromethyl ketones. *Biochemistry* 1989; **28:** 8466–8473.
53. Brodbeck U, Schweikert K, Gentinetta R, Rottenberg M. Fluorinated aldehydes and ketones acting as quasi-substrate inhibitors of acetylcholinesterase. *Biochim Biophys Acta* 1979; **567(2):** 357–369.
54. Dafforn A, Neenan JP, Ash CE *et al.* Acetylcholinesterase inhibition by the ketone transition state analog phenoxyacetone and 1-halo-3-phenoxy-2-propanones. *Biochem Biophys Res Commun* 1982; **104(2):** 597–602.
55. Gelb MH, Svaren JP, Abeles RH. Fluoro ketone inhibitors of hydrolytic enzymes. *Biochemistry* 1985; **24(8):** 1813–1817.
56. Linderman RJ, Leazer J, Roe RM, Venkatesh K, Selinsky BS, London RE. Fluorine ^{19}NMR spectral evidence that 3-octylthio-1,1,1-trifluoropropan-2-one, a potent inhibitor of insect juvenile hormone esterase, functions as a transition state analog. *Pest Biochem Physiol* 1988; **31:** 187–194.
57. Harel M, Quinn DM, Nair HK, Silman I, Sussman JL. The X-ray structure of a transition state analog complex reveals the molecular origins of the catalytic power and substrate specificity of acetylcholinesterase. *J Am Chem Soc* 1996; **118:** 2340–2346.
58. Holmstedt B. Structure–activity relationships of the organophosphorus anticholinesterase agents. In: Koelle GB, ed. *Cholinesterases and Anticholinesterase Agents, Handbuch der Experimentellen Pharmackologie,* Vol 5. Berlin: Springer-Verlag, 1963: 428–485.
59. Fukuto TR. Mechanism of action of organophosphorus and carbamate insecticides. *Environ Hlth Perspect* 1990; **87:** 245–254.
60. Johnson MK. Organophosphorus esters causing delayed neurotoxic effects: mechanism of action and structure activity studies. *Arch Toxicol* 1975; **34(4):** 259–288.
61. Minton NA, Murray VS. A review of organophosphate poisoning. *Med Toxicol Adverse Drug Exp* 1988; **3(5):** 350–375.
62. Heilbronn E. Action of fluoride on cholinesterase. II. In vitro reactivation of cholinesterases inhibited by organophosphorous compounds. *Biochem Pharmacol* 1965; **14(9):** 1363–1373.
63. Bucht G, Puu G. Aging and reactivatability of plaice cholinesterase inhibited by soman and its stereoisomers. *Biochem Pharmacol* 1984; **33(22):** 3573–3577.
64. Milatovic D, Johnson MK. Reactivation of phosphorodiamidated acetylcholinesterase and neuropathy target esterase by treatment of inhibited enzyme with potassium fluoride. *Chem Biol Interact* 1993; **87(1–3):** 425–430.
65. Kiffer D, Minard P. Reactivation by imidazopyridinium oximes of acetylcholinesterase inhibited by organosphosphates. A study with an immobilized enzyme method. *Biochem Pharmacol* 1986; **35(15):** 2527–2533.
66. Langenberg JP, De Jong LP, Otto MF, Benschop HP. Spontaneous and oxime-induced reactivation of acetylcholinesterase inhibited by phosphoramidates. *Arch Toxicol* 1988; **62(4):** 305–310.
67. Balali-Mood M, Shariat M. Treatment of organophosphate poisoning. Experience of nerve agents and acute pesticide poisoning on the effects of oximes. *J Physiol Paris* 1998; **92(5–6):** 375–378.
68. Worek F, Widmann R, Knopff O, Szinicz L. Reactivating potency of obidoxime, pralidoxime, HI 6 and HLo 7 in human erythrocyte acetylcholinesterase inhibited by highly toxic organophosphorus compounds. *Arch Toxicol* 1998; **72(4):** 237–243.
69. Harris LW, Fleisher JH, Clark J, Cliff WJ. Dealkylation and loss of capacity for reactivation of cholinesterase inhibited by sarin. *Science* 1966; **154(747):** 404–407.
70. Fleisher JH, Harris LW. Dealkylation as a mechanism for aging of cholinesterase after poisoning with pinacolyl methylphosphonofluoridate. *Biochem Pharmacol* 1965; **14(5):** 641–650.
71. Wilson BW, Hooper MJ, Hansen ME, Neiberg PS. In: Chamers JE, Levi PE, eds.

Effects of Organophosphates on Cholinesterase Activity. New York: Academic Press, 1992: 107–137.

72. Shafferman A, Ordentlich A, Barak D, Stein D, Ariel N, Velan B. Aging of phosphylated human acetylcholinesterase: catalytic processes mediated by aromatic and polar residues of the active centre. *Biochem J* 1996; **318:** 833–840.
73. Ordentlich A, Barak D, Kronman C *et al.* Exploring the active center of human acetylcholinesterase with stereomers of an organophosphorus inhibitor with two chiral centers. *Biochemistry* 1999; **38(10):** 3055–3066.
74. Harel M, Su CT, Frolow F *et al.* Refined crystal structures of 'aged' and 'non-aged' organophosphoryl conjugates of γ-chymotrypsin. *J Mol Biol* 1991; **221(3):** 909–918.
75. Steinberg N, Grunwald J, Roth E *et al.* Conformational differences between aged and non-aged organophosphoryl conjugates of chymotrypsin. *Prog Clin Biol Res* 1989; **289:** 293–304.
76. Grunwald J, Segall Y, Shirin E *et al.* Aged and non-aged pyrenebutyl-containing organophosphoryl conjugates of chymotrypsin. Preparation and comparison by ^{31}P NMR spectroscopy. *Biochem Pharmacol* 1989; **38(19):** 3157–3168.
77. Kraut J. Serine proteases: structure and mechanism of catalysis. *Annu Rev Biochem* 1977; **46:** 331–358.
78. Ordentlich A, Barak D, Kronman C *et al.* Functional characteristics of the oxyanion hole in human acetylcholinesterase. *J Biol Chem* 1998; **273(31):** 19 509–19 517.
79. Tripathi HL, Dewey WL. Comparison of the effects of diisopropylfluorophosphate, sarin, soman, and tabun on toxicity and brain acetylcholinesterase activity in mice. *J Toxicol Environ Hlth* 1989; **26(4):** 437–446.
80. Casida JE, Fukunaga K. Pesticides: metabolism, degradation, and mode of action. *Science* 1968; **160(826):** 445–450.
81. Dauterman WC. The role of hydrolases in insecticide metabolism and the toxicological significance of the metabolites. *J Toxicol Clin Toxicol* 1982; **19(6–7):** 623–635.
82. Srikanth NS, Seth PK. Alterations in xenobiotic metabolizing enzymes in brain and liver of rats coexposed to endosulfan and malathion. *J Appl Toxicol* 1990; **10(3):** 157–160.
83. Murphy SD. Pesticides. In: Klaassen CD, Amdur MO, Doull J, eds. *Casarett and Doull's Toxicology: The Basic Science of Poisons*, 3rd edn. New York: Macmillan, 1986: 519–581.
84. Barsam PC. The most commonly used miotic—now longer acting. *Ann Ophthalmol* 1974; **6(8):** 809–814.
85. Kisielinski T, Gajewski D, Gidynska T, Owczarczyk H. Antiesterase activity of chlorphenvinphos and phospholine in certain rat tissues after administration by different routes. *Acta Physiol Pol* 1980; **31(3):** 279–288.
86. Bartus RT, Dean RL III, Beer B, Lippa AS. The cholinergic hypothesis of geriatric memory dysfunction. *Science* 1982; **217(4558):** 408–414.
87. Perry EK. The cholinergic hypothesis—ten years on. *Br Med Bull* 1986; **42(1):** 63–69.
88. Shaw FH, Bentley G. Some aspects of the pharmacology of morphine with special reference to its antagonism by 5-aminoacridine (THA) and other chemically related compounds. *Med J Aust* 1949; **2:** 868–874.
89. Heilbronn E. Inhibition of cholinesterase by tetrahydroaminoacrididine. *Acta Chem Scand* 1961; **15:** 1386–1390.
90. Pagala MK, Sandow A. Physostigmine-induced contractures in frog skeletal muscle. *Pflugers Arch* 1976; **363(3):** 223–229.
91. Messamore E, Warpman U, Ogane N, Giacobini E. Cholinesterase inhibitor effects on extracellular acetylcholine in rat cortex. *Neuropharmacology* 1993; **32(8):** 745–750.
92. Giacobini E, Somani S, McIlhany M, Downen M, Hallak M. Pharmacokinetics and pharmacodynamics of physostigmine after intravenous administration in beagle dogs. *Neuropharmacology* 1987; **26(7B):** 831–836.
93. Sharpless NS, Thal LJ. Plasma physostigmine concentrations after oral administration. *Lancet* 1985; **i(8442):** 1397–1398.
94. Thal LJ, Masur DM, Sharpless NS, Fuld PA, Davies P. Acute and chronic effects of oral physostigmine and lecithin in Alzheimer's disease. *Prog Neuropsychopharmacol Biol Psy-*

chiatry 1986; **10(3–5):** 627–636.
95. Becker R, Giacobini E. Mechanisms of cholinesterase inhibition in senile dementia of the Alzheimer's type: clinical, pharmacological, and therapeutic aspect. *Drug Dev Res* 1988; **12:** 163–195.
96. Becker R, Moriearty P, Unni L. The second generation of cholinesterase inhibitors: clinical and pharmacological effects. In: Giacobini E, Becker R, eds. *Cholinergic Basis for Alzheimer Disease*. Boston: Berkäuser, 1991: 263–296.
97. Soncrant TT, Raffaele KC, Asthana S *et al.* Treatment of Alzheimer disease by continuous intravenous infusion of physostigmine. *Alzheimer Dis Assoc Disord* 1995; **9(4):** 223–232.
98. Knapp S, Wardlow ML, Albert K, Waters D, Thal LJ. Correlation between plasma physostigmine concentrations and percentage of acetylcholinesterase inhibition over time after controlled release of physostigmine in volunteer subjects. *Drug Metab Dispos* 1991; **19(2):** 400–404.
99. Pomponi M, Giacobini E, Brufani M. Present state and future development of the therapy of Alzheimer disease. *Aging (Milan)* 1990; **2(2):** 125–153.
100. Brufani M, Castellano C, Marta M *et al.* A long-lasting cholinesterase inhibitor affecting neural and behavioral processes. *Pharmacol Biochem Behav* 1987; **26(3):** 625–629.
101. Cuadra G, Summers K, Giacobini E. Cholinesterase inhibitor effects on neurotransmitters in rat cortex in vivo. *J Pharmacol Exp Ther* 1994; **270(1):** 277–284.
102. De Sarno P, Pomponi M, Giacobini E, Tang XC, Williams E. The effect of heptyl-physostigmine, a new cholinesterase inhibitor, on the central cholinergic system of the rat. *Neurochem Res* 1989; **14(10):** 971–977.
103. Pomponi M, Giardina B, Gatta F, Marta M. Physostigmines and tetrahydroaminoacridine analogs as alternative drugs for the treatment of Alzheimer's disease. *Med Chem Res* 1992; **2:** 306–327.
104. Iversen LL, Bentley G, Dawson S *et al.* Heptylphisostigmine—novel acetylcholinesterase inhibitor: biochemical and behavioral pharmacology. In: Giacobini E, Becker R, eds. *Alzheimer Disease: Therapeutic Strategies.* Boston: Birkäuser, 1991: 297–304.
105. Garrone B, Luparini MR, Tolu L, Magnani M, Landolfi C, Milanese C. Effect of the sub-chronic treatment with the acetyl-cholinesterase inhibitor heptastigmine on central cholinergic transmission and memory impairment in aged rats. *Neurosci Lett* 1998; **245(1):** 53–57.
106. Yu QS, Atack JR, Rapoport SI, Brossi A. Carbamate analogues of (−)-physostigmine: in vitro inhibition of acetyl- and butyryl-cholinesterase. *FEBS Lett* 1988; **234(1):** 127–130.
107. Atack JR, Yu QS, Soncrant TT, Brossi A, Rapoport SI. Comparative inhibitory effects of various physostigmine analogs against acetyl- and butyrylcholinesterases. *J Pharmacol Exp Ther* 1989; **249(1):** 194–202.
108. Canal N, Imbimbo BP. Relationship between pharmacodynamic activity and cognitive effects of eptastigmine in patients with Alzheimer's disease. Eptastigmine Study Group. *Clin Pharmacol Ther* 1996; **60(2):** 218–228.
109. Eptastigmine. *SCRIP* 1999; **2404:** 17–??.
110. Alisi MA, Brufani M, Filocamo L *et al.* Synthesis and structure–activity relationships of new acetylcholinesterase inhibitors: morpholinoalkylcarbamoyloxyeseroline derivatives. *Biorg Med Chem Lett* 1995; **5(18):** 2077–2080.
111. Zhu XD, Cuadra G, Brufani M *et al.* Effects of MF-268, a new cholinesterase inhibitor, on acetylcholine and biogenic amines in rat cortex. *J Neurosci Res* 1996; **43(1):** 120–126.
112. Perola E, Cellai L, Lamba D, Filocamo L, Brufani M. Long chain analogs of physostigmine as potential drugs for Alzheimer's disease: new insights into the mechanism of action in the inhibition of acetyl-cholinesterase. *Biochim Biophys Acta* 1997; **1343(1):** 41–50.
113. Greig NH, Pei XF, Soncrant TT, Ingram DK, Brossi A. Phenserine and ring C hetero-analogues: drug candidates for the treatment of Alzheimer's disease. *Med Res Rev* 1995; **15(1):** 3–31.
114. Brzostowska M, He XS, Greig NH, Rapoport

SI, Brossi A. Phenylcarbamates of (−)-eseroline, (−)-N[1]-noreseroline and (−)-physovenol: selective inhibitors of acetyl and/or butyrylcholinesterase. *Med Chem Res* 1992; **2:** 238–246.
115. Iijima S, Greig NH, Garofalo P *et al.* Phenserine: a physostigmine derivative that is a long-acting inhibitor of cholinesterase and demonstrates a wide dose range for attenuating a scopolamine-induced learning impairment of rats in a 14-unit T-maze. *Psychopharmacol (Berlin)* 1993; **112(4):** 415–420.
116. Yu QS, Pei XF, Holloway HW, Greig NH, Brossi A. Total syntheses and anti-cholinesterase activities of (3a*S*)-*N*(8)-norphysostigmine, (3a*S*)-*N*(8)-norphenserine, their antipodal isomers, and other *N*(8)-substituted analogues. *J Med Chem* 1997; **40(18):** 2895–2901.
117. Ingram DK. Complex maze learning in rodents as a model of age-related memory impairment. *Neurobiol Aging* 1988; **9(5–6):** 475–485.
118. Amstutz R, Enz A, Marzi H, Boelsterly J, Walkingshaw H. Cyclic phenylcarbamate of the myotine-type and their action on acetylcholinesterase. *Helv Chim Acta* 1990; **73:** 739–753.
119. Enz A, Amstutz R, Hofmann A, Gmelin G, Kelly PH. Pharmacological properties of the preferentially centrally acting acetylcholinesterase inhibitor SDN ENA713. In: Kewitz H, Thomson T, Bickel M, eds. *Pharmacological Interventions on Central Cholinergic Mechanisms in Senile Dementia (Alzheimer Disease).* Munich: Zuckscherdt, 1986: 271–277.
120. Enz A, Amstutz R, Boddeke H, Gmelin G, Malanowski J. Brain selective inhibition of acetylcholinesterase: a novel approach to therapy for Alzheimer's disease. *Prog Brain Res* 1993; **98:** 431–438.
121. Anand R, Hartman RD, Hayes PE. An overview of the development of SND ENA713: a brain selective cholinesterase inhibitor. In: Becker R, Giacobini E, eds. *Alzheimer Disease: From Molecular Biology to Therapy.* Boston: Birkäuser, 1997: 239–243.
122. Rivastigmina. *Drug Fut* 1998; **23:** 802–803.
123. van der Staay FJ, Hinz VC, Schmidt BH. Effects of metrifonate, its transformation product dichlorvos, and other organophosphorus and reference cholinesterase inhibitors on Morris water escape behavior in young-adult rats. *J Pharmacol Exp Ther* 1996; **278(2):** 697–708.
124. Wolfenbarger DA, Guerra AA. Toxicity of seven alkyl organophosphorus insecticides to the bollworm, pink and tobacco budworm and persistence of naled in various formulations on cotton leaves. *J Econ Entomol* 1972; **65(5):** 1377–1380.
125. Leland SE Jr, Ridley RK, Dick JW, Slonka GF, Zimmerman GL. Anthelmintic activity of trichlorfon, coumaphos, and naphthalophos against the in vitro grown parastic stages of *Cooperia punctata*. *J Parasitol* 1971; **57(6):** 1190–1196.
126. Becker RE, Colliver J, Elble R *et al.* Effect of metrifonate, a long acting cholinesterase inhibitor, in Alzheimer's disease: report on open trial. *Drug Dev Res* 1990; **49:** 425–434.
127. Hinz VC, Grewig S, Schmidt BH. Metrifonate induces cholinesterase inhibition exclusively via slow release of dichlorvos. *Neurochem Res* 1996; **21(3):** 331–337.
128. Tariot PN. Evaluating response to metrifonate. *J Clin Psychiatry* 1998; **59(Suppl 9):** 33–37.
129. Villen T, Abdi YA, Ericsson O, Gustafsson LL, Sjoqvist F. Determination of metrifonate and dichlorvos in whole blood using gas chromatography and gas chromatography–mass spectrometry. *J Chromatogr* 1990; **529(2):** 309–317.
130. Schmidt BH, Hinz VC, Blokland A, van der Staay F-J, Fanelli RJ. Preclinical pharmacology of metrifonate: a promise for Alzheimer therapy. In: Becker R, Giacobini E, eds. *Alzheimer Disease: From Molecular Biology to Therapy.* Boston: Birkäuser, 1997: 217–221.
131. Unni LK, Hutt V, Imbimbo BP, Becker RE. Kinetics of cholinesterase inhibition by eptastigmine in man. *Eur J Clin Pharmacol* 1991; **41(1):** 83–84.
132. Auteri A, Mosca A, Lattuada N *et al.* Pharmacodynamics and pharmacokinetics of

eptastigmine in elderly subjects. *Eur J Clin Pharmacol* 1993; **45(4)**: 373–376.
133. Bartolucci C, Perola E, Cellai L, Brufani M, Lamba D. 'Back door' opening implied by the crystal structure of carbamoylated acetylcholinesterase. *Biochemistry* 1999; **38:** 5714–5719.
134. Brufani M, Filocamo L, Lappa S, Maggi A. New acetylcholinesterase inhibitors. *Drug Fut* 1997; **22:** 397–410.
135. Velnacrine. *Drug Fut* 1994; **19:** 709.
136. Sugimoto H, Tsuchiya Y, Sugumi H *et al.* Novel piperidine derivatives. Synthesis and anti-acetylcholinesterase activity of 1-benzyl-4-[2-(*N*-benzoylamino)ethyl]piperidine derivatives. *J Med Chem* 1990; **33(7)**: 1880–1887.
137. Kawakami Y, Inoue A, Kawai T *et al.* The rationale for E2020 as a potent acetylcholinesterase inhibitor. *Bioorg Med Chem* 1996; **4(9)**: 1429–1446.
138. Pang YP, Kozikowski AP. Prediction of the binding site of 1-benzyl-4-[5,6-dimethoxy-1-indanon-2-yl)methyl]piperidine in acetylcholinesterase by docking studies with the SYSDOC program. *J Comput Aided Mol Des* 1994; **8(6)**: 683–693.
139. Kryger G, Silman I, Sussman JL. Three-dimensional structure of a complex of E2020 with acetylcholinesterase from *Torpedo californica*. *J Physiol Paris* 1998; **92(3–4)**: 191–194.
140. Cardozo MG, Iimura Y, Sugimoto H, Yamanishi Y, Hopfinger AJ. QSAR analyses of the substituted indanone and benzylpiperidine rings of a series of indanone–benzylpiperidine inhibitors of acetylcholinesterase. *J Med Chem* 1992; **35(3)**: 584–589. [Erratum: *J Med Chem* 1992; **35(25)**: 4767.]
141. Cardozo MG, Kawai T, Iimura Y *et al.* Conformational analyses and molecular-shape comparisons of a series of indanone--benzylpiperidine inhibitors of acetylcholinesterase. *J Med Chem* 1992; **35(3)**: 590–601. [Erratum: *J Med Chem* 1992; **35(25)**: 4768.]
142. Inoue A, Kawai T, Wakita M *et al.* The simulated binding of (±)-2,3-dihydro-5,6-dimethoxy-2-[[1-(phenylmethyl)-4-piperidinyl] methyl]-1*H*-inden-1-one hydrochloride (E2020) and related inhibitors to free and acylated acetylcholinesterases and corresponding structure–activity analyses. *J Med Chem* 1996; **39(22)**: 4460–4470.
143. Villalobos A, Blake JF, Biggers CK *et al.* Novel benzisoxazole derivatives as potent and selective inhibitors of acetylcholinesterase. *J Med Chem* 1994; **37(17)**: 2721–2734.
144. Villalobos A, Butler TW, Chapin DS *et al.* 5,7-Dihydro-3-[2-[1-(phenylmethyl)-4-piperidinyl]ethyl]-6*H*-pyrrolo[3,2-f]-1,2-benzisoxazol-6-one: a potent and centrally-selective inhibitor of acetylcholinesterase with an improved margin of safety. *J Med Chem* 1995; **38(15)**: 2802–2808.
145. Ishihara Y, Hirai K, Miyamoto M, Goto G. Central cholinergic agents. 6. Synthesis and evaluation of 3-[1-(phenylmethyl)-4-piperidinyl]-1-(2,3,4,5-tetrahydro-1*H*-1-benzazepin-8-yl)-1-propanones and their analogs as central selective acetylcholinesterase inhibitors. *J Med Chem* 1994; **37(15)**: 2292–2299.
146. Hirai K, Kato K, Nakayama T *et al.* Neurochemical effects of 3-[1-(phenylmethyl)-4-piperidinyl]-1-(2,3,4,5-tetrahydro-1*H*-1-benzazepin-8-yl)-1-propanone fumarate (TAK147), a novel acetylcholinesterase inhibitor, in rats. *J Pharmacol Exp Ther* 1997; **280(3)**: 1261–1269.
147. Raves ML, Harel M, Pang YP *et al.* Structure of acetylcholinesterase complexed with the nootropic alkaloid (−)-huperzine A. *Nat Struct Biol* 1997; **4(1)**: 57–63.
148. Pang YP, Kozikowski AP. Prediction of the binding sites of huperzine A in acetylcholinesterase by docking studies. *J Comput Aided Mol Des* 1994; **8(6)**: 669–681.

4

Novel roles for cholinesterases in stress and inhibitor responses

Hermona Soreq, David Glick

The first and best known function of acetylcholinesterase (AChE), termination of cholinergic neurotransmission by hydrolysis of acetylcholine (ACh), stimulated the development of organophosphate compounds that were potent cholinesterase (ChE) inhibitors and which found uses as insecticides and chemical warfare agents. More recently, several therapeutic uses have been established for some of these and carbamate anti-ChEs.[1] With the growing recognition of new roles for AChE and butyrylcholinesterase (BuChE), it is appropriate to re-explore the effects of ChE inhibitors and the first steps toward novel approaches to the therapeutic suppression of excess ChE activities.

BuChE operates as a scavenger of anti-ChEs

As a catalyst, BuChE is very similar to AChE. It is less specific for the acetate ester of choline, and it is inhibited by a wider spectrum of compounds, but as a catalyst, it is almost as efficient as AChE.[2] The two human proteins are also similar in sequence and tertiary structure. Their separate identities were established by cloning the two respective genes,[3,4] and locating them on separate chromosomes: *BCHE* to 3q26-*ter*[5,6] and *ACHE* to 7q22[7,8] (Fig. 4.1). Although AChE is the primary ChE of the nervous system and BuChE is present largely in the serum, BuChE was long thought to be a 'back-up' to AChE.

Inconsistent with a role as a terminator of neurotransmission is the highly polymorphic nature of the *BCHE* gene. Over 40 natural mutations have been identified which result in distinct gene products (Fig. 4.2). These mutations generate a range of proteins, including some which are functionally indistinguishable from the typical BuChE, as well as others having no hydrolytic activity whatsoever.[10] That such mutations persist in the population without obvious detriment, even to homozygous carriers, is a good indication that whatever function BuChE performs it is non-essential. However, some BuChE mutations confer a genetic predisposition to adverse responses to AChE inhibitors.[11]

Because of the broad specificity of BuChE, every anti-AChE is also an anti-BuChE. An anti-AChE entering the body will react with BuChE before ever coming into contact with AChE at neuromuscular junctions or brain synapses. The individual is thus protected by the ability of BuChE to absorb AChE inhibitors. Consistent with being a molecular decoy for AChE are also the prominence of BuChE in the serum, its capacity to react quickly with a wide spectrum of compounds, and even the polymorphism of its gene. The polymorphism of *BCHE* has been surveyed extensively, originally by studying the variant characteristic susceptibility to inhibitors of the serum activity[11] and more recently by molecu-

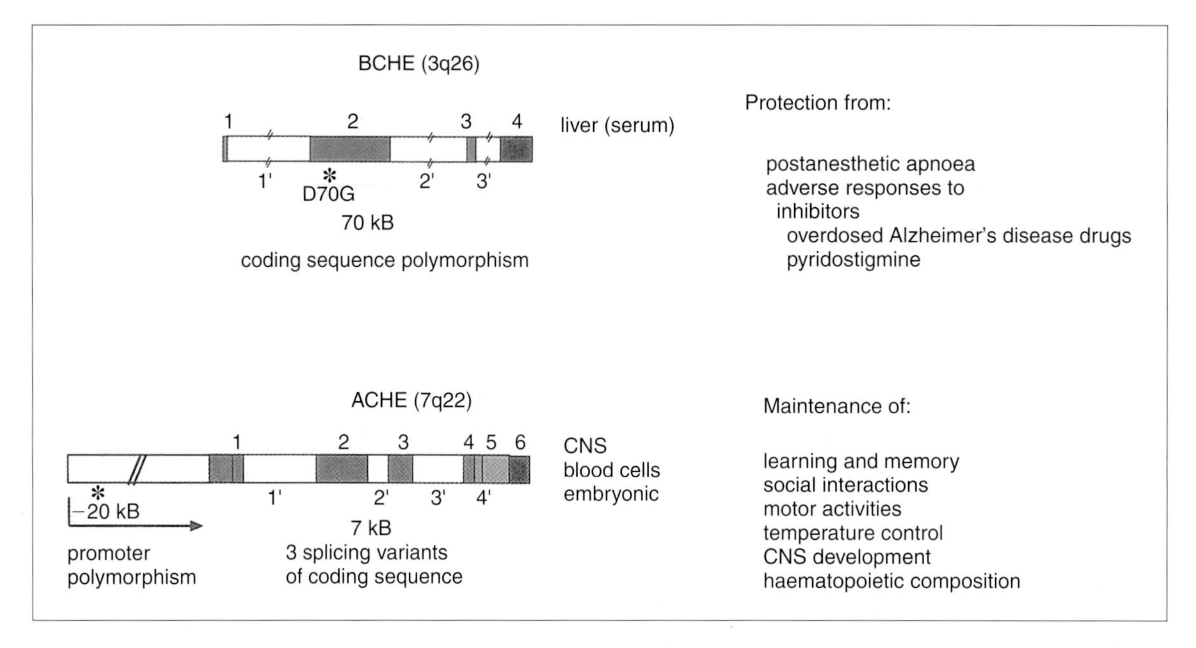

Figure 4.1
The human cholinesterase genes. The BCHE gene, located on chromosome 3q26, is expressed in the liver and its BuChE protein product secreted into the serum. (It also is expressed in the nervous system and, transiently, in many embryonic tissues.[9]) Of the known natural mutations, the most prominent is the D70G ('atypical') substitution (green asterisk). The biological role of BuChE appears to be as a decoy to protect AChE from inhibition. The AChE gene, ACHE, located on chromosome 7q22, is expressed in the nervous system and in developing tissues. Polymorphisms 17 kb upstream from the coding sequence affect expression. Besides its classical role in neurotransmission, it appears to be involved in the plasticity of cholinergic and non-cholinergic neuronal tissues and in developing tissues, such as bone marrow.

lar genotyping.[13] A common variant is 'atypical' BuChE, the product of a single base substitution which replaces aspartate with glycine at the secondary substrate/inhibitor binding site[14–16] (see Fig. 4.2). As a consequence, 'atypical' BuChE has a lower affinity for many inhibitors and unnatural substrates. Under normal conditions, this has no adverse consequences. However, anesthesiologists use the anti-AChE succinylcholine as a muscle relaxant, and for recovery depend upon its slow hydrolysis by BuChE. Homozygous carriers of 'atypical' *BCHE* show a much delayed return to spontaneous breathing following the use of succinylcholine. Less drastic effects, such as slower than normal recovery of spontaneous breathing, may be the consequence of dietary intake of natural anti-AChEs.[19] The very uneven geographical distribution of 'atypical' *BCHE*, with high or low frequencies notable among historically isolated groups, may reflect an evolutionary adaptation to local environment factors. One hypothesis concerns the interplay of the presence of mildly inhibitory

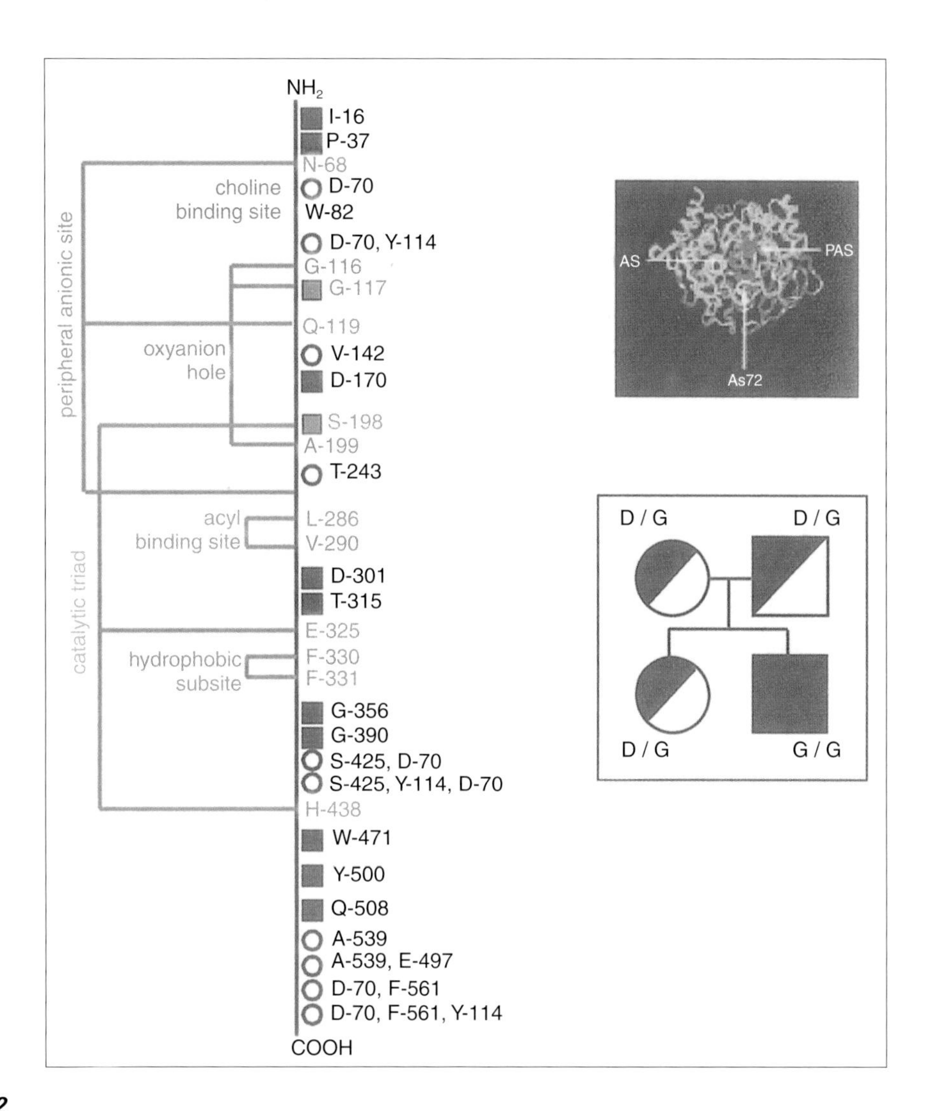

Figure 4.2

Natural mutations of BCHE. Among the sites in BuChE at which mutations have been localized[10,12,17] *are the catalytic triad (green-blue), the acyl binding site (green), the choline binding site (magenta), the oxyanion hole (blue) and the peripheral anionic site (red-orange). The squares indicate mutations that abolish catalytic activity; the circles indicate mutations that result in moderate changes in activity. (a) A ribbon diagram of Torpedo AChE, showing the catalytic triad active site (AS), peripheral anionic site (PAS) and the aspartate residue, which is the cognate of D70 of human BuChE. The end of the helix in the upper left is the site of attachment of the carboxy-terminal residue, which is indistinct in the X-ray crystallography model.*[18] *(b) The pedigree of a soldier who carries two copies of the D70G 'atypical' BCHE mutation (filled square), and who displayed post-anaesthesia apnoea under succinylcholine administration, as well as deep depression, insomnia and massive weight loss following pyridostigmine treatment during the Gulf War.*[11] *In the failure of the 'atypical' enzyme to protect AChE from this anti-AChE, this case history strikingly illustrates the role of the typical BuChE as an anti-AChE scavenger.*

alkaloids in *Solanaceae* food plants (potatoes, tomatoes, aubergines and peppers) and the appearance of BuChE in the placenta, to explain the co-occurrence of a high frequency of the 'atypical' allele and the traditional consumption of these foods; because 'atypical' BuChE has a low affinity for these alkaloids, it survives to play other protective roles during pregnancy.[20,21]

Multi-organ indications of non-catalytic roles

The occurrence of AChE on the surface of erythrocytes is a paradox of long standing: there, it has no known cholinergic role. Early hints at non-classical roles for AChE and BuChE came from their presence in non-neuronal tissues such as the blood cells,[22,23] developing avian cartilage[24] and developing oocytes and sperm,[25,26] where no catalytic role for AChE is likely. Also, within the nervous system there is not always a correlation between the occurrence of AChE and other cholinergic proteins.[27–29] Furthermore, non-neuronal brain cells, the meninges,[30] blood vessel endothelium[31] and glia[32] also express AChE. Genomic studies have placed AChE in other unexpected contexts. The human *ACHE* gene undergoes massive amplification under diverse conditions: leukaemias,[33] ovarian carcinomas,[34] thrombocytopenia[35] and exposure to organophosphate ChE inhibitors.[36] These first indications of unexpected long-term danger in exposure to ChE inhibitors initiated a systematic search for cells and tissues in which *CHE* genes are expressed. As expected, expression of *ACHE*, i.e. the presence of the corresponding mRNA, was prominent in mammalian brain, but surprisingly AChE and BuChE mRNA were observed also in developing human oocytes,[26] which correlates with the report of an inherited amplification of the *BCHE* gene following exposure to insecticides. Human *CHE* expression was also observed in primary carcinomas,[37] placental chorionic villi,[38] developing blood cells[39] and embryonic bone.[40] The ubiquitous expression of human *CHE*s in developing tissues reinforced the hypothesis that these proteins are involved in function(s), beyond ACh hydrolysis. Also, cytological and functional studies of neurons indicated a correlation of AChE with the morphology of neurites, which was not easily explained solely by the cholinergic function of these structures.[41–43]

Alternative splicing leads to AChE variants with distinct structural functions

The single *ACHE* gene may give rise to different protein products by alternative splicing in the coding region of the original transcript. Production of carboxy-terminal (C-terminal) variant protein isoforms was observed in embryonic tissues[44] and in tumour cell lines.[32] Among the consequences of this alternative splicing are synaptic or epidermal accumulation of C-terminal distinct AChE isoforms in *Xenopus* tadpoles,[45] modulation of process extension in rat glioma cells[46] and induction of neurite growth in cultured *Xenopus* motor neurons.[47] Most of these functions of AChE are independent of its catalytic ability, as they survive insertion of a seven-residue polypeptide sequence into the active site.

A schematic diagram of alternative splicing in the coding region of human AChE mRNA is shown in Fig. 4.3. In all three *ACHE* mRNAs the common core, exons 2, 3 and 4, is sufficient for catalytic activity of the protein. Alternate splicing gives rise to AChE isoforms with different C-termini; these confer characteristic hydrodynamic properties, capacities for multimerization and/or attachment to

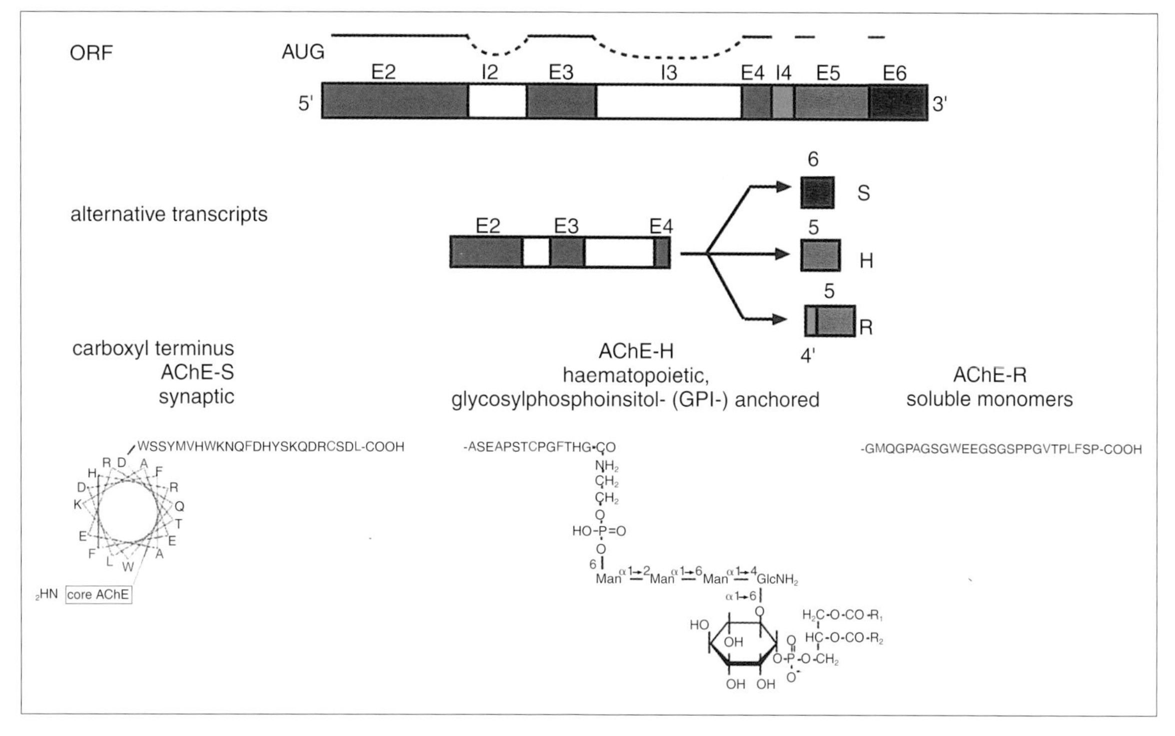

Figure 4.3
Natural AChE splicing variants. The three natural AChE mRNA variants produced by splicing in the coding region yield proteins with characteristic C-terminal sequences, encoded by the open-reading frame (ORF), of pseudointorn I4, exons E5 and E6.[47] Each differs in its developmental and tissue distributions. Synaptic AChE-S is encoded by the transcript 3′-terminated with E6. AChE-S has a C-terminal peptide predicted by the peptide structure programme of the GCG software package (University of Wisconsin) to have a helical amphipathic sequence before the final 23 residues. The haematopoietic AChE-H protein carries a 14-residue C-terminus that is covalently linked to a glycosylphosphoinositol (GPI) anchor.[48] The readthrough variant, AChE-R, includes a hydrophilic 26-residue C-terminus with no predicted secondary structure and/or option for membrane anchor. Non-polar residues are in red.

membranes and tissue and cellular distribution patterns. The major brain and muscle form, AChE-S (S, for synaptic), is encoded by mRNA carrying the core exons plus exon 6; the haematopoietic form, AChE-H, derives from mRNA carrying exon 5; a third mRNA species, reads through intron 4 directly into exon 5, hence the name, 'readthrough', or AChE-R. The AChE-R isoform has only recently emerged as a physiologically significant member of the AChE family. When expressed in *Xenopus*, human AChE-R is localized to secretory epidermal cells, and does not accumulate at neuromuscular junctions. Therefore, AChE-R represents a soluble, secretable, non-synaptic form of AChE.

Readthrough AChE accumulates following stress

In vivo and in cultured hippocampal brain slices, long-lasting accumulation of readthrough AChE is induced in response to acute stress and to anti-ChE exposure[49] (Fig. 4.4). In the short term, excess of AChE-R would be advantageous, as it can suppress the initial excitation state by enhancing ACh hydrolysis. However in the long term, this accumulation may be harmful. Many of the delayed consequences of inhibitor exposure, notably cognitive impairments and neuromo-

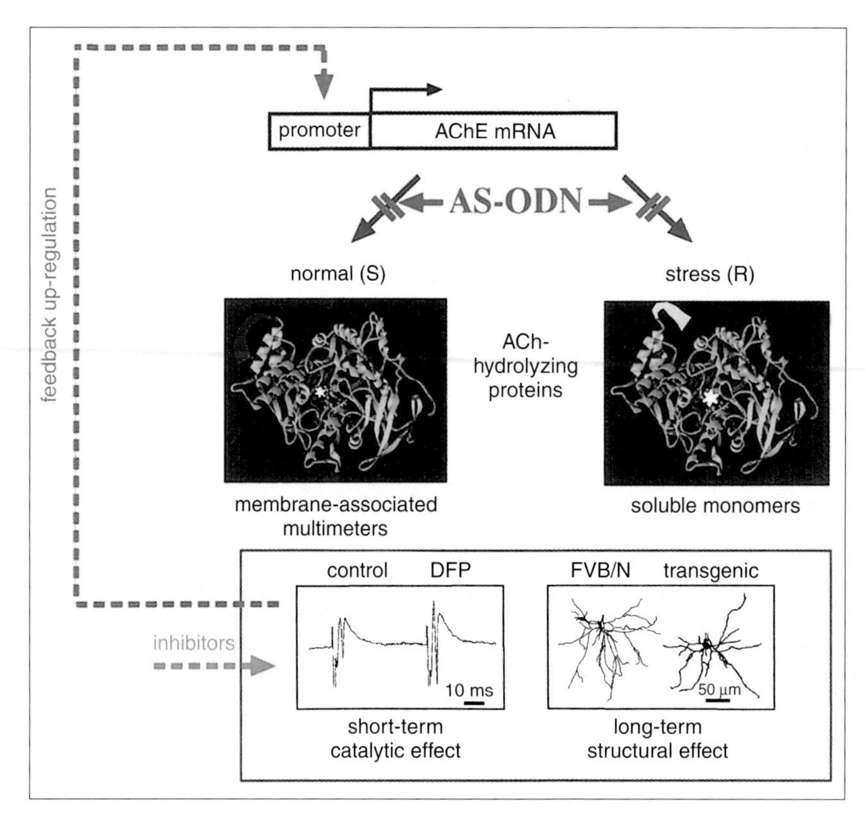

Figure 4.4
Feedback regulation of ACHE expression. Inhibition of synaptic AChE-S, for instance by an organophosphate, causes changes in neurotransmission, evident as increases in the electrophysiological excitation due to inhibition of membrane-associated AChE-S multimers (short-term) (lower left inset). At the transcriptional control level, such excitation leads to c-fos and/or egr-mediated increase in ACHE gene expression (arrow pointing right) and causes accumulation of extrasynaptic AChE-R soluble monomers. AChE accumulation is associated with long-term structural effects on neuron morphology and plasticity. This is evident, among other pathologies, as thinning of the dendritic trees of pyramidal neurons in the somatosensory cortex of AChE transgenic mice (lower right inset). This feedback, and the expression of all variants, may be blocked by antisense oligodeoxynucleotides (AS-ODNs) targeted to the ACHE gene (see Fig. 4.6), as such agents prevent AChE accumulation, reducing the catalytic and structural effects associated with AChE overproduction.

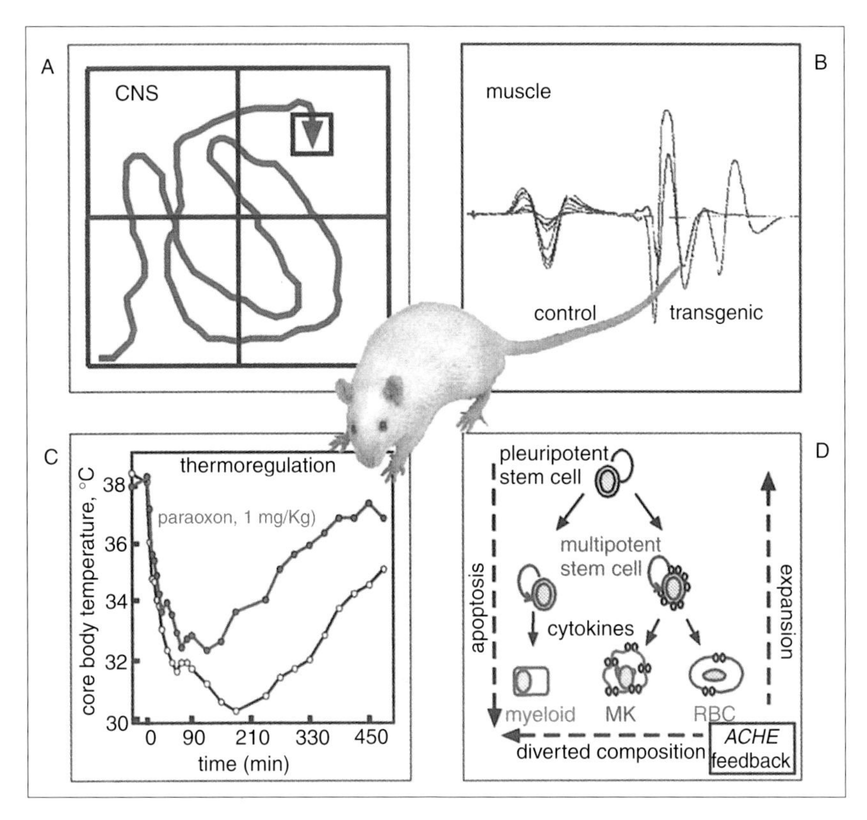

Figure 4.5
Multi-organ consequences of AChE overproduction. AChE-transgenic mice, which overexpress AChEs in their brain neurons, have been characterized as suffering cognitive impairments, evident as failure to perform the Morris water maze tasks (CNS: a).[56] *They also display progressive neuromotor deterioration, evident as rapid fatigue in their electromyographic response (muscle; b).*[51] *At the level of drug responses, these mice fail to adjust their body temperature under paraoxon exposure (thermoregulation: c).*[50] *In cultured primary haematopoietic cells, AChE overproduction is associated with enhanced expansion of multipotent stem cells and diversion of their differentiation toward myeloid lineages*[57] *(D Grisaru, unpublished observations).*

tor deficits, are strikingly similar to effects noted in transgenic mice that express human AChE (hAChE) in their CNS[50,51] (Fig. 4.5 (a,b)). What these experimental models have in common is elevated levels of AChE-R, due, in the case of exposure to anti-ChEs, to long-term up-regulation of *ACHE* expression, and in the case of *ACHE*-transgenic animals, to increased gene dose. Under normal conditions, the blood–brain barrier is practically impermeable to many AChE inhibitors, especially those that are not lipophilic.[52] However, under acute psychological stress, there is an efficient penetrance by inhibitors and other large compounds of the blood–brain barrier.[53] This would lead to AChE-R accumulation through the above-described feedback response. However, our transgenic animal studies suggested that elevated levels of AChE are harmful, and indeed, this is supported by findings in a study of closed-head-injury animals (E Shohami, unpublished observations). The extent of neuronal survival and recovery (evaluated from a panel of physical performance tasks to which injured animals were subjected) were both decreased in AChE-S-transgenic mice as compared to controls. Stress-induced accumulation of AChE-R was also observed in blood. To investigate this issue, the haematopoietic response to stress was mimicked in vitro by submitting human umbilical stem cells to hydrocortisone levels which are characteristic of stress and noting the accumulation of both AChE-R and its C-terminal peptide. Surprisingly, a synthetic peptide representing the 26 C-terminal residues of AChE-R improved the survival of cultured human stem cells, potentiated their AChE-R mRNA levels and enhanced their ex vivo expansion by early acting cytokines (D Grisaru, unpublished observations). These findings identify the C-terminal peptide of AChE-R as an autoregulatory stress-responsive haematopoietic element promoting the myeloid and megakarocytic expansion characteristic of acute and chronic stress responses.

Novel transgenic animal models reveal cortical neuropathologies

Besides the extensively studied hAChE-S-transgenic mouse, additional transgenic strains have been created:[54] two transgenics for human AChE-R, and another which expresses a control variant of AChE-S that is catalytically inactive because of a seven-residue peptide inserted near the active site (IN-AChE-S). Immunohistochemical detection of neurofilaments revealed in the brains of these mice pathological 'corkscrew' patterns of curled axons with sinusoidal regularity at the somatosensory cortex. Cumulative length measurements of these processes demonstrated a significantly more severe pathology in brains of AChE-S and IN-AChE-S transgenics than in AChE-R transgenics or non-transgenic controls. Corkscrew pathology is known to occur in degenerating cortical pyramidal neurons following exposure to phencyclidine (PCP), which blocks NMDA receptors of cortical γ-aminobutyric acid releasing (GABAergic) interneurons.[55] However, GABAergic cell counts and morphological properties of transgenic somatosensory cortices were indistinguishable from controls, suggesting that the corkscrew phenotype reflects acutely imbalanced cholinergic inputs to pyramidal neurons due to AChE-S overproduction. The neuropathologies of IN-AChE-S mice indicate that deleterious consequences of AChE increases, induced by AChE inhibitors, are primarily due to the structural properties of AChE and may therefore occur also in the presence of the inhibitor-blocked enzyme. At the same time, the limited pathology in AChE-R transgenics indicates that this soluble, extrasynaptic protein is designed to cause minimal damage under conditions of its stress-induced overproduction.

Very possibly AChE mediates other stress-related responses that are currently being investigated. Closed-head injury increases the risk of late-onset neuropathologies in humans.[58] It also promotes cholinergic hyperexcitation and facilitates initiation of the feedback loop of AChE overproduction (E Shohami, unpublished observations). This denotes a convergent outcome of acute psychological stress, chemical inhibition and traumatic injury. Together with the demonstration that a long-term excess of AChE promotes neurodeterioration, this suggests that elevated expression of *ACHE* due to various initial causes, may mediate delayed neurodegenerative deterioration. This further implies that genetic variabilities which affect the activity of the *CHE* genes or their protein products represent important risk factors that can determine susceptibility to neurodeterioration.

Implications for anti-AChE therapies

AChE inhibitors are increasingly used in therapy. They had long been used to treat glaucoma (the organophosphate, tetraisopropylpyrophosphoramide) and some symptoms of myasthenia gravis (the carbamate, physostigmine).[1] More recently, and much more extensively, they have been used to protect soldiers from the anticipated use of chemical warfare agents (pyridostigmine) and as Alzheimer disease (AD) drugs (tacrine, donepezil, rivastigmine, etc.). If our findings in experimental animals are applicable to humans, and we fear they are, the short-term benefits of AChE inhibitor therapies will have to be weighed against the long-term dangers of neuromuscular and cognitive deterioration.[59] In fact, the danger is posed not only by the use of AChE inhibitors, but also by any condition that results in long-term up-regulation of AChE, such as seen in the closed-head injury and stressed animals.

The use of AChE inhibitors for treatment of neurodegenerative diseases is a rational approach to cholinergic imbalances.[60,61] Deficient cholinergic neurotransmission, it was reasoned, can be enhanced by partially blocking the hydrolysis of the neurotransmitter through the inhibition of AChE. (However, the presence of AChE in the pathological AD plaques[62] and its potential function in their formation,[63] complicate this neat picture.) Until very recently, the only established function of AChE was termination of cholinergic neurotransmission. Therefore, all of the consequences of its inhibition might be evaluated by the efficiency of inhibition of AChE enzymatic activity and observation of cholinergic processes, such as cognition in AD patients. Now that additional functions of AChE are established, the effects of AChE inhibitors must be more broadly considered. This is especially so as one of these newly recognized effects is the up-regulation of *ACHE* expression and the other concerns changes in neurite growth. We see abnormal responses to anti-AChE as well as detrimental effects of AChE overproduction in experimental animals,[54,56] and suspect that they will be recognized in humans, as well. In line with this is the finding that in anti-AChE treatment of AD patients, the amelioration of cognitive decline is both limited and temporary.

Toward genome-based antisense therapy

That the morphological effects of AChE are unrelated to the catalytic capacity of AChE warns us that the detrimental consequences of elevated AChE will not necessarily be alleviated by AChE inhibitors. Moreover, the feedback response to such inhibition increases the amount of the AChE protein in the brain of treated patients (A Nordberg, personal communication). Therefore, we have investigated the effects of blocking ChE production by specifically interfering with *ACHE* expression. Antisense oligodeoxynucleotides (AS-ODNs) targeted to AChE mRNA suppressed the development of cultured platelet-forming megakaryocytes.[57] Conversely, the subsequent increases in AChE-R were associated with induced progenitor cell expansion and suppressed haematopoietic apoptosis[64] (see Fig. 4.5(d)). An antisense cRNA, capable of

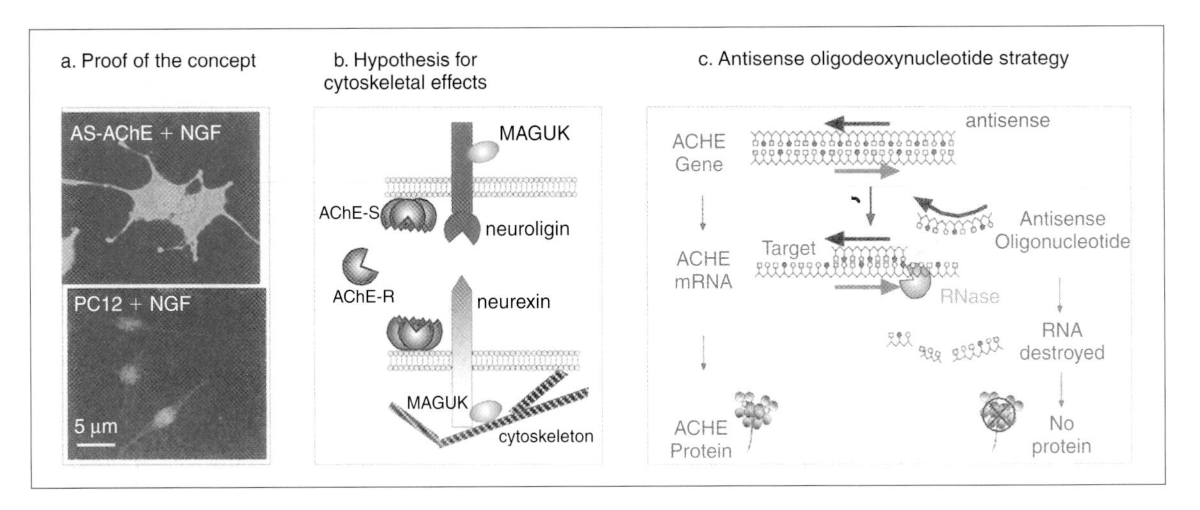

Figure 4.6
The antisense approach to suppression of AChE production. (a) Proof of the concept. When treated with nerve growth factor (NGF), PC12 cells extend neurites and express the AChE protein, detected on the cell surface with an antibody directed against the core AChE protein and visualized with a fluorescein-labelled secondary antibody (top). Transfection with a DNA plasmid encoding antisense AS-AChE cRNA to exon 6 yielded a stable phenotype with 80% lowered AChE levels and failure to extend neurites under NGF (bottom). Retransfection of the AS-AChE cells with vectors encoding catalytically active or inactive AChE or the AChE-homologous protein, neuroligin, rescued the neurite extension capacity of AS-AChE cells.[65] This proves that antisense suppression can deplete neurons of their AChE protein, that such depletion reversibly impairs neurite growth and that in affecting this growth process, AChE competes with neuroligin. (b) Hypothesis for cytoskeletal effects. AChE-S multimers, but not AChE-R monomers, associate with the plasma membrane through their amphipathic C-terminal tail and/or structural subunits (e.g. ColQ tails[67]). Proteins with single transmembrane domains and with AChE-like extracellular domains, such as neuroligin, can transduce signals into the cell which expresses them through interactions of their cytoplasmic carboxy-terminal domains with membrane-associated guanylate kinase (MAGUK) proteins. Extracellular interaction of neuroligin with neurexin[68] would further transduce signals, through yet other MAGUK proteins, into the other cell. MAGUK proteins are known to modulate the properties of cytoskeletal elements, thus potentially affecting neurite growth.[69,70] Therefore, competition between AChE-S and/or AChE-R and neuroligin (or, possibly, other AChE-like proteins), can explain the involvement of AChE with neuritogenesis, synapse plasticity under drug exposure and stress and perhaps the disruption of blood–brain barrier under stress. (c) Antisense oligodeoxynucleotide strategy. The 'sense' strand (rightward arrow) of the ACHE gene is transcribed to yield the AChE mRNA product. Chemically protected 'antisense' oligonucleotides are directed inversely and form DNA–mRNA hybrids with their target AChE mRNA.[66] These hybrids are destroyed by induced RNase activities, preventing production of the AChE protein and reducing both its catalytic and its non-catalytic activities.[71]

suppressing AChE activity, suppressed neurite growth in mammalian neuroendocrine PC12 cells cultured with nerve growth factor.[65] Neurite growth was rescued by retransfecting these cells (Fig. 4.6(a)) with the AChE-homologous protein, neuroligin, which was the basis for our theory of the redundant function of these two proteins (Fig. 4.6(b)).

Following characterization and optimization of AS-ODNs in cultured cells,[66] AS-AChEs are being used in vivo for improving recovery from injury and for enhancing short-term memory in the cognitively impaired transgenic mice with excess human AChE-S in their neurons.

Intracerebroventricular injection of 1 μg/kg doses of AS-AChE ODNs (Fig. 4.6(c)) reduces AChE-R levels and improves short-term memory for at least 24 h at similar efficiency with the AD drug tacrine, which is used in 1000-fold higher doses and at closely spaced intervals.[72] In developing the next generation of antisense agents, we have synthesized selective ribozymes for catalytic destruction of AChE mRNA transcripts.[73] If successful, such agents may lead to a novel approach for therapeutic suppression of AChE levels in those diseases where conventional inhibitors are now being used.

Discussion

Accumulated findings implicate genetic polymorphism and recently described regulatory processes that control expression of the *CHE* genes in unexpected responses to ChE inhibitors. In particular, induction of a prolonged and exaggerated *ACHE* expression, associated with both immediate and long-term changes in ACh metabolism following acute psychological stress, closed-head injury (itself, a known AD risk factor) and exposure to ChE inhibitors. All of these insults elicit rapid, transient excitation of cholinergic neurons in the CNS. Such excitation activates a feedback response that dramatically up-regulates AChE production, mediated by early immediate transcription factors such as c-fos and egr[74] and dependent on increases in free intracellular calcium. Together with parallel down-regulation of the genes encoding the ACh-synthesizing and ACh-packaging proteins choline acetyltransferase and vesicular ACh transporter, this leads to delayed yet persistent suppression of electrophysiological activity in the hippocampus.[75] This response causes, within 1 h and for at least several days, a beneficial calming of brain hyperactivity by suppression of the increased ACh levels. However, studies with AChE-transgenic mice suggest that excesses of the AChE protein in neurons can also cause late-onset limitation of dendrite branching and depletion of dendritic spines (i.e. impaired networks) in cortical pyramidal neurons. This leads to progressively impaired performance in tests of memory and muscle strength. These findings suggest that at least some of the delayed phenomena induced by AChE excess are caused by the AChE protein per se, independently of its hydrolytic activity.

The biomedical and environmental implications of hAChE research indicate that genomic polymorphisms in the coding sequence and/or promoters of the *ACHE* and *BCHE* genes (and possible additional loci) may modulate individual responses to ChE inhibitors in a complex and yet not fully predictable manner, affecting both the nervous and the haematopoietic systems. Recent cosmid sequencing and genotyping efforts have revealed novel polymorphisms in the *ACHE* upstream promoter sequence.[76] Such polymorphisms can alter *ACHE* expression and/or properties, affecting both short- and long-term manifestations of cholinergic functions in a manner that may increase the risk for neurodegenerative disease due to physical, chemical or psychological insults.

Acknowledgements

This research has been supported by grants from the US Army Medical Research and Development Command (DAMD 17-97-1-

7007), the Israel Science Foundation (590/97), the US–Israel Binational Science Foundation (96-00110) and Ester Neuroscience, Ltd.

References

1. Taylor P. Agents acting at the neuromuscular junction and autonomic ganglia. In: Hardman JG, Limbird LE, Molinoff PB, Ruddon RW, eds. *Goodman and Gilman's The Pharmacological Basis of Therapeutics*. New York: McGraw-Hill, 1996: 177–197.
2. Chatonnet A, Lockridge O. Comparison of butyrylcholinesterase and acetylcholinesterase. *Biochem J* 1989; **260:** 625–634.
3. Prody CA, Zevin-Sonkin D, Gnatt A, Goldberg O, Soreq H. Isolation and characterization of full-length cDNA clones coding for cholinesterase from fetal human tissues. *Proc Natl Acad Sci USA* 1987; **84:** 3555–3559.
4. Soreq H, Ben-Aziz R, Prody CA *et al.* Molecular cloning and construction of the coding region for human acetylcholinesterase reveals a G + C-rich attenuating structure. *Proc Natl Acad Sci USA* 1990; **87:** 9688–9692.
5. Gnatt A, Prody CA, Zamir R, Lieman-Hurwitz J, Zakut H, Soreq H. Expression of alternatively terminated unusual human butyrylcholinesterase messenger RNA transcripts, mapping to chromosome 3q26-*ter*, in nervous system tumors. *Cancer Res* 1990; **50:** 1983–1987.
6. Allderdice PW, Gardner HA, Galutira D, Lockridge O, LaDu BN, McAlpine PJ. The cloned butyrylcholinesterase (BChE) gene maps to a single chromosome site, 3q26. *Genomics* 1991; **11:** 452–454.
7. Ehrlich G, Viegas Pequignot E, Ginzberg D, Sindel L, Soreq H, Zakut H. Mapping the human acetylcholinesterase gene to chromosome 7q22 by fluorescent in situ hybridization coupled with selective PCR amplification from a somatic hybrid cell panel and chromosome-sorted DNA libraries. *Genomics* 1992; **13:** 1192–1197.
8. Getman DK, Eubanks JH, Camp S, Evans GA, Taylor P. The human gene encoding acetylcholinesterase is located on the long arm of chromosome 7. *Am J Hum Genet* 1992; **51:** 170–177.
9. Soreq H, Zakut H. *Human Cholinesterases and Anticholinesterases*. San Diego: Academic Press, 1993.
10. Primo-Parmo SL, Bartels CF, Wiersema B, van der Spek AF, Innis JW, La Du BN. Characterization of 12 silent alleles of the human butyrylcholinesterase (BCHE) gene. *Am J Hum Genet* 1996; **58:** 52–64.
11. Loewenstein Lichenstein Y, Schwarz M, Glick D, Norgaard Pedersen B, Zakut H, Soreq H. Genetic predisposition to adverse consequences of anti-cholinesterases in 'atypical' BCHE carriers. *Nat Med* 1995; **1:** 1082–1085.
12. Whittaker M. *Cholinesterase*. Basel: Karger, 1986.
13. La Du BN. Identification of human serum cholinesterase variants using the polymerase chain reaction amplification technique. *Trends Pharmacol Sci* 1989; **10:** 309–313.
14. McGuire MC, Nogueira CP, Bartels CF *et al.* Identification of the structural mutation responsible for the dibucaine-resistant (atypical) variant form of human serum cholinesterase. *Proc Natl Acad Sci USA* 1989; **86:** 953–957.
15. Neville LF, Gnatt A, Loewenstein Y, Soreq H. Aspartate-70 to glycine substitution confers resistance to naturally occurring and synthetic anionic-site ligands on in-ovo produced human butyrylcholinesterase. *J Neurosci Res* 1990; **27:** 452–460.
16. Neville LF, Gnatt A, Loewenstein Y, Seidman S, Ehrlich G, Soreq H. Intramolecular relationships in cholinesterases revealed by oocyte expression of site-directed and natural variants of human *BCHE*. *EMBO J* 1992; **11:** 1641–1649.
17. Schwarz M, Glick D, Loewenstein Y, Soreq H. Engineering of human cholinesterases explains and predicts diverse consequences of administration of various drugs and poisons. *Pharmacol Ther* 1995; **67:** 283–322.
18. Sussman JL, Harel M, Frolow F *et al.* Atomic structure of acetylcholinesterase from *Torpedo californica*: a prototypic acetylcholine-binding protein. *Science* 1991; **253:** 872–879.
19. Krasowski MD, McGehee DS, Moss J. Natural inhibitors of cholinesterases: implications for adverse drug reactions. *Can J Anaesth* 1997; **44:** 525–534.

20. Ehrlich G, Ginzberg D, Loewenstein Y *et al.* Population diversity and distinct haplotype frequencies associated with *ACHE* and *BCHE* genes of Israeli Jews from trans-Caucasian Georgia and from Europe. *Genomics* 1994; **22:** 288–295.
21. Sternfeld M, Rachmilewitz J, Loewenstein-Lichtenstein Y *et al.* Normal and atypical butyrylcholinesterases in placental development, function, and malfunction. *Cell Mol Neurobiol* 1997; **17:** 315–332.
22. Paoletti F, Mocali A, Vannucchi AM. Acetylcholinesterase in murine erythroleukemia (Friend) cells: evidence for megakarocyte-like expression and potential growth-regulatory role of enzyme activity. *Blood* 1992; **79:** 2873–2879.
23. Patinkin D, Lev Lehman E, Zakut H, Eckstein F, Soreq H. Antisense inhibition of butyrylcholinesterase gene expression predicts adverse hematopoietic consequences to cholinesterase inhibitors. *Cell Mol Neurobiol* 1994; **14:** 459–473.
24. Alber R, Sporns O, Weikert T, Willbold E, Layer PG. Cholinesterases and peanut agglutinin binding related to cell proliferation and axonal growth in embryonic chick limbs. *Anat Embryol Berlin* 1994; **190:** 429–438.
25. Beeri R, Gnatt A, Lapidot Lifson Y *et al.* Testicular amplification and impaired transmission of human butyrylcholinesterase cDNA in transgenic mice. *Hum Reprod* 1994; **9:** 284–292.
26. Soreq H, Malinger G, Zakut H. Expression of cholinesterase genes in human oocytes revealed by in-situ hybridization. *Hum Reprod* 1987; **2:** 689–693.
27. Levey AI, Wainer BH, Rye DB, Mufson EJ, Mesulam MM. Choline acetyltransferase-immunoreactive neurons intrinsic to rodent cortex and distinction from acetylcholinesterase-positive neurons. *Neuroscience* 1984; **13:** 341–353.
28. Mesulam MM, Geula C. Acetylcholinesterase-rich neurons of the human cerebral cortex: cytoarchitectonic and ontogenetic patterns of distribution. *J Comp Neurol* 1991; **306:** 193–220.
29. Appleyard ME. Secreted acetylcholinesterase: non-classical aspects of a classical enzyme. *Trends Neurosci* 1992; **15:** 485–490.
30. Razon N, Soreq H, Roth E, Bartal A, Silman I. Characterization of activities and forms of cholinesterases in human primary brain tumors. *Exp Neurol* 1984; **84:** 681–695.
31. Pakaski M, Kasa P. Glial cells in coculture can increase the acetylcholinesterase activity in human brain endothelial cells. *Neurochem Int* 1992; **21:** 129–133.
32. Karpel R, Ben Aziz Aloya R, Sternfeld M *et al.* Expression of three alternative acetylcholinesterase messenger RNAs in human tumor cell lines of different tissue origins. *Exp Cell Res* 1994; **210:** 268–277.
33. Lapidot-Lifson Y, Prody CA, Ginzberg D, Meytes D, Zakut H, Soreq H. Coamplification of human acetylcholinesterase and butyrylcholinesterase genes in blood cells: correlation with various leukemias and abnormal megakaryocytopoiesis. *Proc Natl Acad Sci USA* 1989; **86:** 4715–4719.
34. Zakut H, Ehrlich G, Ayalon A *et al.* Acetylcholinesterase and butyrylcholinesterase genes coamplify in primary ovarian carcinomas. *J Clin Invest* 1990; **86:** 900–908.
35. Zakut H, Lapidot-Lifson Y, Beeri R, Ballin A, Soreq H. In vivo gene amplification in non-cancerous cells: cholinesterase genes and oncogenes amplify in thrombocytopenia associated with lupus erythematosus. *Mutat Res* 1992; **276:** 275–284.
36. Prody CA, Dreyfus P, Zamir R, Zakut H, Soreq H. De novo amplification within a 'silent' human cholinesterase gene in a family subjected to prolonged exposure to organophosphorous insecticides. *Proc Natl Acad Sci USA* 1989; **86:** 690–694.
37. Zakut H, Even L, Birkenfeld S, Malinger G, Zisling R, Soreq H. Modified properties of serum cholinesterases in primary carcinomas. *Cancer* 1988; **61:** 727–737.
38. Zakut H, Lieman Hurwitz J, Zamir R, Sindell L, Ginzberg D, Soreq H. Chorionic villus cDNA library displays expression of butyrylcholinesterase: putative genetic disposition for ecological danger. *Prenat Diagn* 1991; **11:** 597–607.
39. Lev-Lehman E, Deutsch V, Eldor A, Soreq H. Immature human megakaryocytes produce nuclear-associated acetylcholinesterase. *Blood* 1997; **89:** 3644–3653.

40. Grisaru D, Lev-Lehman E, Shapira M *et al.* Human osteogenesis involves differentiation-dependent increases in the morphogenically active 3' alternative splicing variant of acetylcholinesterase. *Mol Cell Biol* 1999; **19:** 788–795.
41. Llinas RR, Greenfield SA. On-line visualization of dendritic release of acetylcholinesterase from mammalian substantia nigra neurons. *Proc Natl Acad Sci USA* 1987; **84:** 3047–3050.
42. Dupree JL, Maynor EN, Bigbee JW. Inverse correlation of acetylcholinesterase (AChE) activity with the presence of neurofilament inclusions in dorsal root ganglion neurons cultured in the presence of a reversible inhibitor of AChE. *Neurosci Lett* 1995; **197:** 37–40.
43. Koenigsberger C, Chiappa S, Brimijoin S. Neurite differentiation is modulated in neuroblastoma cells engineered for altered acetylcholinesterase expression. *J Neurochem* 1997; **69:** 1389–1397.
44. Massoulie J, Pezzementi L, Bon S, Krejci E, Vallette FM. Molecular and cellular biology of cholinesterases. *Prog Neurobiol* 1993; **41:** 31–91.
45. Seidman S, Sternfeld M, Ben Aziz Aloya R, Timberg R, Kaufer Nachum D, Soreq H. Synaptic and epidermal accumulations of human acetylcholinesterase are encoded by alternative 3'-terminal exons. *Mol Cell Biol* 1995; **15:** 2993–3002.
46. Karpel R, Sternfeld M, Ginzberg D, Guhl E, Graessmann A, Soreq H. Overexpression of alternative human acetylcholinesterase forms modulates process extensions in cultured glioma cells. *J Neurochem* 1996; **66:** 114–123.
47. Sternfeld M, Ming G, Song H *et al.* Acetylcholinesterase enhances neurite growth and synapse development through alternative contributions of its hydrolytic capacity, core protein, and variable C termini. *J Neurosci* 1998; **18:** 1240–1249.
48. Futerman AH, Low MG, Ackermann KE, Sherman WR, Silman I. Identification of covalently bound inositol in the hydrophobic membrane-anchoring domain of Torpedo acetylcholinesterase. *Biochem Biophys Res Commun* 1985; **129:** 312–317.
49. Kaufer D, Friedman A, Seidman S, Soreq H. Acute stress facilitates long-lasting changes in cholinergic gene expression. *Nature* 1998; **393:** 373–377.
50. Beeri R, Andres C, Lev Lehman E *et al.* Transgenic expression of human acetylcholinesterase induces progressive cognitive deterioration in mice. *Curr Biol* 1995; **5:** 1063–1071.
51. Andres C, Beeri R, Friedman A *et al.* Acetylcholinesterase-transgenic mice display embryonic modulations in spinal cord choline acetyltransferase and neurexin Iβ gene expression followed by late-onset neuromotor deterioration. *Proc Natl Acad Sci USA* 1997; **94:** 8173–8178.
52. Mollgard K, Dziegielewska KM, Saunders NR, Zakut H, Soreq H. Synthesis and localization of plasma proteins in the developing human brain. Integrity of the fetal blood–brain barrier to endogenous proteins of hepatic origin. *Dev Biol* 1988; **128:** 207–221.
53. Friedman A, Kaufer D, Shemer J, Hendler I, Soreq H, Tur Kaspa I. Pyridostigmine brain penetration under stress enhances neuronal excitability and induces early immediate transcriptional response. *Nat Med* 1996; **2:** 1383–1385.
54. Sternfeld M, Patric, JD, Soreq H. Position effect variegations and brain-specific silencing in transgenic mice overexpressing human acetylcholinesterase variants. *J Physiol (Paris)* 1998; **92:** 249–255.
55. Shoham S, Sternfeld M, Milay T, Patrick JW, Soreq H. Transgenic human AChE variants display different capacities of inducing 'corkscrew-like' neuronal processes in mouse somatosensory cortex. *Neurosci Lett* 1998; **51(Suppl):** S38.
56. Beeri R, Le Novere N, Mervis R *et al.* Enhanced hemicholinium binding and attenuated dendrite branching in cognitively impaired acetylcholinesterase-transgenic mice. *J Neurochem* 1997; **69:** 2441–2451.
57. Soreq H, Patinkin D, Lev-Lehman E *et al.* Antisense oligonucleotide inhibition of acetylcholinesterase gene expression induces progenitor cell expansion and suppresses hematopoietic apoptosis ex vivo. *Proc Natl Acad Sci USA* 1994; **91:** 7907–7911.
58. Gennarelli TA, Graham DI. Neuropathology of the head injuries. *Semin Clin Neuropsychiatry* 1998; **3:** 160–175.

59. Grisaru D, Sternfeld M, Eldor A, Glick D, Soreq H. Structural roles of acetylcholinesterase variants in biology and pathology. *Eur J Biochem* 1999; **264:** 672–686.
60. Coyle JT, Price DL, DeLong MR. Alzheimer's disease: a disorder of cortical cholinergic innervation. *Science* 1983; **219:** 1184–1190.
61. Giacobini E. Cholinergic foundations of Alzheimer's disease therapy. *J Physiol Paris* 1998; **92:** 283–287.
62. Wright CI, Guela C, Mesulam MM. Protease inhibitors and indoleamines selectively inhibit cholinesterases in the histopathologic structures of Alzheimer disease. *Proc Natl Acad Sci USA* 1993; **90:** 683–686.
63. Inestrosa NC, Alvarez A, Perez CA *et al.* Acetylcholinesterase accelerates assembly of amyloid-β-peptides into Alzheimer's fibrils: possible role of the peripheral site of the enzyme. *Neuron* 1996; **16:** 881–891.
64. Ehrlich G, Patinkin D, Ginzberg D, Zakut H, Eckstein F, Soreq H. Use of partially phosphorothioated 'antisense' oligodeoxynucleotides for sequence-dependent modulation of hematopoiesis in culture. *Antisense Res Dev* 1994; **4:** 173–183.
65. Grifman M, Galyam N, Seidman S, Soreq H. Functional redundancy of acetylcholinesterase and neuroligin in mammalian neuritogenesis. *Proc Natl Acad Sci USA* 1998; **95:** 13 935–13 940.
66. Grifman M, Soreq H. Differentiation intensifies the susceptibility of pheochromocytoma cells to antisense oligodeoxynucleotide-dependent suppression of acetylcholinesterase activity. *Antisense Nucleic Acid Drug Dev* 1997; **7:** 351–359.
67. Donger C, Krejci E, Serradell AP *et al.* Mutation in the human acetylcholinesterase-associated collagen gene, COLQ, is responsible for congenital myasthenic syndrome with end-plate acetylcholinesterase deficiency (Type Ic). *Am J Hum Genet* 1998; **63:** 967–975.
68. Ichtchenko K, Nguyen T, Sudhof TC. Structures, alternative splicing, and neurexin binding of multiple neuroligins. *J Biol Chem* 1996; **271:** 2676–2682.
69. Hata Y, Butz S, Sudhof TC. CASK: a novel dlg/PSD95 homolog with an N-terminal calmodulin-dependent protein kinase domain identified by interaction with neurexins. *J Neurosci* 1996; **16:** 2488–2494.
70. Irie M, Hata Y, Takeuchi M *et al.* Binding of neuroligins to PSD-95. *Science* 1997; **277:** 1511–1515.
71. Seidman S, Eckstein F, Grifman M, Soreq H. Antisense technologies have a future fighting neurodegenerative diseases. *Antisense Nucleic Acid Drug Dev* 1999; **9:** 333–340.
72. Seidman S, Cohen O, Ginsberg D *et al.* Multilevel approaches to AChE-induced impairments in learning and memory. In: Doctor BP, Taylor P, Quinn DM, Rotundo RL, Gentry MK, eds. *Structure and Function of Cholinesterases and Related Proteins.* New York: Plenum Press, 1998: 183–184.
73. Birikh KR, Berlin YA, Soreq H, Eckstein F. Probing accessible sites for ribozymes on human acetylcholinesterase RNA. *RNA* 1997; **3:** 429–437.
74. von der Kammer H, Mayhaus M, Albrecht C, Enderich J, Wegner M, Nitsch RM. Muscarinic acetylcholine receptors activate expression of the EGR gene family of transcription factors. *J Biol Chem* 1998; **273:** 14 538–14 544.
75. Kaufer D, Friedman A, Soreq H. The vicious circle: long-lasting transcriptional modulation of cholinergic neurotransmission following stress and anticholinesterase exposure. *Neuroscientist* 1999; **5:** 173–183.
76. Shapira M, Korner M, Bosgraaf L, Tur-Kaspa I, Soreq H. The human ACHE locus includes a polymorphic enhancer domain 17 kb upstream from the transcription start site. In: Doctor BP, Taylor P, Quinn DM, Rotundo RL, Gentry MK, eds. *Structure and Function of Cholinesterases and Related Proteins.* New York: Plenum Press, 1998: 111.

5

The genes encoding the cholinesterases: structure, evolutionary relationships and regulation of their expression

Palmer Taylor, Z David Luo, Shelley Camp

Introduction

The first cholinesterase (ChE) sequence was obtained from the cloning of a cDNA encoding acetylcholinesterase (AChE) from *Torpedo californica* using degenerate oligonucleotides encoding amino acid sequences from isolated AChE peptides. The sequence showed no global homology with any of the other serine hydrolases known at the time, despite similarity in catalytic parameters and a common pentapeptide sequence around the active-center serine.[1] Rather, sequence identity was evident between AChE and the carboxyl-terminal (C-terminal) region of thyroglobulin.[1,2] That an enzyme designed to rapidly hydrolyze a neurotransmitter showed structural similarities to the precursor of the thyroid hormones, thyroxin (T_3) and T_4, provided an indication that the ChEs were part of a larger superfamily of proteins. Thus, both hydrolase and non-hydrolase functions might be subserved by a common structural matrix.

Soon after *Torpedo* AChE was cloned, the *Drosophila* ChE gene was characterized from genetic studies and the sequence of the gene determined.[3] This was followed by butyrylcholinesterase (BuChE) sequences obtained from amino acid sequencing[4] and from molecular cloning.[5,6] Mammalian AChEs proved somewhat more intractable, but in 1990 mouse, bovine, and human AChE sequences were reported.[7–9] Since then sequences for a variety of vertebrate and invertebrate species, including arthropods, reptiles, and nematodes, have been published.

Hydrolases with distinctive substrate specificities from microbes (*Streptomyces*, *Pseudomonas*), fungi (*Geotrichum* and *Candida*), insects other than *Drosophila*, plants (*Arabidopsis*), *Caenorhabditis elegans*, and *Dictyostelium* have been reported.[10–18] Other proteins serving related functions fall in the superfamily.[10]

Analysis of the common features of three-dimensional structure of AChE[19] and a homologous lipase[20] indicates that these proteins have a common fold termed the α,β hydrolase fold.[21] Several other proteins related in sequence, the structures of which have more recently been determined, also show these common features in three-dimensional structure. A database containing a list of 497 sequences related to the ChEs, of which 54 are ChEs based on their catalytic activity or close sequence identity, may be found on the ESTER web site (http://www.ensam.inra.fr/cholinesterase/). The subclassifications of catalytic activity include: acetylhydrolases, ChEs, carboxylesterases, carboxypeptidases, cutinases, dienelactone hydrolases, endo- and exopeptidases, haloalkane dehydrogenases, haloperoxidases, hemagglutinins, lipases, lyases, and vitallogenins.

In addition, proteins of the tactin family (entactin, glutactin, gliotactin, and neurotactin) are also homologous to the ChEs but, like thyroglobulin, these proteins lack apparent hydrolase activity or the features of sequence that give rise to a full catalytic triad.[22–25] Another protein with similarities in sequence to AChE, neuroligin, has been isolated from mouse.[26,27] Neuroligin lacks apparent catalytic activity as a hydrolase, but its interacting partner, neurexin, is well characterized. In fact, it was the interaction of neuroligin with neurexin 1-β that led to its isolation.

A comparison of sequences of members of the family shows neuroligin to be slightly more closely related to the ChEs than are the tactins.[27] Homology modeling has revealed that neuroligin and AChE have electrostatic dipoles with similar orientations in the two molecules.[28] Moreover, both molecules appear to have a similar internal core structure and exhibit departures in structure at the tips of the loops.[29] Common EF hand motifs are found in the two molecules, and these may form the basis of the Ca^{2+} dependent neuroligin–neurexin interactions.[29] The fact that members of the α,β hydrolase fold family are involved in the formation or maintenance of heterologous cell contacts lends credence to the possibility that ChEs may serve a non-hydrolytic function in development. However, since proteins with similarities in structure may confer overlapping functions at excessive concentrations, it may prove difficult to establish a physiologic function from overexpression of the gene. Figure 5.1 shows the similarities and differences in structure of key members of the α,β hydrolase fold family. This listing and the more comprehensive presentation in the ESTER database reveal that the α-β hydrolase fold defines a functionally eclectic group of proteins, the functions of which are not solely restricted to hydrolytic catalysis.

Key functional and structural components

Since the initial cloning of the ChE relied heavily on generated amino acid sequence, shortly thereafter disulfide bond patterns were established for AChE[30] and BuChE.[31] Labeling with diisopropylphosphorofluoridate (DFP) identified the active-center serine;[32] the histidine in the active-center triad was established by site-specific mutagenesis,[33] while for the third participant, glutamate, the crystal structure was required for identification.[19] The amino acid order of the catalytic triad in this family, using the initial *Torpedo* numbering system is S200, H440, E327. This order differs from the trypsin and other families of serine hydrolases and constitutes another argument against divergent evolution amongst the serine hydrolases. Corresponding residues in the triad can be found in all family members possessing hydrolase activity, whereas those members with projected other functions typically lack one or more of the residues in the catalytic triad.

Three disulfide loops are found in several proteins in the family (all ChEs, neuroligin, and the *Dictyostelium* proteins); others contain two disulfide loops, while certain esterases contain only the amino-terminal (N-terminal) disulfide loop. The third or most C-terminal loop, in addition to containing the catalytic histidine in AChE, forms part of the four-helix bundle responsible for intersubunit contacts.[19] In many, but not all, of the alternatively spliced ChE gene products, an additional cysteine is found near the very C-terminus. It forms disulfide bonds with other catalytic or structural subunits in the formation of homo-

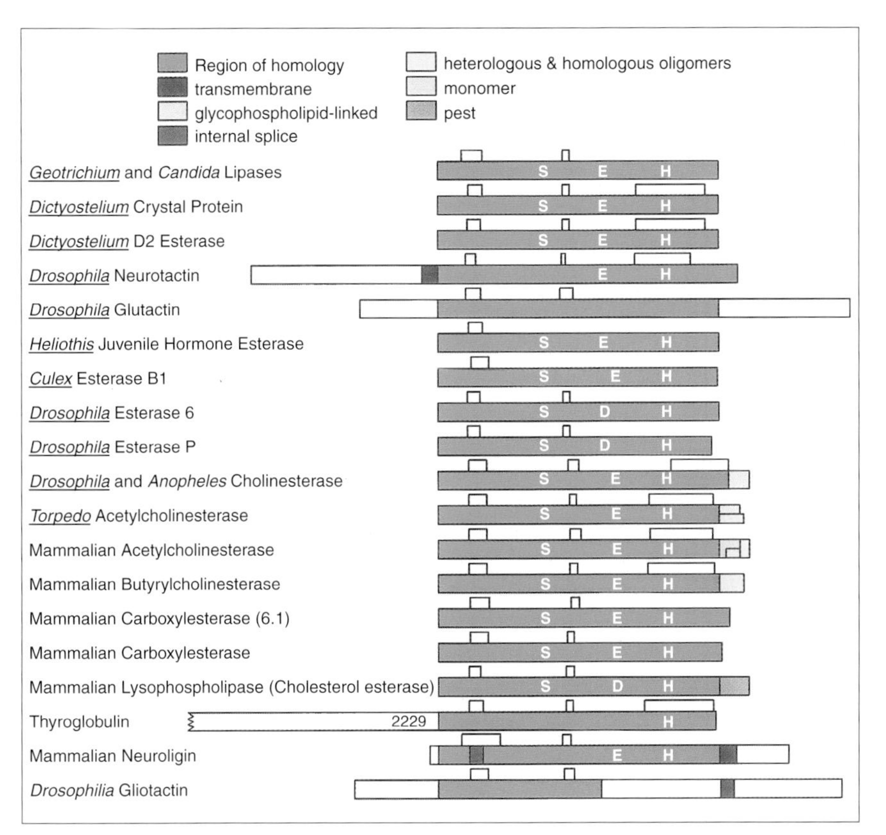

Figure 5.1
Structural relationships between the α,β hydrolase fold family of proteins. The serines, histidines, and glutamates (or aspartates) in homologous positions to S200, H440, and E327 in Torpedo AChE are shown. Presumed intrasubunit disulfide bonds are shown by the bracketed loops above the sequences. The homologous or invariant regions are shown in green.

meric or heteromeric oligomers. In some family members, a seventh cysteine is found at variable locations in the sequence.

Gene organization in relation to protein structure

Relating gene structure to that of the gene products provides additional insight into the potential factors that control gene expression. Typically, ChEs have been defined as AChEs (EC 3.1.1.7) and BuChEs (EC 3.1.1.8). The latter have a broad selectivity with respect to the size of the acyl group, whereas for AChE a marked reduction in catalytic rate is seen at the propionyl to butyryl juncture for acyl moiety substitutions.[34] Over several decades, selective inhibitors of AChE and BuChE have been identified.[34–37] Inhibitor specificity is based on the larger acyl pocket of BuChE for tetraisopropyl pyrophosphoramide (iso-OMPA) inhibition, the absence of a tyrosine or phenylalanine in the choline subsite of BuChE for ethopropazine inhibition, and a patch of aromatic residues near the mouth of the active-center gorge conferring inhibition of AChE by fasciculin, BW284C51, and propidium.[37–40]

Genomic clones of *Drosophila* ChE,[41] *Caenorhabditis elegans* ChEs,[42] *Torpedo* AChE,[43] snake AChE,[44] mouse AChE,[45]

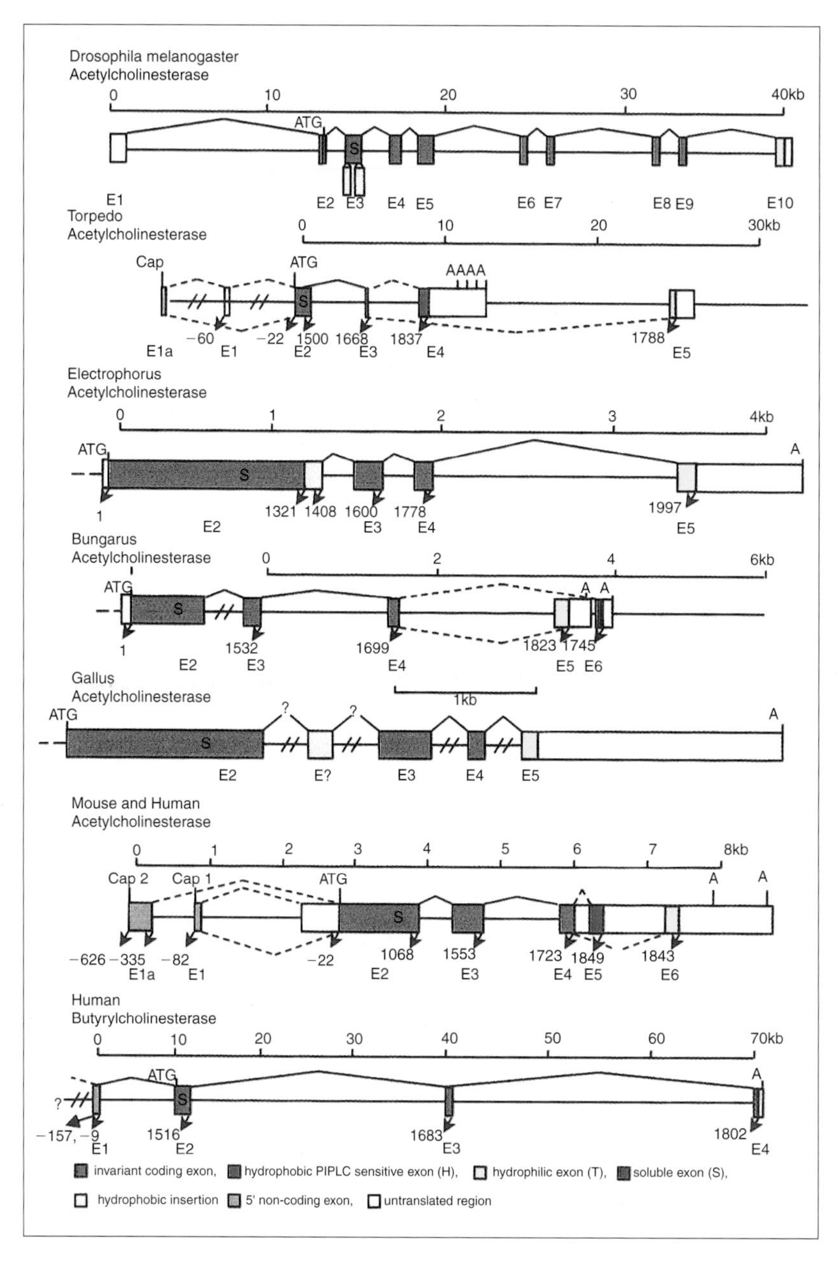

Figure 5.2
Structure of ChE genes from higher organisms. The exons are numbered sequentially starting with the most proximal Cap site (i.e. the start site for transcription). Alternate Cap sites are designated as 1a, etc. Splicing between invariant exons is shown by a continuous line; alternatively spliced exons are connected by dashed lines. The retained intron in the mammalian gene after exon 4 is shown by an open box. Numbering considers only base pairs in the exons beginning at the ATG start site for translation. The active-center serine (S) is noted in its encoding exon. Translational stop signal boundaries and polyadenylation signals (A) are also noted. Intron lengths are approximate, since not all have been completely sequenced. The individual genes differ substantially in the range of sequence covered. Data are derived from the ESTHER website (http://meleze.ensam.inra.fr/cholinesterase/java/genes. html), the GenomeNet www server (http://www. genome.ad.jp) and references 41–50, 66, 82, 83, 87 and 91.

Electrophorus AChE,[46] human AChE,[47,48] and human BuChE[49] have been isolated (Fig. 5.2). *C. elegans* ChE is encoded by four genes, the structures of which are not completely resolved. The *Drosophila* gene has a complex organization of exons, whereas the organization of the *Torpedo* AChE gene and mammalian AChE and BuChE genes are surprisingly simple (see Fig. 5.2). For example, the *Torpedo* gene has six identified exons, two of which are upstream of the translation start site.[50] In the open-reading frame, two invariant and two alternatively spliced exons are found. The second invariant exon, exon 3 in Fig. 5.2, splices to one of two downstream exons, giving rise to the two alternatively spliced forms of *Torpedo* AChE. One splice gives rise to a glycophospholipid linkage suitable for tethering the enzyme to the plasma membrane. The other splice yields a tryptophan amphiphilic tetramerization domain which associates with a proline-rich attachment domain on a structural subunit.[51] Both spliced forms contain a cysteine capable of disulfide linking to an identical catalytic subunit or a structural subunit. A similar situation exists for the mammalian enzyme except that the invariant region is divided into three exons. Alternative splicing occurs at approximately the same position in the linear sequence for the *Torpedo*, reptilian, and mammalian genes.

Evolutionary relationships among the ChEs

The basic structural motif, the α,β hydrolase fold, can be found in proteins from prokaryotic cells, some of which have hydrolase activity. Early evidence revealed ChE-like activity in *Pseudomonas* and other bacterial species.[52] AChE is also found in species and tissues devoid of cholinergic nerves.[53] Thus, acetylcholine and its degradative enzyme may serve functions other than neurotransmission in these primitive species.

The invertebrate *Drosophila melanogaster* and other insects express a single ChE, the substrate specificity of which falls between that of mammalian AChE and BuChE. In fact, the region determined by homology modeling to be the acyl pocket has a single phenylalanine instead of two phenylalanines found in AChE or none found in BuChE. The invertebrate nematode *C. elegans* has four ChE genes encoding proteins closely related in sequence. These four genes encode putative acyl pockets also containing a single phenylalanine.[42]

The cephalochordate invertebrate *Amphioxus* is a close relative to the vertebrates and has two ChE genes: one gene product has AChE-like properties, whereas the other has catalytic properties similar to BuChE or the carboxylesterases.[54,55] Active-center sequences for the two ChEs in two species of *Amphioxus* have been elucidated.[55] The ChE containing the phenylalanine in the active center is selective for catalyzing AChE hydrolysis and is inhibited by many of the classical ChE inhibitors, while the second ChE shows no acetyl to butyryl selectivity and is resistant to the inhibitors. The two enzymes appear to differ in acyl pocket dimensions and in the disposition of aromatic residues in the active center gorge. This gene duplication in *Amphioxus* may bear ancestral relationships to vertebrate AChE and BuChE or represent a parallel gene duplication to that found in vertebrates.

Agnathan or jawless vertebrates have only a single ChE; its amino acid sequence in the acyl pocket and substrate specificity resemble that of AChE. The jawed vertebrates have both AChE and BuChE genes conferring distinctive activities, although the large departure in substrate specificity between the two enzymes occurs only as we ascend the evolutionary scale in the vertebrates to birds and mammals. Pezzementi *et al.*[55] have developed some intriguing models of the relationships of the ChEs from various species. A comprehensive analysis of this extensive family awaits the acquisition of complete sequences for the species believed to reside at critical stages of evolution.

Relationships between the gene structures of the ChEs

Regulation of gene expression

An overriding consideration in AChE expression is that all electric fish, mammalian, and avian AChEs are encoded by single genes[43–48,56] located at chromosomal position 7q23 in man[57,58] and at the distal position of chromosome 5 in mouse.[59] BuChE is also encoded by a single gene located 3q26 in man.[60,61] The absence of a functional BuChE leads to an apparently normal phenotype unless drugs such as succinylcholine are administered.[35,62] A therapeutic dose of succinylcholine accounts for a large fraction of the drug being hydrolyzed in plasma prior to entry to its extravascular site of action. In the absence of plasma hydrolysis, the dose is sufficient to cause excessive paralysis and prolonged apnea. This evidence, along with differing BuChE mutations in South American populations eating foodstuffs containing toxic esters, would suggest that BuChE functions primarily to catalyze hydrolysis of toxic esters in the diet.[63] By contrast, humans devoid of AChE are not known, and a transgenic mouse lacking the capacity to express AChE, and thus devoid of the enzyme in all tissues, experiences severe tremors. Despite a near normal appearance at birth, weight gain in the AChE knockout mouse is delayed, and the animals do not live beyond postnatal day 21.[64]

If we consider that the two families of cholinergic receptors subserved by AChE exhibit considerable diversity in subtype (i.e. five muscarinic receptor genes and 14 nicotinic receptor subunit genes that can be expressed in various combinations to form a pentameric nicotinic receptor), it is noteworthy that an enzyme expressed from a single gene could partner effectively with such a diversity in receptor subtypes. To do so may require extraordinary complexity in the expression pattern of AChE.

The regulation of gene expression can occur at the level of transcription, splicing of the gene, stabilization of the mRNA, and subsequent mRNA translation. Processing, assembly and stability of the translation product relative to the rate of translation will also govern steady-state levels of the protein. Of the events between transcription and translation, only alternative splicing will govern directly the diversity in the gene product in terms of the molecular forms produced or observed at steady state (Fig. 5.3). Thus, alternative splicing is critical to achieving the distinct distributions of the AChE species arising from a single gene.

Regions of alternative mRNA splicing

Two regions of alternative splicing have been identified in the AChE gene (see Fig. 5.2). They are found in the 5' untranslated region and in the 3' region just in front of the stop codon. Alternative splicing in the 5' untranslated region allows for the gene to have different translational control, while alternative splicing in the 3' region of the open-reading frame of the gene gives rise to distinct C-termini. Intervening these regions is an invariant region encompassing most of the open-reading frame.

Control of transcription and its sites of initiation

Primer extension and mRNA protection led to the identification of the primary Cap sites for the initiation of transcription in the *Torpedo*,[50] *C. elegans*,[65] mouse,[45,66] and human[45,47,48] genomic clones. Upstream of the Cap sites, a plethora of candidate consensus sequences for association of transcription factors can be

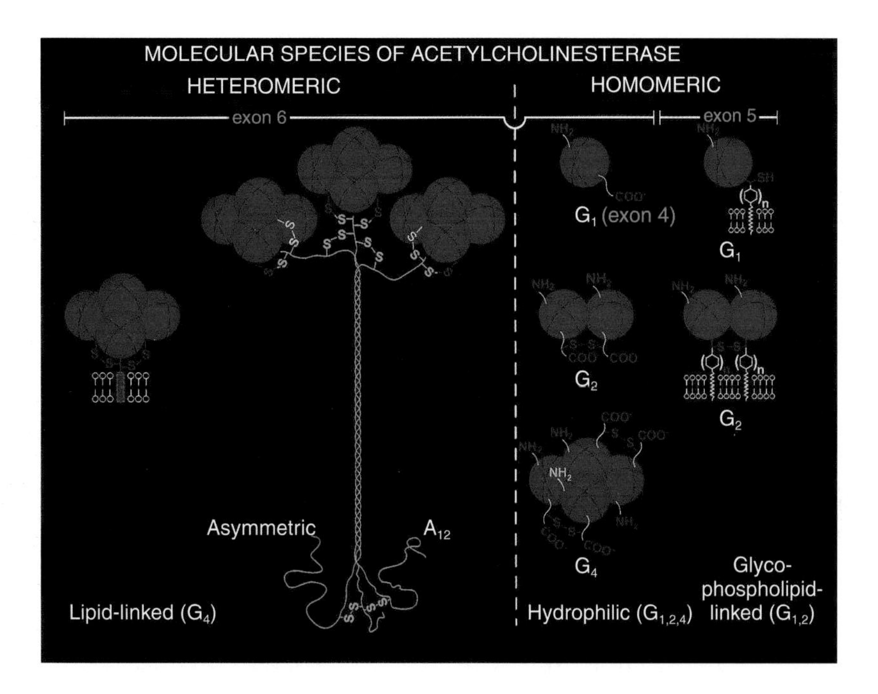

Figure 5.3
Molecular species of mammalian AChE. The species are divided into two classes: a heteromeric class in which the catalytic subunits exist as tetramers and are disulfide bonded to either a lipid-linked subunit or a triple helix of collagen-containing subunits through a proline-rich attachment domain.[51,87] *The homomeric class exists as monomers, dimers, and tetramers and can be divided into the hydrophilic and amphiphilic forms. Hydrophilic forms may be formed by retaining the intron between exons 4 and 5 and by splicing exon 4 to exon 6. For the gene product of the latter splice to be hydrophilic, it must assemble into a tetramer; otherwise, an amphipathic helix is exposed at the C-terminus of each subunit. The amphiphilic forms result from the exon 4 to 6 splice, where the assembled forms are only monomeric and dimeric, and from the exon 4 splice to exon 5, which encodes a signal peptide for glycophospholipid attachment at the C-terminus of the nascent peptide. Thus, the initial determinant for the multiplicity of molecular forms of AChEs arises from alternative mRNA processing.*

found. However, given the sequence redundancy in their recognition sequences and the multiplicity of potential *trans*-acting factors, their significance can only be ascertained from the binding of endogenous factors and, more importantly, their capacity to control expression. To date there is a paucity of data for establishing the factors influencing AChE gene expression in specific tissues or during differentiation.

In the *Torpedo* AChE gene, two short exons are found in the 5' untranslated region. Two closely spaced candidate Cap sites are found at −143 and −138 base pairs from the start methionine signal. Twenty-one base pairs in the 5' direction from the Cap site a repeating

TATA sequence is found, and further upstream consensus sequences for Sp1, AP1, and AP2 transcription factors are detected. When this candidate promoter region is linked to a reporter gene, enhanced transcription is evident in C2–C12 myoblasts, but no further increase is seen upon differentiation of the myoblasts into myotubes. When AP1 (c-Jun) and AP2 expression vectors are co-transfected with the reporter gene into HepG2 hepatoma or C2–C12 muscle cells, marked increases in transcription are evident (up to 18-fold). Further increases can be achieved with agents that stimulate protein kinases A and C. AP1 and AP2 binding can also be detected on the candidate consensus sequences by gel shift and footprinting assays.[50]

The putative promoter and enhancer sequences, Cap site, and the exon–intron boundaries in the promoter and 5' untranslated region are remarkably similar in the mouse[45,66] and human[45,47,48] AChE genes. However, they differ markedly from these regions in the *Torpedo* AChE gene. The mammalian AChE gene lacks a TATA box, the absence of which is often encountered in housekeeping genes that maintain constitutive activity. Moreover, the potential transcription factors, their positions with respect to the CAP site and their apparent capacity to activate gene expression in the mammalian AChE genes differ substantially from the *Torpedo* gene.

The binding of transcription factors, such as Sp1, AP1, and Myo D, to the mammalian genes can be demonstrated by band shift and footprinting experiments.[48,66,67] Moreover, interactions between the transcription factors are evident.[48,66,67] However, co-transfection of the AChE gene with genes expressing putative *trans*-acting factors only yields modest transcriptional activation. For example, both the AChE gene and the genes expressing the α and γ subunits of the nicotinic AChE receptor contain multiple E boxes, which are activated by the muscle determination genes. However, in co-transfection experiments, only the receptor is transcriptionally activated by Myo D and myogenin.[67] Position of the *cis* elements may be important here, since the E box elements in the receptor subunits are closer to the Cap site. To date, little evidence has accrued showing that transcription factors directly control AChE gene expression in early differentiation steps in muscle and nerve. Whether transcriptional activation is modest in all mammalian systems or the right combinations of factors and cell systems have yet to be found remain as open questions. However, the recent considerations discussed below indicate that transcription cannot be disregarded.

A unique feature of the mammalian AChE gene is its relatively small size. The entire gene containing the promoter region and extending through the second polyadenylation signal can be found in an 8.5 kb *Hin*dIII fragment and easily cloned into vectors lacking functional components for mammalian expression. Camp and Taylor[68] have transfected the entire mouse gene into C2–C12 myoblasts and have found the transfected gene to be regulated upon muscle differentiation in a fashion similar to the endogenous gene in these cells. Upon dissecting the gene they found that the first intron (the segment between exons 1 and 2 and residing in the 5' untranslated region) is required to achieve appreciable specific AChE expression associated with myoblast to myotube differentiation. Delimiting this region further shows that a region of 257 base pairs in the 2 kb intron is responsible; this region of the intron is uniquely conserved between mouse and human AChE genes. Since this region of the intron is ultimately spliced from the gene, and it increases expression even when inverted, it is likely that it functions in an enhancer capacity. Moreover, this enhancer capacity seems only to function when linked to the endogeneous AChE promoter, since increased expression is not achieved when viral promoters, such as SV-40 and CMV, are substituted (S Camp, unpublished).

Jasmin and colleagues[69] examined AChE gene expression at a different stage of muscle differentiation by injecting plasmids containing portions of rat genomic clones directly into muscle. This approach enabled them to examine AChE gene expression in junctional and extra-junctional areas of innervated muscle. They found that this first intronic region enhanced expression, particularly in junctional areas of the intact muscle. Multiple N-boxes may enhance transcription and a critical N-box sequence lies in the conserved sequence region between 1057 and 1314 base pairs identified by Camp and Taylor.[68] Thus, this intronic region, when acting with the endogenous promoter, may enhance transcription at several stages in muscle differentiation and play a role in the focal expression of the gene in innervated muscle.

Stabilization of AChE mRNA associated with muscle differentiation

Early studies with C2–C12 muscle cells showed a parallel increase in AChE and nicotinic acetylcholine receptor (nAChR) subunit expression during myoblast to myotube differentiation when examined either as levels of the respective mRNAs or gene products.[70] However, the underlying basis for the increased expression of these two genes appears to differ. Reporter gene studies showed increased transcription of the nAChR α or γ subunit genes with little change in AChE gene transcription. Run-on transcription also showed enhanced nAChR subunit gene transcription with little change for the AChE gene. Finally, studies employing cycloheximide as an inhibitor of translation, or 5,6 dichloro-1β-D-ribofuranosyl benzimidazole (DRB), as an inhibitor of transcription, show during differentiation a superinduction of the AChE mRNA, but not of the nAChR subunit mRNAs. Thus, control of gene expression differs for these two postsynaptic proteins, despite a parallel overall expression pattern.[70]

Superinduction produced by inhibitors of transcription or protein synthesis complicates the measurement of mRNA half-life in myoblasts and myotubes, since any block of transcription is a global one. Hence, block preludes the expression of rapidly turning over proteins that may control stability of the mRNA. Nevertheless, substantive evidence has accrued that mRNA stabilization is a major factor responsible for the enhanced expression of the AChE gene during the transition from myoblasts to myotubes. Although differentiation is slower and requires specific cofactors, the increase in AChE expression found upon retinoic acid elicited differentiation of P19 cells from neuroblasts to neurons also involves mRNA stabilization.[71]

During differentiation of hematopoietic cells of erythroid and megakaryocyte lineages, large increases in AChE activity have been observed. Studies on erythroleukemia cells that are allowed to differentiate in response to dimethylsulfoxide (DMSO) showed enhanced AChE mRNA levels and a decreased rate of mRNA degradation.[72] Hence, this differentiation process again appears to be associated with stabilization of the AChE mRNA.

Enhanced AChE gene expression in muscle requires the mobilization of Ca^{2+} through L-type Ca^{2+} channels in the cell membrane and ryanodine sensitive Ca^{2+} channels in the T-tubule, since both the dihydropyridines and ryanodines will block the enhanced expression in differentiating cells.[73] Moreover, mice deficient in the L-type Ca^{2+} channel show greatly diminished AChE expression, whether examined in the nullizygote at birth or in cultured cells produced from embryonic animals.[74] Calcineurin, a protein phosphatase, appears to destabilize the AChE mRNA during differentiation of muscle cells. Cyclophyllin association with calcineurin results in calcineurin inhibition, and this inhibition can be elicited by the immunosuppressant agent, cyclosporine, which forms a complex with its receptor, cyclophyllin. Cyclosporine treatment of C2–C12 cells during the myoblast to myotube conversion results in a three-fold enhancement of differentiation-induced AChE expression.[75]

Several trophic factors have been reported to influence AChE gene expression. TrK A is a high-affinity neurotropin receptor that is responsive to nerve growth factor and responsible for development of sympathetic paravertebral neurons and chromaffin tissue. A deficiency in TrK A influences preganglionic innervation as well as AChE expression in the adrenal medulla and sympathetic preganglionic neurons.[76] Related neurotropins, TrK B and TrK C, influence AChE gene expression in motor neurons after axotomy.[77] Calcitonin gene related peptide (CGRP) will decrease AChE expression in differentiating C2–C12 myotubes, cells of mouse origin.[78] Others have shown this peptide to increase AChE expression in cultured chick myotubes. Hence, distinct differences in AChE expression in response to trophic factors are evident in these species.[79] AChE expression has long been known to differ following denervation in murine and avian species.[80]

Both innervation of muscle[81] and subsequent synaptic activity[82] influence the distribution and level of AChE mRNA. Production of mRNA appears asynchronous and restricted to focal patches of innervation.[81] Fast muscle maintains higher mRNA levels, and mRNA expression is dramatically lowered by low frequency stimulation of fast muscle.[82] In osteosarcoma cells, steroid osteogenic factors, such as vitamin D_3 and 17β-estradiol, enhance AChE expression.[83] This study raises the intriguing issue of whether AChE itself influences proliferation or differentiation in osteogenesis.

Alternative mRNA processing

Regions of alternative mRNA processing have been identified in the reptilian, mammalian and *Torpedo* AChE genes,[43–45,66,84–88] whereas alternative splicing has yet to be found in mammalian BuChE. In mammalian and *Torpedo* AChE, the first alternative splice is in the 5' untranslated region and indicates usage of more than a single Cap site. In addition, the acceptor site which splices out intron I in the mammalian gene may have alternative positions.[45,66,84,85]

Two or more Cap sites exist for the mammalian gene, and they splice into an invariant region which encodes the open-reading frame of the gene. Early studies showed the proximal Cap site to be approximately 1.6 kb upstream of the methionine start site.[45,66] Protection studies also indicate that an upstream exon may also be used in brain.[66] Subsequently, Brimijoin and colleagues[85] further characterized and sequenced this upstream exon and its promoter in mouse. A candidate Cap site was found at −626 base pairs to the methonine start site (see Fig. 5.2). The first exon extends to position −335 where it splices to the same acceptor site as the proximal promoter (−22 base pairs). The upstream promoter region also shows the earmarks of a regulated transcriptional site;[85] however, factors controlling expression from this site have not been identified. It is not clear whether the alternative sites and their respective promoters influence tissue-specific expression or are regulated during differentiation. Conditions where alternative Cap site usage can be regulated would prove informative in this regard.

The second region of alternative splicing is found near the end of the open-reading frame and gives rise to important divergences in structure of the C-terminus of the gene product. Three alternative sequences have been identified in mammalian species, two in snake and *Torpedo*, and none in *Drosophila* and chicken. The most prevalent splice in mammalian AChE is from the invariant exon 4 to exon 6 giving rise to a C-terminus with diverse characteristics (see Fig. 5.2). First, exon 6 encodes for a sulfhydryl group positioned for intersubunit disulfide bonding, and a disulfide linkage may form between monomers to form a dimer between the catalytic subunits or with structural subunits to form a heteromeric species (see Fig. 5.3). Second, the exon encodes a recognition site for a proline-rich attachment domain. This site serves to associate with proline-rich regions in the structural subunits, allowing a heterologous association between catalytic and structural subunits.[51,89] Third, the encoded sequence folds into an amphipathic helix with the hydrophobic region directed to the surface. In the monomeric and dimeric forms of AChE, this region is exposed, producing a propensity for these forms to associate with membranes or other hydrophobic surfaces. Formation of the tetramer occludes these hydrophobic surfaces, yielding a hydrophilic species.

The region, encoded by exon 6 in mammals and consisting of 40 amino acids, has been termed the tryptophan amphiphilic tetramerization domain.[89] Upon formation of a tetramer this domain associates with the proline-rich attachment domain of the structural subunit. Crystallographic studies of murine AChE show tetramer formation in the unit crystal.[90] The tetramer contains a 35° tilt from a square planar configuration and provides an initial indication of how AChE tetramers may be organized.

Splicing from the exon 4 donor to exon 5 gives rise to a translated sequence that serves as a signal sequence for glycophospholipid attachment. Just N-terminal to the signal sequence is a cysteine allowing the enzyme to form a stable dimer. Glycophospholipid attachment results in the enzyme being tethered to the outer leaflet of the plasma membrane. Treatment with a phosphatidyinositol specific phospholipase C results in the release of this form of AChE as a soluble entity.

The third splice alternative at the 3' end is reading through the intronic sequence between exons 4 and 5. With the retained intron an additional 30 amino acids are encoded before reaching a termination signal. This sequence is hydrophilic, cationic, and lacks a disulfide bond for oligomerization. Thus a soluble monomer is formed with this splice.

Another splice alternative has been found in the snake where an exon found in the 3' direction from the exon encoding the oligomeric forms of AChE serves as an alternative acceptor to exon 4.[46] This exon also encodes a soluble form of AChE that is found in the venomous gland. Snakes appear to lack the splice to an exon in the position of mammalian exon 5 that encodes a glycophospholipid linked form of AChE. This exon has also not been found in avian species.

Control of alternative mRNA processing

Alternative mRNA processing of the newly transcribed mRNA from the gene appears under strict control by cellular factors, although the factors influencing this process remain uncharacterized. For example, in mammalian

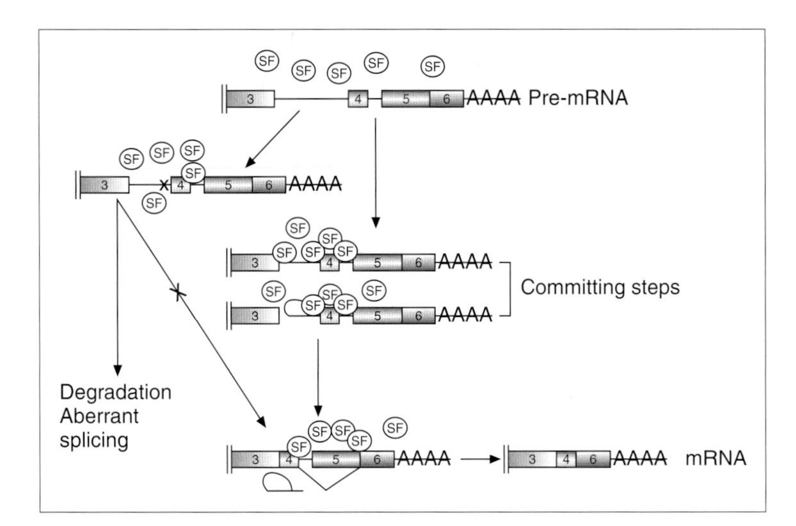

Figure 5.4
Influence of neighboring exons on alternative splicing preferences of the mammalian AChE gene. Only the alternative splice region and its upstream intron and exon are shown. The numbers indicate the order of the exons. SF denotes the splicing factors. Control of splicing by the upstream or downstream exons has been termed 'exon definition'.[92]

skeletal muscle and brain, the exon 4 to 6 splice is predominant, whereas the exon 4 to 5 splice is predominant in certain hematopoietic cells. Splicing to both alternative exons is found in the *Torpedo* electric organ, yielding both glycophospholipid-linked as well as tetrameric species. Retention of intron 4 gives rise to a minor abundance mRNA species and gene product in certain mammalian hematopoietic cells[84] and in embryonic muscle.[88] Typically, in alternatively spliced regions, there is no clear consensus for a strong acceptor, and it is believed that factors influencing splicing efficacy then can tip the balance.

However, regions of the mRNA extending beyond the alternative spliced region (i.e. in the case of the AChE gene, outside of the region from the donor site in exon 4 to the acceptor in exon 6) also appear to be determinants in splice selection.[91] Removal of introns II and III, located between exons 2 and 3 and exons 3 and 4, respectively (see Fig. 5.2), abolishes the preference for splicing to exon 6 seen in differentiating C2–C12 myotubes. Rather, apparently random splicing occurs where all three alternatives, the exon 4 to 6 splice, the exon 4 to 5 splice, and retention of intron IV, are found in near equal abundance.[91] Moreover, analysis of the relative abundance of the products of these transfected genes indicates that the spliced mRNAs are translated with similar efficiencies. Deletion of only a single upstream intron causes the proportion of mRNA species to approach, but not replicate, that of endogenous or transfected unmodified gene. Deletions within the introns and branch point mutations indicate that the upstream splicing process and not the secondary structure of the gene as conferred by the intron, is the determining factor in splice selection (Fig. 5.4). Moreover, deletions in intron IV, which exists between the donor and alternative acceptor sites, modify the efficiency of splicing of the upstream introns. Thus, splicing of the AChE pre-mRNA during myogenesis occurs in an ordered sequence in which the flanking introns influence splicing efficiency and alternative splice preference. Hence, splicing factors on both sides of the exon appear to come into contact or interact with each other, a concept previously described as exon definition.[92]

References

1. Schumacher M, Camp S, Maulet Y *et al.* Primary structure of *Torpedo californica* acetylcholinesterase deduced from cDNA sequences. *Nature* 1986; **319:** 407–409.
2. Swillens S, Ludgate M, Mercken L *et al.* Analysis of sequence and structure homologies between thyroglobulin and acetylcholinesterase: possible functional and clinical significance. *Biochem Biophys Res Commun* 1986; **137:** 142–148.
3. Hall LMC, Spierer P. The *Ace* locus of *Drosophilia melanogaster*: structural gene for acetylcholinesterase with an unusual 5' leader. *EMBO J* 1986; **5:** 2949–2954.
4. Lockridge O, Bartels CF, Vaughan TA *et al.* Complete amino acid sequence of human serum cholinesterase. *J Biol Chem* 1987; **262:** 549–557.
5. Prody CA, Zevin-Sonkin D, Gnatt A *et al.* Isolation and characterization of full-length cDNA clones coding for cholinesterase from fetal human tissues. *Proc Natl Acad Sci USA* 1987; **84:** 3555–3559.
6. McTiernan C, Adkins S, Chatonnet A *et al.* Brain cDNA clone for human cholinesterase. *Proc Natl Acad Sci USA* 1987; **84:** 6682–6686.
7. Rachinsky TL, Camp S, Li Y, Ekström TJ, Newton M, Taylor P. Molecular cloning of mouse acetylcholinesterase: tissue distribution of alternatively spliced mRNA species. *Neuron* 1990; **5:** 317–327.
8. Doctor BP, Chapman TC, Christner CE *et al.* Complete amino acid sequence of fetal bovine serum acetylcholinesterase and its comparison in various regions with other cholinesterases. *FEBS Lett* 1990; **266:** 123–127.
9. Soreq H, Ben-Aziz R, Prody CA *et al.* Molecular cloning and construction of the coding region for human acetylcholinesterase reveals a G+C-rich attenuating structure. *Proc Natl Acad Sci USA* 1990; **87:** 9688–9692.
10. Derewenda Z, Wei Y, Derewenda U. Structure–function relationships in high molecular weight PAF-acetylhydrolases from the studies of microbial α/β hydrolase. In: Doctor BP, Taylor P, Quinn DM, Rotundo RL, Gentry MK, eds. *Structure and Function of the Cholinesterases and Related Proteins.* New York: Plenum, 1998: 309–314.
11. Collet C, Nielsen KM, Russell RJ, Karl M, Oakeshott JG, Richmond RC. Molecular analysis of duplicated esterase genes in *Drosophila melanogaster. Mol Biol Evol* 1990; **7:** 9–28.
12. Bomblies L, Biegelmann E, Doring V *et al.* Membrane-enclosed crystals in *Dictyostelium discoideum* cells, consisting of developmentally regulated proteins with sequence similarities to known esterases. *J Cell Biol* 1990; **110:** 669–679.
13. Long RM, Satoh H, Martin BM, Kimura S, Gonzalez FJ, Pohl LR. Rat liver carboxylesterase: cDNA cloning, sequencing, and evidence for a multigene family. *Biochem Biophys Res Commun* 1988; **156:** 866–873.
14. Mouches C, Pauplin Y, Agarwal M *et al.* Characterization of amplification core and esterase B1 gene responsible for insecticide resistance in Culex. *Proc Natl Acad Sci USA* 1990; **87:** 2574–2578.
15. Hanzlik TN, Abdel-Aal YAI, Harshman LG, Hammock BD. Isolation and sequencing of cDNA clones coding for juvenile hormone esterase from *Heliothis virescens.* Evidence for a catalytic mechanism for the serine carboxylesterases different from that of the serine proteases. *J Biol Chem* 1989; **264:** 12 419–12 425.
16. Pathak D, Ngai KL, Ollis D. X-ray crystallographic structure of dienelactone hydrolase at 2.8 Å. *J Mol Biol* 1991; **204:** 435–445.
17. Han JH, Stratowa C, Rutter WJ. Isolation of full-length putative rat lysophospholipase cDNA using improved methods for mRNA isolation and cDNA cloning. *Biochemistry* 1987; **26:** 1617–1625.
18. Franken SM, Rozeboom HJ, Kalk KH, Dijkstra BW. Crystal structure of haloalkane dehalogenase: an enzyme to detoxify halogenated alkanes. *EMBO J* 1991; **10:** 1297–1302.
19. Sussman JL, Harel M, Frolow F *et al.* Atomic structure of acetylcholinesterase from *Torpedo californica*: a prototypic acetylcholine-binding protein. *Science* 1991; **253:** 872–879.
20. Cygler M, Schrag J, Sussman JL *et al.* Relationship between sequence conservation and three-dimensional structure in a large family of esterases, lipases, and related proteins. *Protein Sci* 1993; **2:** 366–382.

21. Ollis DL, Cheah E, Cygler M *et al.* The α/β hydrolase fold. *Protein Eng* 1992; **5:** 197–211.
22. De la Escalera S, Backamp E-O, Moya F, Piovant M, Jimenez F. Characterization and gene cloning of neurotactin, a *Drosophila* transmembrane protein related to cholinesterases. *EMBO J* 1990; **9:** 3593–3601.
23. Olsen PF, Fessler LI, Nelson RE, Sterne RE, Campbell AG, Fessler JH. Glutactin a novel *Drosophila* basement membrane-related glycoprotein with sequence similar to serine esterases. *EMBO J* 1990; **9:** 1219–1227.
24. Hortsch M, Patel NH, Bieber AJ, Traguina ZR, Goodman CS. *Drosophila* neurotactin, a surface glycoprotein with homology to serine esterases, is dynamically expressed during embryogenesis. *Development* 1990; **110:** 1327–1340.
25. Auld VJ, Fetter RD, Broadic K, Goodman CS. Gliotactin, a novel transmembrane protein on peripheral glia, is required to form the blood nerve barrier in *Drosophila. Cell* 1995; **81:** 757–767.
26. Nguyen T, Sudhof TC. Binding properties of neuroligin 1 and neurexin 1β reveal function as heterophilic cell adhesion molecules. *J Biol Chem* 1997; **272:** 26 032–26 039.
27. Ichtchenko K, Hata Y, Nguyen F *et al.* Neuroligin 1: a splice site-specific ligand for β-neurexins. *Cell* 1995; **81:** 435–443.
28. Bottis SA, Felder CE, Sussman JL, Silman I. Electrotactins: a class of adhesion proteins with conserved electrostatic and structural motifs. *Protein Eng* 1998; **11:** 415–420.
29. Tsigelny I, Shindyalov IN, Bourne PE, Sudhof TC, Taylor P. Common EF hand motifs in cholinesterases and neuroligins suggest a role for Ca^{2+} binding in cell surface associations. *Protein Sci* 1999; in press.
30. MacPhee-Quigley K, Vedvick TS, Taylor P. Profile of disulfide bonds in acetylcholinesterase. *J Biol Chem* 1986; **261:** 13 565–13 570.
31. Lockridge O, Adkins S, La Du BN. Location of disulfide bonds within the sequence of human serum cholinesterase. *J Biol Chem* 1987; **262:** 12 945–12 952.
32. MacPhee-Quigley K, Taylor P, Taylor SS. Primary structures of the catalytic subunits from two molecular forms of acetylcholinesterase: a comparison of NH_2-terminal and active center sequences. *J Biol Chem* 1985; **260:** 12 185–12 189.
33. Gibney G, Camp S, Dionne M, MacPhee-Quigley K. Mutagenesis of essential functional residues in acetylcholinesterase. *Proc Natl Acad Sci USA* 1990; **87:** 7546–7550.
34. Augustinsson KB. Cholinesterases: a study in comparative enzymology. *Acta Physiol Scand* 1948-; **52(Suppl 15):** 1–182.
35. Silver A. *The Biology of Cholinesterases*. Amsterdam: North-Holland, 1974.
36. Aldridge WN, Reiner E. *Enzyme Inhibitors as Substrates*. Amsterdam: Elsevier, 1972.
37. Radić Z, Pickering N, Vellom DC, Camp S, Taylor P. Three distinct domains in the cholinesterase molecule confer selectivity for acetyl- and butyrylcholinesterase inhibitors. *Biochemistry* 1993; **32:** 12 074–12 084.
38. Ceveransky C, Dajas F, Harvey AL, Karlsson E. Fasciculins, acetylcholinesterase toxins from mamba venoms: biochemistry and pharmacology. In: Harvey AL, ed. *Snake Toxins*. New York: Pergamon, 1991: 303–321.
39. Radic Z, Duran R, Vellom DC, Li Y, Cevenansky C, Taylor P. Site of fasciculin interaction with acetylcholinesterase. *J Biol Chem* 1994; **269:** 11 233–11 239.
40. Eastman J, Wilson EJ, Cervenansky C, Rosenberry TL. Fasciculin 2 binds to the peripheral site on acetylcholinesterase and inhibits substrate hydrolysis by slowing a step involving proton transfer during enzyme acylation. *J Biol Chem* 1995; **270:** 19 694–19 701.
41. Fournier D, Karch F, Bridge J-M, Hall LMC, Berge JB, Spierer P. *Drosophila melanogaster* acetylcholinesterase gene: structure, evolution and mutations. *J Mol Biol* 1989; **210:** 15–22.
42. Grauso M, Culetto E, Combes D, Feyon Y, Toutant JP, Aspagaus M. Existence of four acetylcholinesterase genes in nematodes *Caenorhabditis elegans* and *Caenorhabditis briggsae. FEBS Lett* 1998; **424:** 65–71.
43. Maulet Y, Camp S, Gibney G, Rachinsky T, Ekström T, Taylor P. Single gene encodes glycophospholipid-anchored and asymmetric acetylcholinesterase forms: alternative coding exons contain inverted repeat sequences. *Neuron* 1990; **4:** 289–301.
44. Cousin X, Bon S, Massoulie J, Bon C. Identifi-

cation of a novel type of alternatively spliced exon from the acetylcholinesterase gene of *Bungarus fasciatus*. *J Biol Chem* 1998; **273:** 9812–9820.
45. Li Y, Camp S, Rachinsky TL, Getman D, Taylor P. Gene structure of mammalian acetylcholinesterase. Alternative exons dictate tissue specific expression. *J Biol Chem* 1991; **266:** 23 083–23 090.
46. Simon S, Massoulie J. Cloning and expression of acetylcholinesterase from *Electrophorus*: splicing pattern of 3' exons in vivo and in transfected mammalian cells. *J Biol Chem* 1997; **272:** 33 045–33 055.
47. BenAziz-Aloya R, Seidman S, Timberg R, Sternfeld M, Zakut H, Soreq H. Expression of a human acetylcholinesterase promoter–reporter construct in developing neuromuscular junctions of *Xenopus* embryos. *Proc Natl Acad Sci USA* 1993; **90:** 2471–2475.
48. Getman DK, Mutero A, Inoue K, Taylor P. Transcription factor repression and activation of the human acetylcholinesterase gene. *J Biol Chem* 1995; **270:** 23 511–23 519.
49. Arpagaus M, Kott M, Vatsis KP, Bartels CF, La Du BN, Lockridge P. Structure of the gene for human butyrylcholinesterase. Evidence for a single copy. *Biochemistry* 1990; **29:** 124–131.
50. Ekström TJ, Klump WM, Getman D, Karin M, Taylor P. Promoter elements and transcriptional regulation of the acetylcholinesterase gene. *DNA Cell Biol* 1989; **12:** 63–72.
51. Bon S, Coussen F, Massoulie J. Quaternary associations of acetylcholinesterase. II. The polyproline attachment domain of the collagen tail. *J Biol Chem* 1997; **272:** 3016–3021.
52. Searle BW, Goldstein A. Mutation to neostigmine resistance in a cholinesterase-containing Pseudomonas. *J Bacteriol* 1962; **83:** 789–799.
53. Rama-Sastry BV, Sadavongvivad C. Non-neuronal acetylcholine. *Pharmacol Rev* 1979; **30:** 65–132.
54. Sutherland D, McClellan DS, Milner D *et al.* Two cholinesterase genes and activities are present in amphioxus. *J Exp Zool* 1997; **277:** 213–229.
55. Pezzementi L, Sutherland D, Sanders M *et al.* Structure and function of cholinesterases from agnathans and cephalochordates. In: Doctor BP, Taylor P, Quinn DM, Rotundo RL, Gentry MK, eds. *Structure and Function of Cholinesterases and Related Proteins.* New York: Plenum, 1998: 105–110.
56. Rotundo RL, Gomex AM, Fernandez-Valle C *et al.* Allelic variants of acetylcholinesterase forms in avian nerves and muscle and encoded by a single gene. *Proc Natl Acad Sci USA* 1988; **52:** 7121–7125.
57. Getman DK, Eubanks J, Evans G, Taylor P. Assignment of the human acetylcholinesterase gene to chromosome 7q22. *Am J Hum Genet* 1992; **50:** 170–177.
58. Ehrlich G, Viegas-Pequignot E, Ginzberg D. Mapping the human acetylcholinesterase gene to chromosome 7q22 by fluorescent in situ hybridization coupled with selective PCR amplification from a somatic hybrid cell panel and chromosome-sorted DNA libraries. *Genomics* 1992; **13:** 1192–1197.
59. Rachinsky TL, Crenshaw EB III, Taylor P. Assignment of the gene for acetylcholinesterase to distal mouse-chromosome 5. *Genomics* 1992; **14:** 511–514.
60. Gaughan G, Park H, Priddle J, Craig I. Refinement of the localization of human butyrylcholinesterase to chromosome 3q26.1–q26.2 using a PCR-derived probe. *Genomics* 1991; **11:** 455–458.
61. Allderdice PW, Gardner HAR, Galutira D *et al.* The cloned butyryl-cholinesterase (BCHE) gene maps to a single chromosome site, 3q26. *Genomics* 1991; **11:** 452–454.
62. La Du B, Lockridge O. Molecular biology of human serum cholinesterase. *Fed Proc* 1986; **46:** 2965–2974.
63. Loewenstein Y, Gnatt A, Neville LF, Zakut H, Soreq H. Structure–function relationship studies in human cholinesterase reveal genomic origins for individual variations in cholinergic drug responses. *Prog Neuropsychopharmacol Biol Psychiatry* 1993; **17:** 905–926.
64. Xie W, Stribley JA, Chatonnet A *et al.* Postnatal developmental decay in mice lacking acetylcholinesterase. Submitted for publication (1999).
65. Culetto E, Combes D, Fedon Y, Roig A, Toutant JP, Arpagaus M. Structure and promoter activity of the 5' flanking region of ace-1, the gene encoding acetylcholinesterase of class A in *Caenorhabditis elegans*. *J Mol Biol* 1999; **290:** 951–966.

66. Li Y, Camp S, Rachinsky T, Bongiorno C, Taylor P. Promoter elements and transcriptional control of the mouse acetylcholinesterase gene. *J Biol Chem* 1993; **268:** 3563–3572.
67. Mutero A, Camp S, Taylor P. Promoter elements of the mouse *acetylcholinesterase* gene: transcriptional regulation during muscle differentiation. *J Biol Chem* 1995; **270:** 1866–1872.
68. Camp S, Taylor P, Intronic elements appear essential for the differentiation-specific expression of acetylcholinesterase in C2C12 myotubes. In: Doctor BP, Taylor P, Quinn DM, Rotundo RL, Gentry MK, eds. *Structure and Function of Cholinesterases and Related Proteins*. New York: Plenum Press, 1998: 51–57.
69. Chan RYY, Boudreau-Larivierra C, Angus LM, Markal FA, Jasmin BJ. An intronic enhancer containing an N-box motif is required for synapse and tissue-specific expression of the acetylcholinesterase gene in skeletal muscle fibers. *Proc Natl Acad Sci USA* 1999; **96:** 4627–4632.
70. Fuentes ME, Taylor P. Control of acetylcholinesterase gene expression during myogenesis. *Neuron* 1993; **10:** 679–687.
71. Coleman BA, Taylor P. Regulation of acetylcholinesterase expression during neuronal differentiation. *J Biol Chem* 1996; **271:** 4410–4416.
72. Chan RYY, Adatia FA, Frupa AM, Jasmin BL. Increased expression of acetylcholinesterase T and R transcripts during hematopoietic differentiation is accompanied by parallel elevations in levels of their respective molecular forms. *J Biol Chem* 1998; **273:** 9727–9733.
73. Luo Z, Fuentes ME, Taylor P. Regulation of acetylcholinesterase mRNA stability by calcium during differentiation from myoblasts to myotubes. *J Biol Chem* 1994; **269:** 27 216–27 223.
74. Luo ZD, Pincon-Raymond M, Taylor P. Acetylcholinesterase and nicotinic acetylcholine receptor expression diverge in muscular dysgenic mice lacking the l-type calcium channel. *J Neurochem* 1996; **67:** 111–118.
75. Luo ZD, Camp S, Wang Y, Werlen G, Chien KR, Taylor P. Calcineurin enhances acetylcholinesterase mRNA stability during C2–C12 muscle cell differentiation. *Mol Pharmacol* 1999; in press.
76. Schoben A, Minichiella L, Keller M *et al.* Reduced acetylcholinesterase activity in adrenal medulla and loss of sympathetic preganglionic neurons in TrkA-deficient but not TrKB-deficient mice. *J Neurosci* 1997; **17:** 891–903.
77. Fernandes KJL, Kobayashi NR, Jasmin BJ, Tetzlaff W. Acetylcholinesterase gene expression in axotomized rat facial motoneuron is differentially regulated by neurotropins: correlation with TrkB and TrkC mRNA levels and isoforms. *J Neurosci* 1998; **18:** 9936–9947.
78. Boudreau-Lariviere C, Jasmin BJ. Calcitonin gene-related peptide decreases expression of acetylcholinesterase in mammalian myotubes. *FEBS Lett* 1999; **444:** 22–26.
79. Choi RC, Yung LY, Dong TT, Wan DC, Wong YH, Tsim KW. Calcitonin gene related peptide induced acetylcholinesterase synthesis in cultured chick myotubes is mediated by cAMP. *J Neurochem* 1998; **71:** 152–160.
80. Massoulie J, Pezzementi L, Bon S, Krejci E, Vallette FM. Molecular and cellular biology of the cholinesterases. *Proc Neurosci* 1993; **41:** 31–91.
81. Grubić Z, Komel R, Walker WF, Miranda AF. Myoblast fusion and innervation with rat motor nerve alter distribution of acetylcholinesterase and its mRNA in cultures of human muscle. *Neuron* 1995; **14:** 317–327.
82. Sketelj J, Crue-Finderle N, Struckelj B, Trontelj JV, Pette D. Acetylcholinesterase mRNA level and synaptic activity in rat muscles depend on nerve-induced pattern of muscle activation. *J Neurosci* 1998; **18:** 1944–1952.
83. Grisaru D, Lev-Lehman E, Shapira M *et al.* Human osteogenesis involves differentiation dependent increases in the morphogenically active 3' alternative splicing variant of acetylcholinesterase. *Mol Cell Biol* 1999; **19:** 788–795.
84. Li Y, Camp S, Taylor P. Tissue specific expression and alternative mRNA processing of the mammalian acetylcholinesterase gene. *J Biol Chem* 1993; **268:** 5790–5797.
85. Atanasova E, Chiappa S, Wieben E, Brimijoin

S. Novel messenger RNA and alternative promoter for murine acetylcholinesterase. *J Biol Chem* 1999; **274:** 21 078–21 084.
86. Gibney G, McPhee-Quigley K, Thompson B *et al.* Divergence in primary structure between the molecular forms of acetylcholinesterase. *J Biol Chem* 1988; **263:** 1140–1145.
87. Sikorav JL, Duval N, Anselmet A *et al.* Complex alternative splicing of acetylcholinesterase in *Torpedo* electric organ; primary structure of the precurser of the glycolipid anchored form. *EMBO J* 1988; **7:** 2983–2993.
88. Legay C, Hatchet M, Massoulie J, Changeux JP. Developmental regulation of acetylcholinesterase transcripts in mouse diaphragm: alternative splicing and focalization. *Eur J Neurosci* 1995; **7:** 1803–1809.
89. Simon S, Krejci E, Massoulie J. A four-to-one association between peptide motifs: four C-terminal domain (PRAD) is the secretory pathway. *EMBO J* 1998; **17:** 6178–6187.
90. Bourne Y, Taylor P, Bougis P, Marchot P. Crystal structure of mouse acetylcholinesterase. *J Biol Chem* 1999; **274:** 2963–2970.
91. Luo ZD, Camp S, Mutero A, Taylor P. Splicing of 5' introns dictates alternative splice selection of acetylcholinesterase pre-mRNA and specific expression during myogenesis. *J Biol Chem* 1998; **273:** 28 486–28 495.
92. Berget SM. Exon recognition in vertebrate splicing. *J Biol Chem* 1998; **270:** 2411–2424.

6

Molecular forms and anchoring of acetylcholinesterase

Jean Massoulié

Introduction

The vertebrates possess two cholinesterase (ChE) genes, encoding acetylcholinesterase (AChE) and butyrylcholinesterase (BuChE). Each of these enzymes exists under multiple molecular forms which correspond to various localizations: they include soluble forms in the plasma or in certain snake venoms, glycolipid-anchored forms on the surface of blood cells, collagen-tailed forms at neuromuscular junctions, and hydrophobic-tailed forms in the brain. This diversity of molecular forms arises at the levels of alternative splicing and post-translational processing. In this chapter, after a brief description of the catalytic domain, we focus on the origin and role of the non-catalytic carboxy-terminal (C-terminal) domains in the generation of ChE forms and on the properties of these various molecular forms. For more complete reviews, see Massoulié *et al.*[1–3]

ChE forms may be distinguished on the basis of their oligomeric state, which can be deduced from their hydrodynamic properties, usually their sedimentation in sucrose gradients, and on the basis of their hydrophobic interactions: amphiphilic molecules are operationally distinguished from non-amphiphilic molecules by the fact that detergent micelles (Triton X-100, Brij-96) modify their sedimentation and their electrophoretic migration under non-denaturing conditions.[4]

The catalytic domain

AChE or BuChE mRNAs encode a secretion signal peptide, that is removed during biosynthesis, a catalytic domain of about 500 residues, and short C-terminal peptides of about 40 residues (these are discussed later). The catalytic domain possesses a variable number of *N*-glycosylation sites (as discussed later) and three intracatenary disulphide bonds.

The various molecular forms of AChE possess the same catalytic domain. The three-dimensional structure of this domain has been established by X-ray crystallography in the case of *Torpedo* AChE.[5] The catalytic site contains an active serine which is part of a catalytic triad, Glu-His-Ser. The active site is not directly accessible at the surface of the protein, but can apparently be reached only through a narrow 'catalytic gorge'. Considering that AChE is one of the fastest enzymes known, this is surprising because it seems to limit the traffic of substrates and products.

ChEs are inhibited by various compounds, acting in different manners, some of which are used therapeutically in myasthenia gravis, glaucoma and Alzheimer's disease. The various inhibitors of AChE can be classified according to their binding sites. Inhibitors that bind to the active site itself are either non-covalent ligands, such as edrophonium and tacrine, or pseudo-substrates that phosphorylate, sulphonylate or carbamylate the active serine. These covalent inhibitors include, respectively, metrifonate, phenylmethylsulphonyl fluoride and physostigmine and its derivatives. Other inhibitors, such as curare, propidium, gallamine and the snake peptide toxin fasciculin do not penetrate into the active site, but bind at a peripheral site located at the entrance of the catalytic gorge. Finally, some inhibitors bind simultaneously to the active and peripheral sites, such as the bis-quaternary compound BW284C51. Both the active and the peripheral sites contain aromatic residues that contribute strongly to the binding of ligands. AChE and BuChE differ by the volume of the acyl pocket that accommodates the acyl group of the substrate, explaining that AChE hydrolyses acetylcholine (ACh) but not butyrylcholine, while BuChE can hydrolyse the larger substrate as well as a variety of other esters. This difference also explains why BuChE is sensitive to bulky inhibitors, like tetramon*iso*propyl pyrophosphortetramide (iso-OMPA), while AChE is not.

The structure of AChE complexes with several inhibitors has been studied by crystallography. Fasciculin was found to occlude the opening of the catalytic gorge essentially completely.[6,7] E2020 (Aricept) occupies the entire gorge, interacting with aromatic groups such as W279 in the peripheral site and F330 in the acyl pocket: this explains its specificity for AChE, relative to BuChE which does not possess these residues.[8,9] (*Note*: By convention, the numbering of AChE residues corresponds to that of *Torpedo* AChE.)

The distribution of charged residues in ChEs creates a strong electrostatic dipole, which is aligned with the axis of the catalytic gorge: it has been suggested that the resulting local electrostatic field facilitates access of the charged substrate, ACh, towards the active site, but this is very controversial. It has also been suggested that intramolecular movements may open a passage between the active site and the external medium. This 'back door', which is closed in the crystalline state, would allow the exit of products, thus resolving the paradox created by the traffic limitations through the catalytic gorge. However, its existence has not been established.

Apart from their esteratic activity, ChEs have been proposed to play a variety of catalytic or non-catalytic roles (see references in Massoulié *et al.*[3]). The possibility that they might possess a proteolytic activity was considered, but seems to be ruled out.[11] In contrast, the hypothesis that ChEs might be involved in heterophilic protein–protein interactions is supported by their homology with adhesion proteins such as neuroactin in *Drosophila*,[12] or neuroligin in mammals.[13] The electric dipole of ChEs and homologous proteins might play a role in heterophilic interactions, so that this family of molecules has been named 'electrotactins'.[14] As discussed below, physicochemical interactions with AChE, probably involving its peripheral site, may accelerate the assembly of Aβ peptides into amyloid fibrils in the brain.[15]

Alternative splicing and multiplicity of C-terminal domains

The coding region of vertebrate ChEs comprises a variable number of exons corresponding to the catalytic domain, and either one exon or several alternative exons, corresponding to short carboxy-terminal domains. The catalytic domain is well conserved and is sufficient to generate an active ChE: when the C-terminal domain is deleted, the enzyme forms soluble active monomers.[16] Because the numbers of exons differ among ChE genes, we refer to the regions encoding carboxy-terminal

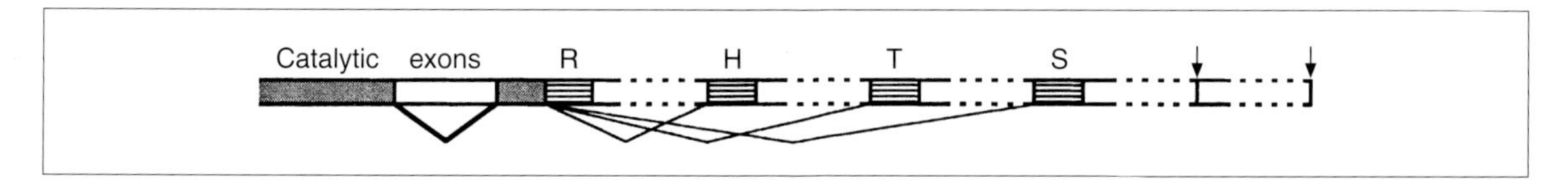

Figure 6.1
Schematic representation of the 3′ region of a vertebrate AChE gene. This composite scheme illustrates all known possible splice choices, but they have not been simultaneously observed in any single species: Torpedo and mammalian AChE genes generate R, H and T transcripts; Bungarus AChE gene generates T and S transcripts; and teleost fish (Electrophorus, Danio) and avian AChE genes generate only T transcripts. Electrophorus constructs, however, when expressed in mammalian cells, reveal abnormal splice donor and acceptor sites, producing non-functional transcripts. The shaded boxes indicate common coding exons, corresponding to the catalytic domain; the hatched boxes represent coding regions corresponding to the C-terminal peptides; empty boxes represent non-coding regions. The thick lines indicate splicing of a constitutive intron, between exons corresponding to the catalytic domain; thin lines indicate splicing possibilities within the 3′ region of the gene. The arrows indicate the possible existence of multiple polyadenylation sites.

domains by functional names, that identify their equivalence between different species. The different C-terminal domains are thus called H (hydrophobic), R (readthrough), S (soluble or snake) and T (tailed).[1,2,10] They correspond to alternative splicing that may occur in the 3' region of AChE genes, as shown schematically in Fig. 6.1. The regions encoding these domains do not always correspond to exons in the strict sense: for example, R is usually considered as intronic but is retained in some transcripts, and T is retained in the 3'-untranslated region of transcripts of type H.[17,18] However, H and T are called 'alternative exons'.

The 3' regions of vertebrate ChE genes do not possess all the different splice choices, and in fact, BuChE genes and AChE genes of many species appear to contain a single 3' coding exon, of type T. Thus, the T peptide constitutes the only C-terminal domain that is common to all vertebrate ChEs.

H 'exons' exist in the *Torpedo* AChE gene and in mammalian AChE genes, upstream of the T 'exons'. They have not been found in the bony fishes *Danio* (zebrafish)[19] and *Electrophorus* (electric eel),[20] in the snake *Bungarus fasciatus*,[21] or in the chicken (A Anselmet, unpublished results).

The S 'exon' was found in the *Bungarus* AChE gene, downstream of the T 'exon'.[21] It encodes a short hydrophilic peptide, generating a monomeric soluble AChE species which is abundantly secreted in the venom of this Elapid snake.

R transcripts correspond to mature mRNAs, since constitutive introns interrupting the coding sequence of the catalytic domain are spliced out, but they retain the 'intronic region' preceding 'exon H'. Such transcripts have been observed in *Torpedo*[22] and in mammals.[17,18,23] In mammals, R transcripts are relatively abundant in embryonic liver and muscle and they seem to be specifically induced by stress in the mouse brain.[24] The differentiation of lymphocytes is accompanied by an increase in expression of R and T transcripts.[25] When produced

Subunit		*Peptide sequence of C-terminal region*	*Molecular forms*	*Tissue distribution*
R	*Torpedo* AChE[50]	GNVFAFHMQKVRTPAKTYHFGVIVAHLLLLSLPTASDVPRLASSKWWAHSCPLCSRRCWESWGRIL		– R transcripts in *Torpedo* electric organ; in mouse embryonic muscle; expression level increased by stress in mouse brain
	Rat AChE[17]	GRRGVGKQGMHKAARVGRTGERKGGKHRM		
	Mouse AChE[18]	GRRMEWGEQGMHKAARVGRRGERWGAKHRV		
	Human AChE[21]	GMQGPAGSGWEEGSGSPPGVTPLFSP		
H	*Torpedo* AChE[49,50]	ACDGELS**S**SSTSSSKG*IIFYVLFSILYLIFY*	GPI-G_2^a (type I)	– nervous tissue in insects – muscles, nervous tissue and electric organs in *Torpedo* – haematopoietic cells in mammals
	Rat AChE[17,18]	ATEVPCTCPSPAH**G**EAAPRPGPALSLS*LLFFLFLLHSGLRWL*		
T	*Torpedo* AChE[22,49]	ETIDEAERQ**W**KTEFHR**W**-SS**Y**MMH**W**KNQ**F**DQ**Y**_ _ _ _RHENCAEL	G_1^a, G_2^a (type II)	– muscles and nervous tissues in all vertebrates
	Electrophorus AChE[20]	ENIDDAERQ**W**KAEFHR**W**-SS**Y**MMH**W**KNQ**F**DH**Y**– – – –SKQERCTNL	G_4^{na}, G_2^a	
	Bungarus AChE[27]	DNIEEAERQ**W**KLEFHL**W**-SA**Y**MMH**W**KSQ**F**DH**Y**– – – –NKQDRCSEL	A_4, A_8, A_{12}	
	Quail AChE[83]	GPTEDAER-**W**RLEFHR**W**-SS**Y**MGR**W**RTQ**F**EH**Y**– – – –SRQQPCATL		
	Rat AChE[115]	DTLDEAERQ**W**RAEFHR**W**-SS**Y**MVH**W**KNQ**F**DH**Y**– – – –SKQDRCSDL		
	Human BChE[114]	GNIDEAEWE**W**KAGFHR**W**-NN**Y**MMD**W**KNQ**F**ND**Y**T– – – –SKKESCVGL		
S	*Bungarus* AChE[27]	VDPPRADRRRRSARA	G_1^{na}	– venom glands and other tissues in Elapid snakes

*The underlined regions are removed from the mature proteins; the last residue of the mature protein (ω) is shown in bold type, and the C-terminal region is shown in italics. A C-terminal part of the T peptide may be removed upon secretion, yielding amphiphilic or non-amphiphilic forms, but the sites of cleavage have not been determined. Conserved aromatic residues in T peptides are shown in bold letters; in H and T peptides, cysteine residues that may establish intercatenary disulphide bonds are shown as C. For additional sequences, see the ESTHER server.[27]

Table 6.1
*Different types of C-terminal peptides in cholinesterases**

by transfected cells, using an appropriate vector, the mouse R subunits form soluble monomers,[26] but these molecules have not yet been unambiguously identified in vivo. It should be noted, however, that these 'intronic' sequences are not well conserved, even among mammals. In particular, the deduced peptides differ in their length, and this casts some doubt on their functional significance (Table 6.1). Moreover, R transcripts have been demonstrated in mouse erythroid cells, but not in human cells, and the human gene did not produce them in transfected murine erythroleukaemia cells, indicating that this type of transcript depends on the gene structure.[26] Thus, the mammalian AChE genes generate two major transcripts, H and T. In the following sections, we examine the processing of the corresponding subunits, AChE-H and AChE-T.

The H subunits: addition of the GPI anchor

H transcripts have been characterized in *Torpedo* and in mammalian AChE genes, but not in the other vertebrate phyla. They also exist in various invertebrates, including insects and nematodes (*Caenorhabditis elegans*).[28] H 'exons' encode C-terminal domains which contain a signal for addition of a glycophosphatidylinositol (GPI) allowing the anchoring of AChE at the outer surface of cellular membranes. This signal consists of a cleavage/addition site, called the ω site, to which the GPI anchor is added by a transamidation reaction, located about ten residues upstream of a C-terminal hydrophobic region, which is removed during processing in the endoplasmic reticulum, very shortly after completion of the polypeptide chain.[29,30] The primary structure of GPI-addition signals is in fact extremely flexible, and there is no sequence homology between the GPI-addition signals of *Torpedo* and mammalian AChEs. Considering this lack of homology and the absence of H exons in vertebrate classes, which are phylogenetically intermediate between mammals and *Torpedo*, a primitive cartilaginous fish, it seems likely that H exons have disappeared and reappeared independently several times during evolution.

We have analysed the peptidic environment of the ω site in the rat H sequence, by site-directed mutagenesis (S Bon, F Coussen, J Massoulié, unpublished results). This allowed us to identify the ω site as a glycine residue, G588, in agreement with direct analyses of peptide fragments.[31,32] We also found that mutations that reduce the efficiency of GPI addition generally also reduce the level of AChE activity, probably because unprocessed precursors are retained in intracellular compartments and degraded.[33,34] It is interesting that some mutations improved GPI addition and increased AChE activity in transfected COS cells, indicating that the H sequence has not evolved towards an optimal production of GPI-anchored AChE. On the other hand, the wild type sequence allows a relatively high level of AChE secretion, which seems to result from abortive transamidation:[35] cleavage at the ω site would be followed by hydrolysis instead of GPI addition, producing soluble non-amphiphilic dimers (G_2^{na}). It is possible that such secretion of AChE dimers may play a physiological role.

In addition to the GPI-addition signal, the *Torpedo* and mammalian H peptides contain one or two cysteine residues, which allow the formation of dimers, so that the AChE-H subunits generate GPI-anchored dimers (G_2^{a}).

The T subunits: oligomeric organizations; associations with structural subunits

The AChE-T subunits generate a series of homomeric and heteromeric molecular forms.

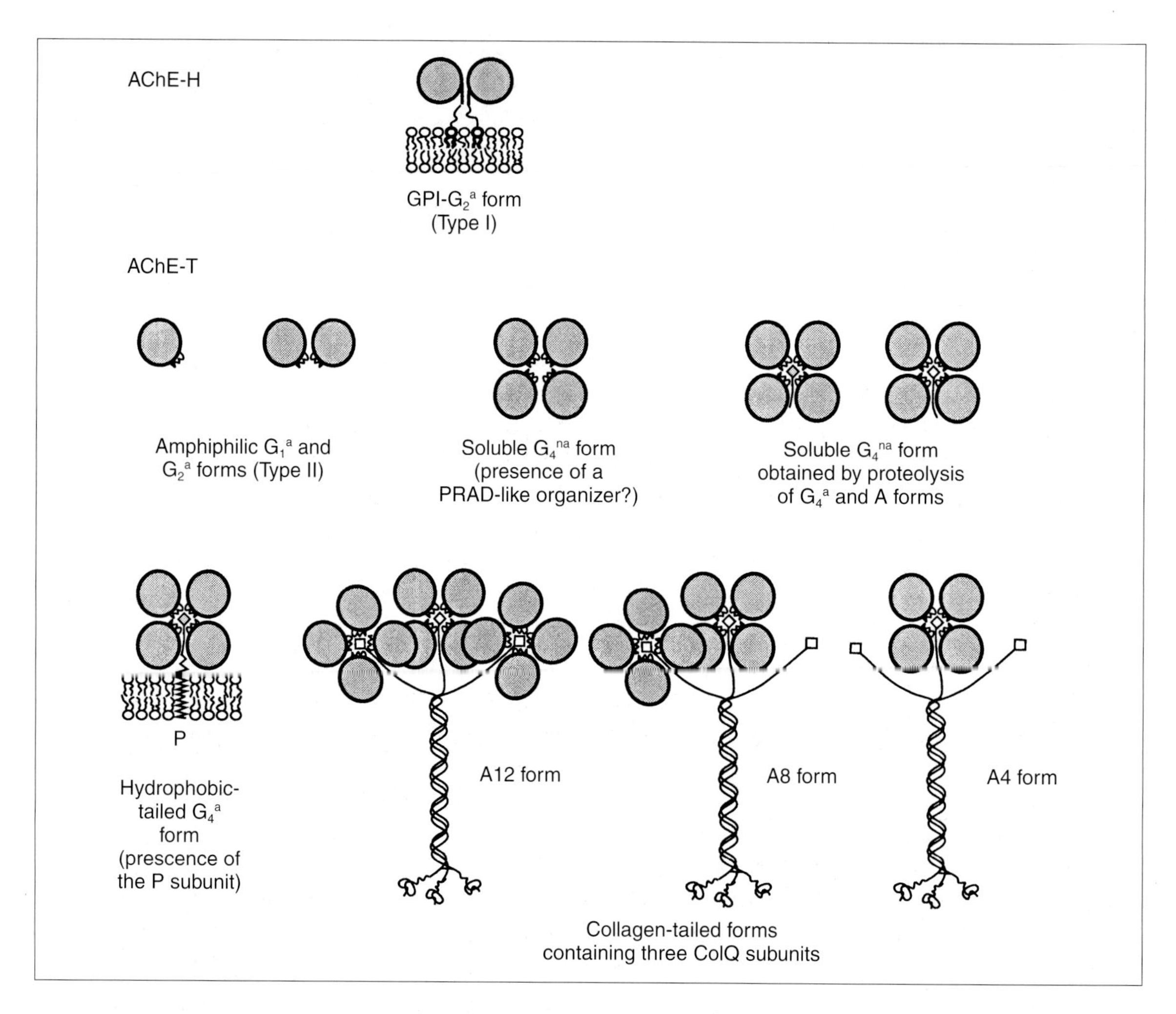

Figure 6.2
Quaternary structures of AChE forms. AChE-H subunits only generate GPI-anchored dimers, while AChE-T subunits generate a wide diversity of molecular forms. In A forms, AChE-T tetramers are organized around the proline rich attachment domain (PRAD), represented as a white square; the quaternary organization of membrane-anchored G_4^a, that contain a hydrophobic P subunit is probably similar. Proteolysis of hydrophobic-tailed and collagen-tailed forms produces soluble G_4^{na} forms that contain a PRAD or PRAD-like 'organizer'; the existence of homomeric G_4^{na} tetramers that do not possess an organizer is debated.

The homomeric forms include monomers, dimers and tetramers; the heteromeric forms are collagen-tailed and hydrophobic-tailed molecules, in which a tetramer of AChE-T subunits is associated with a collagen subunit, or with an hydrophobic protein, respectively. In these heteromeric molecules, two AChE-T subunits are linked by a disulphide bond to each other, and the other two are linked to the structural protein, through cysteine residues located near the C-terminus of the T peptides (Fig. 6.2). The assembly of collagen-tailed, and probably also of hydrophobic-tailed molecules, occurs in the endoplasmic reticulum.[1] Non-amphiphilic tetramers, G_4^{na}, can be released by digestion of the collagen-tailed molecules by trypsin, or by collagenase that specifically cleaves the collagen tail.

C-terminal peptides of type T exist for all vertebrate ChEs. They are also extremely well conserved, even between AChE and BuChE. In addition to the presence of a free cysteine that may form intercatenary bonds, these peptides of 40 or 41 residues contain a series of conserved aromatic residues (see Table 6.1). On the basis of spectroscopic data, we proposed several years ago that the T peptide may assume an α-helical conformation, in which the aromatic side-chains would form a hydrophobic patch.[3] This amphiphilic α-helix would explain the interaction of monomers and dimers of AChE-T subunits with detergent and lipid micelles. These molecules have been characterized as amphiphilic forms of type II (G_1^a and G_2^a), as opposed to GPI-anchored amphiphilic forms of type I, which differ in their solubilization and aggregation properties.[36–38] A more elaborate theoretical model has been proposed for the structure of the T peptide, according to which the amphiphilic α-helix bends upon itself, forming an hydrophobic interaction domain.[39] However, the three-dimensional structure of AChE subunits, and in particular of the T peptide, have not yet been resolved by X-ray crystallography.

We have recently renamed the T peptide as the 'tryptophan amphiphilic tetramerization' domain, or WAT domain, because this C-terminal region endows AChE-T subunits with the capacity to form not only dimers, but also tetramers which may be homotetramers or tetramers attached to a structural protein.[40] We showed that the WAT domain associates with the proline-rich attachment domain (PRAD) of the collagen Q (ColQ), generating the collagen-tailed forms; WAT domains may in fact organize as tetramers with the PRAD, even in the absence of the catalytic domain. The PRAD represents a small, 17-residue peptide motif containing two cysteines and eight proline residues, located in the amino-terminal (N-terminal) non-collagenous region of ColQ.[41] A combination of deletions and mutations showed that it is sufficient for the binding of an AChE-T tetramer and that the cysteine residues may stabilize the quaternary interactions, but are not necessary; the essential feature is only the presence of a succession of proline residues. In fact, synthetic polyproline was found to induce the tetramerization of AChE-T subunits efficiently. The interaction between the collagen ColQ and the AChE-T subunits, which conditions the assembly of collagen-tailed AChE forms, thus relies on the association of one PRAD with four WAT domains, which forms the core of the quaternary assembly. This association does not require a cell-specific biosynthetic capacity, since it occurs in the COS cells and in *Xenopus* oocytes, as well as in myogenic C2C12 cells.

AChE-T subunits can also associate with a hydrophobic subunit, called P, forming membrane-anchored tetramers which constitute the major AChE species in mammalian brain and also exist in muscles. The biochemical structure of this tetrameric form has been analysed in the case of the bovine brain.[42,43] Electrophoresis under denaturing conditions, with and without reduction, showed that it is composed of two disulphide-linked dimers, one of which contains the 20 kDa hydrophobic P protein. The C-terminal cysteine of the WAT domain forms a disulphide bond with the P protein.[44] Since this organization appears extremely similar to that of ColQ-linked tetramers, it is tempting to assume that the assembly of hydrophobic-tailed AChE tetramers is similar to that of collagen-tailed forms. However, the AChE-associated hydrophobic P subunit has not yet been cloned, and we do not know whether it contains a proline-rich PRAD-like motif, like the AChE-associated ColQ collagen. In fact, we do not know whether the brain and muscle membrane-bound tetramers contain the same hydrophobic P subunits, or distinct anchoring subunits. It is possible that AChE-T subunits, and more generally WAT-containing proteins interact with a number of different 'organizers', such as PRAD, which assemble them into tetramers.

Rat AChE-T subunits, when overexpressed in foreign cells such as COS cells or *Xenopus* oocytes, may produce unstable multimers sedimenting around 13 S.[40,45] This 13 S species has also been observed in brain extracts from rat embryos.[46] These oligomers might correspond to hexamers of catalytic subunits:[45] they dissociate readily, especially in the presence of detergent (Triton X-100). In COS cells, AChE-T subunits also form amphiphilic tetramers. In *Xenopus* oocytes, tetramers are only detected in trace amounts.

The quaternary organizations of amphiphilic and non-amphiphilic tetramers are not known. They may represent homotetramers in which the hydrophobic regions of the WAT domains are more or less exposed, and their formation depends on the species. The fact that tetramers do not form efficiently in *Xenopus* oocytes[40] suggests that most tetrameric species contain an organizer. For example, inactivation of the ColQ gene in mice leads to the disappearance of the soluble G_4^{na} form that is normally found in neonatal muscles,[47] indicating that the assembly of these molecules depends on the presence of the PRAD. (*Note*: Hydrophobic-tailed G_4^{a} molecules are still present in these mice.) If other organizers exist, they could determine the amphiphilic or non-amphiphilic character of G_4 molecules. Careful biochemical analyses of the tetrameric forms of plasma BuChE have not revealed the presence of an associated protein, but a putative organizer could have escaped detection if it is a small peptide.

Tissue-specific expression of H and T subunits in vertebrates

In *Torpedo*, GPI-anchored AChE dimers represent the only form of enzyme in the dorsal muscles, indicating that they may be functional in neuromuscular junctions (S Bon, unpublished results). They also represent about two-thirds of total AChE activity in the *Torpedo* electric organ; the remaining third corresponds to collagen-tailed molecules, consisting of AChE-T subunits.[48] Both types of molecule are produced in the electromotor neurons of the electric lobes, as well as in the postsynaptic electroplaques. The synthesis of AChE-H and AChE-T subunits in electroplaques was demonstrated by the cloning of the two types of a DNA from electric organs[22,49,50] and by cultures of embryonic electroplaque prisms (S Bon, unpublished results).

In mammals, H transcripts are produced in embryonic tissues (muscle and liver), but in the adult they seem to be restricted to blood cells (erythrocytes and lymphocytes). Their presence on the surface of blood cells may be correlated with cellular differentiation, as suggested by antisense experiments: blockade of AChE synthesis in progenitor cells suppressed megakaryocyte formation and erythropoiesis, while increasing cell expansion and reducing apoptosis.[51,52] Antisense oligonucleotides against BuChE also reduced differentiation of megakaryocytes.[53,54]

Adult mammalian muscles and nervous tissue only contain the T subunits of AChE, together with a minor proportion of BuChE. Post-translational processing may be thought to introduce a cell-dependent bias in favour of one type of subunit: for example, the rat basoleukaemia cell line (RBL) produces over ten times more AChE activity when transfected with H than with T subunits, whereas COS cells produce similar levels of activity in both cases.[55] This seems appropriate, since blood cells normally express H transcripts, and probably results from the fact that T peptides act as a retention/degradation signal. Tissue-specific processing and degradation might explain why transfection of *Xenopus* embryos results in a restricted expression of AChE-T in muscles and brain, even when directed by a non-specific viral promoter (cytomegalovirus (CMV)).[56,57] Other AChE subunits might thus be expressed, as in embryonic tissues, but eliminated at a post-translational stage. However, analysis of transcripts by reverse transcriptase polymerase chain reaction (RT-PCR) shows that this is not the case: only T transcripts can be detected in postnatal rat muscle.[23] In addition, transfection experiments in the myogenic cell line C_2C_{12}, and in muscle in vivo showed that AChE-H subunits produce a higher AChE activity than AChE-T subunits.[58] The choice of AChE-T subunits in adult tissues thus relies exclusively on the specificity of splicing, which eliminates the R and H sequences.

Paradoxically, AChE-T subunits appear to be degraded in myogenic cells. This hypothesis is confirmed by the fact that the yield of active AChE is increased by co-expression with an associated protein, such as the collagen-tailed one, ColQ, or a fragment of ColQ which facilitates the oligomerization and export of AChE-T subunits.[58] This observation has also been made in COS cells, where co-transfection with an N-terminal fragment of ColQ (Q_N) 'rescues' a fraction of AChE-T subunits.[45] Thus, the presence of an exposed T peptide may trigger the degradation of AChE subunits.[59,60] This may be caused in part by some resemblance between the C-terminal tetrapeptide (CAEL in *Torpedo*, CSDL in rat) with the KDEL reticulum retention peptide,[61] and by the presence of a free cysteine, which is also thought to act as a retention signal.[62] The formation of an intercatenary disulphide-bond would at least partially mask those signals and allow secretion, explaining why the ratio of dimers to monomers is higher in the culture medium than in extracts from COS cells expressing AChE-T subunits.

Physiological assembly and significance of heteromeric AChE forms in mammalian muscles

Adult mammalian muscles express AChE-T subunits and contain G_1^a, G_2^a, G_4^a, G_4^{na} and collagen-tailed forms, A_4, A_8 and A_{12}, the major forms being usually G_1^a, G_4^a and A_{12}. The G_1^a and G_2^a forms at least partly represent precursors of the more complex molecules, and there is no clear evidence that they are exposed at the cell surface and play a physiological role. The probable localization of AChE forms in a mammalian neuromuscular junction is represented schematically in Fig. 6.3, based on the localizations observed in the rat fast muscles.

The rat fast and slow muscles differ in their content and distribution of collagen-tailed AChE forms. Fast muscle contains almost

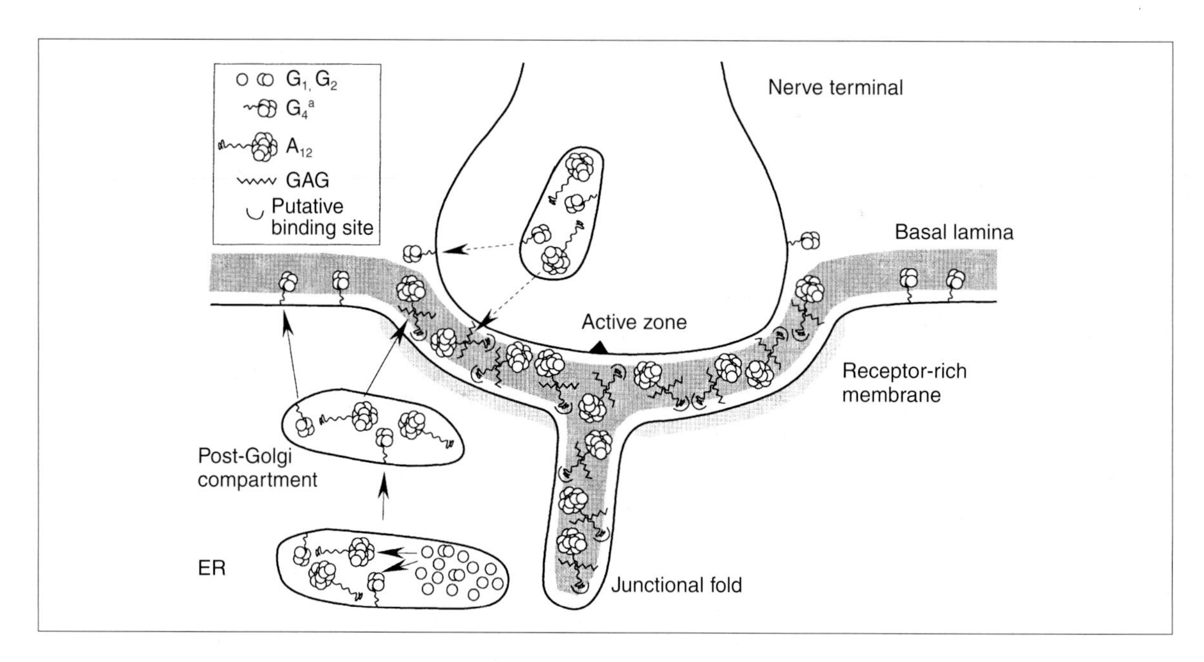

Figure 6.3
Schematic localization of AChE forms at the neuromuscular junction (based on rat fast muscles). The receptor-rich region of the postsynaptic membrane is indicated. The different molecular species of AChE are shown at their major localizations (not to scale). AChE monomers and dimers are assembled into collagen-tailed molecules in the endoplasmic reticulum (ER), and this is probably also the case for hydrophobic-tailed molecules. Collagen-tailed molecules, shown here as the A_{12} form, are then transported to the synaptic basal lamina through post-Golgi compartments, and associate with glycosaminoglycans (GAGs), such as perlecan, and possibly to unidentified synaptic binding sites, corresponding to 'molecular parking lots'.[63] Note that the length of the collagen-tailed molecule (about 50 nm) is close to the width of the synaptic cleft.[64] The A_{12} molecules are represented as randomly oriented in the synaptic cleft, but they are more probably integrated in an ordered supramolecular organization. Hydrophobic-tailed tetramers, G_4^a, are transported to the 'peri-junctional zone'.[65] Dashed arrows indicate the possible contribution of the presynaptic motor neuron to these localizations, through the axonal flow. The relative contributions of the muscle fibre and the motor neuron to synaptic AChE are still debated and may vary from muscle to muscle. An active zone, where synaptic vesicles containing ACh may be released, is shown in the nerve terminal, opposed to the opening of a postsynaptic junctional fold.

exclusively the A_{12} molecular form, in which the three ColQ chains forming a triple-helical collagen are attached to an AChE-T tetramer, and this collagen-tailed form is exclusively localized at the endplates.[66,67] This localization may be partly due to biosynthesis in motoneurons and transport to the presynaptic nerve terminals, but at least some of these molecules are produced in the muscle fibres, as shown by ectopic reinnervation and electric stimulation of denervated muscles.[68,69] In contrast, the slow soleus muscle contains a large proportion of 'unsaturated' collagen-tailed molecules in which only one (A_4) or two (A_8) ColQ chains are attached to an AChE-T tetramer, and these molecules are distributed all along the muscle fibre. We recently analysed the expression of AChE-T and ColQ in rat muscles and found that these distributions may be explained by the relative levels of AChE-T and ColQ transcripts in muscles.[70] In soleus, the ColQ transcripts are uniformly expressed all along the fibres. In the fast muscles, ColQ transcripts are less abundant, especially when compared to AChE-T transcripts, and they are restricted to the endplate region. Thus, collagen-tailed molecules are only produced in this region, in the presence of an excess of AChE-T subunits, explaining why the A_{12} form is largely predominant. By injecting appropriate amounts of AChE-T and ColQ mRNAs in *Xenopus* oocytes, we could in fact reproduce the patterns of A_{12}, A_8 and A_4 forms observed in the fast and slow muscles. It thus appears that the ratio of AChE-T and ColQ subunits determines the level of occupancy of the three PRADs in the triple-helical collagen tail, and thus defines the pattern of the collagen-tailed forms.

At the neuromuscular junction, collagen-tailed AChE molecules interact with polyanionic components of the extracellular matrix such as glycosaminoglycans (GAGs),[71–73] and in particular with perlecan.[74] These interactions with sulphated GAGs probably involve two heparin-binding sites, located within the triple-helical domain of ColQ.[75,76] In addition, the non-collagenous C-terminal domain probably interacts with other binding sites which are necessary for the functional localization of the enzyme: in a family presenting a congenital myasthenic syndrome (CMS-1c), the patients are homozygous for a point mutation in this domain; their muscles contain collagen-tailed forms but they are not concentrated at neuromuscular junctions.[77] This is in excellent agreement with the hypothesis that specific binding sites in the synaptic basal lamina constitute 'molecular parking lots' for collagen-tailed molecules. Following its binding, the collagen-tailed enzyme seems to become covalently bound to the synaptic basal lamina; it cannot be solubilized by heparin or at a high salt concentration, but catalytic tetramers can be detached by collagenase.[63,78,79]

The physiological importance of collagen-tailed AChE forms in muscles has recently been illustrated by the identification of several mutations in the ColQ genes of CMS-1c patients, which suppress the formation of collagen-tailed molecules.[80] The role of collagen-tailed AChE has been demonstrated experimentally by inactivating the ColQ gene in mice.[47] The $ColQ^{-/-}$ mice are born and may survive for several weeks, although they suffer from rather severe muscular dysfunction, including tremor and weakness. Their muscles show no histochemically detectable accumulation of AChE at neuromuscular junctions, in agreement with the fact that the normal accumulation of AChE at neuromuscular junctions consists mostly of collagen-tailed molecules and can be removed by collagenase.[81] Moreover, electrophysiological recordings show that miniature endplate potentials (MEPPs) are not affected by AChE inhibition (e.g. by the toxin fasciculin), in contrast to those of $ColQ^{+/+}$ muscles. When administered intraperitoneally, the toxin paralyses normal or heterozygous $ColQ^{+/-}$ mice but has no effect on $ColQ^{-/-}$ mice. The life time of these MEPPs appears rather normal, suggesting that adaptive changes have allowed the neuromuscular junction to function without an active hydrolysis of ACh by AChE. Other forms of AChE are also present in muscle, in particular G_1^a, G_2^a and G_4^a. However, they do not form histochemically detectable accumulations at neuromuscular junctions and their inhibition has no detectable effect on MEPPs, indicating that their localization does not allow them to efficiently contribute to synaptic function under these experimental conditions.

In addition to AChE, neuromuscular junctions contain a minor complement of BuChE, which is barely detectable, either biochemically and histochemically. The junctional BuChE appears to present two components. One component corresponds to collagen-tailed forms: it is removed by collagenase and suppressed in the ColQ$^{-/-}$ mice, since both ChEs may be associated with the same collagen tail. The second component appears to be membrane bound, since it is removed by the detergent Triton X-100, and is probably presynaptic, because it disappears shortly after denervation.[47] It will be interesting to assess the functional importance of this enzyme in normal and ColQ$^{-/-}$ muscles.

The analysis of neuromuscular transmission in ColQ$^{-/-}$ mice illustrates the fact that AChE function strictly depends on the precise anchoring and localization of the enzyme. It confirms the idea that collagen-tailed and hydrophobic-tailed molecules may respond to distinct physiological requirements.

The importance of collagen-tailed AChE forms in muscles does not exclude a possible function for other molecular forms, in particular the membrane-bound tetramers, G_4^a, which seem to be preferentially located in a 'perijunctional' zone.[65] The level of these molecules is modified in dystrophic muscles and is specifically controlled by exercise in fast muscles.[82,83] This suggests that the hydrophobic-tailed G_4^a form plays an adaptive role in muscle activity, presumably distinct from that of collagen-tailed forms. The presence of an excess of AChE-T subunits, relative to ColQ, at the neuromuscular junction of fast muscles, may explain that exercise specifically affects the level of membrane-bound tetramers (G_4^a), without any marked change in the A_{12} form: muscle activity may control the expression of a hydrophobic anchoring protein, which recruits some of the excess AChE-T subunits into G_4^a molecules.

AChE forms in the central nervous system

The major forms of AChE in the central nervous system of birds and mammals are monomers, dimers and tetramers of AChE-T subunits. As in muscles, it seems likely that monomers and dimers (G_1^a and G_2^a) mostly represent intracellular precursors of the mature heteromeric collagen-tailed and hydrophobic-tailed molecules.

Collagen-tailed forms of AChE are abundant in the central nervous systems of primitive vertebrates, e.g. in the electric lobes of *Torpedo* and in the brains of Teleost fishes and frogs.[67,84] In contrast, they are barely detectable in the brains of higher vertebrates.

An analysis of AChE forms during development of the quail showed the presence of non-amphiphilic and amphiphilic tetramers: the latter molecules, which incorporate a hydrophobic anchoring P protein, become predominant during development, representing over 85% of the total AChE activity.[85] A similar evolution was observed in rat: in the brain of 15-day old embryos, the major form of AChE is G_1^a, with minor proportions of G_2^a and G_4 (probably G_4^a and G_4^{na}); the proportion of G_4^a increases to about 80% in the adult.[46] The hydrophobic-tailed G_4^a form is very slightly soluble or insoluble without detergent, so that an aqueous extract contains essentially G_4^{na} while a detergent extract contains the bulk of G_4^a.

The G_1 form of embryonic mouse brain was found to differ from the adult form by a higher K_m value, a lower affinity for the peripheral-site inhibitors fasciculin and propidium, and it did not bind to an acridinium affinity column.[86] In contrast, the G_4 form appeared identical at embryonic and adult stages.

Glycosylation, maturation and catalytic differences between molecular forms

The catalytic domains of AChE and BuChE carry several N-linked glycans (four for *Torpedo* AChE and nine for human BuChE). The structure of the glycan chains may vary, depending on the cell type and on the molecular species. For example, *Torpedo* electric organs contain glycolipid-anchored dimers and collagen-tailed molecules (see below), and only the former possess a carbohydrate epitope recognized by a family of monoclonal antibodies including Elec-39 and HNK-1.[87] This suggests that these molecular forms are partially segregated during their processing in the secretory pathway, since they undergo different glycosylation processes. Similarly, the collagen-tailed forms of mammalian neuromuscular junctions are recognized by lectins from *Vicia villosa* (VVA) and *Dolichos biflorus* (DBA) that do not bind the globular forms.[88,89] Differences in lectin binding were also observed between the globular forms, G_1, G_2 and G_4, in bovine muscle and superior cervical ganglia: in particular, wheat germ agglutinin (WGA) binds the G_4 form, but not the G_1 form.[90] This may reflect a different state of maturation of the glycan chains, suggesting that the G_1 form carries oligomannosidic glycans and is contained in early parts of the secretory pathway, while the more mature G_4 form has acquired *N*-acetyllactosaminic glycans, in agreement with a predominantly extracellular localization. Lectins also reveal differences between enzymes produced in different cells, e.g. between bovine lymphocytes and erythrocytes.[90]

In general, the various AChE forms are catalytically equivalent, indicating that their oligomeric state has no direct influence on their catalytic activity.[91] However, covalent inhibitors may show a preferential inhibition of some molecular forms relative to others when administered systemically, because of their accessibility. For example, phenylmethylsulphonyl fluoride showed the same efficiency on all solubilized AChE forms, but when injected in vivo it was found to inhibit the rat brain G_4^a form more efficiently than G_1, in agreement with the fact that G_4^a is mostly extracellular while G_1 is mostly intracellular.[92]

Molecular forms of AChE can also present intrinsic kinetic differences, related to their glycosylation. For example, the GPI-anchored AChEs (G_2^a) from bovine lymphocytes and erythrocytes cannot be distinguished by their sedimentation coefficient or by anti-AChE antibodies, but the K_m value of the former is about twice that of the latter.[90] The lymphocyte enzyme is recognized by concanavalin A and is inhibited by agglutinin from *Ulex europeus*, while the corresponding erythrocyte enzyme is not affected by these lectins: the inhibition by *Ulex europeus* agglutinin shows that glycan interactions can affect the catalytic activity, and it is therefore likely that the observed difference in K_m values is caused by the different maturation of carbohydrate chains in the two types of cell.

Differences in glycosylation may thus explain the reported differences in the sensitivity of solubilized, isolated G_1 and G_4 AChE forms to various inhibitors. For example, ethopropazine inhibits G_1 AChE from rat muscle more efficiently than G_4.[93] A similar observation was made with heptylphysostigmine in the case of rat brain[94] and human brain.[95–97] SDZ ENA 713 (rivastigmine,

Exelon) was also more potent on G_1 than on G_4. Tacrine and physostigmine inhibited both forms equally well, and edrophonium had a slightly higher affinity for G_4 than for G_1. There was no significant difference between the corresponding molecular forms from normal and Alzheimer disease (AD) brains.

The differences observed in the affinity of G_1 and G_4 for inhibitors did not exceed tenfold, but concerned molecules of therapeutic interest, and have therefore been considered to provide some selectivity. It should be recalled, however, that G_1 is probably not physiologically active, either in muscle or in brain.

AChE forms in normal and Alzheimer human brain

In adult human brain, as in other mammals, AChE is mainly present as the membrane-bound G_4^a form, and the more soluble G_1 form, with minor contributions of G_2 and other forms.[98] During development, AChE becomes progressively less soluble without detergent (about 25% at birth, and 15% in the adult). Surprisingly, the adult proportion of G_4 (70–75% of the total activity) is reached as early as 11 weeks in the human embryo, in contrast with the fact that this mature form only becomes predominant postnatally in the rat.[46] It is likely that the membrane-bound G_4^{na} form progressively replaces the soluble G_4^a form during postnatal maturation.

In AD, the total AChE activity was found to be lower than in normal brains, mainly because of a decrease in the membrane-bound G_4^a form, principally in the frontal cortex, with a concomitant increase in the relative percentage of the soluble G_4^{na} form and of the G_2^a and G_1^a forms.[96,97,99–102] Since the G_2^a and G_1^a forms probably represent intracellular precursors of the mature physiologically active G_4^a species, these changes resemble a regression to an embryonic state.[99] They are variable according to the brain area, and this is possibly related to the localization of ChE-expressing cell bodies and terminals. Although the level of collagen-tailed forms is very low in normal brain, it was reported to increase three- to four-fold in the cortex of AD patients, reaching about 15% of the total AChE activity.[102]

Both AChE and BuChE were found to be associated with amyloid plaques and neurofibrillary tangles,[103,104] and this may correspond, at least partially, to collagen-tailed forms, since AChE could be solubilized from these structures as a G_4 form, by collagenase and trypsin.[105] This is consistent with the presence of perlecan in the plaques,[106] a heparan sulphate proteoglycan which has been shown to bind collagen tailed AChE at neuromuscular junctions.[74] In fact, sulphated glycosaminoglycans were found to accelerate the aggregation of Aβ peptides into amyloid deposits.[106,107] It is interesting that AChE itself accelerates Aβ aggregation in vitro, probably through an interaction with its peripheral site,[15,108,109] and may thus contribute to the formation of senile plaques. This may be related to the fact that AChE exerts a toxic effect on glial neuronal cells in culture, inducing apoptosis by a mechanism that does not seem to require its catalytic activity.[110]

The plaque-associated AChE was found to differ from the normal brain enzyme in its catalytic properties: traditional AChE inhibitors (BW284C51, tacrine and physostigmine) were more efficient on the enzyme of normal axons and cell bodies, while ChEs of plaques and tangles were selectively inhibited by serotonin and other compounds.[104] AChE forms stable complexes with Aβ peptides, which present similar modifications, including resistance to low pH, a higher K_m, reduced inhibition by excess substrate, and reduced sensitivity to inhibitors that bind to the active site (edrophonium and tacrine), to the peripheral site (propidium and gallamine) or to both sites simultaneously (BW284C51).[111,112] These differences may be due to glycosylation, as discussed previously, to the microenvironment of the enzymes, or to the occupancy of its peripheral site.

Inhibition of AChE has been found to exert a beneficial therapeutic effect in some Alzheimer patients. It might therefore be useful specifically to reduce the physiological efficiency of brain AChE, if possible in a controlled fashion, by modifying the functional localization of the mature G_4^a molecules. Cloning of the P subunit will not only fill an important gap in the understanding of AChE structure, but it will also allow us to understand the functional anchoring of AChE in brain at the molecular level. Molecular genetics of the P subunits may also offer new tools for interfering with this anchoring, either by an antisense or by a dominant negative strategy. It is not inconceivable that administration of appropriately targeted peptides, such as the WAT domains, may compete with the assembly of the brain membrane-bound tetramers.

In this chapter, we have emphasized the role of the C-terminal domains in the functional localization of ChE for ensuring acetylcholine hydrolysis in cholinergic transmission. These domains may also participate in non-classical functions of ChEs, as suggested for example by the fact that AChE subunits containing the WAT domain (AChE-T) stimulates neurite extension in transfected glioma cells, while other AChE variants had no such effect.[113]

Acknowledgements

I thank Alain Anselmet, Suzanne Bon, Eric Krejci and Claire Legay for useful discussions. This work was supported by grants from the Centre National de la Recherche Scientifique, the Association Française contre les Myopathies, the Direction des Forces et de la Prospective, and the European Community.

References

1. Massoulié J, Anselmet A, Bon S *et al.* Diversity and processing of acetylcholinesterase. In: Doctor BP, Quinn DM, Rotundo RL, Taylor P, eds. *Structure and Function of Cholinesterases and Related Proteins.* New York: Plenum, 1998: 3–24.
2. Massoulié J, Anselmet A, Bon S *et al.* Acetylcholinesterase: C-terminal domains, molecular forms and functional localization. *J Physiol (Paris)* 1998; **92:** 183–190.
3. Massoulié J, Pezzementi L, Bon S, Krejci E, Vallette FM. Molecular and cellular biology of cholinesterases. *Prog Neurobiol* 1993; **41:** 31–91.
4. Toutant JP. An evaluation of the hydrophobic interactions of chick muscle acetylcholinesterase by charge shift electrophoresis and gradient centrifugation. *Neurochem Int* 1986; **9:** 111–119.
5. Sussman JL, Harel M, Frolow F *et al.* Atomic structure of acetylcholinesterase from *Torpedo californica*: a prototypic acetylcholine-binding protein. *Science* 1991; **253:** 872–879.
6. Bourne Y, Taylor P, Marchot P. Acetylcholinesterase inhibition by fasciculin: crystal structure of the complex. *Cell* 1995; **83:** 503–512.
7. Harel M, Kleywegt GJ, Ravelli RBG, Silman I, Sussman JL. Crystal structure of an acetylcholinesterase–fasciculin complex: interaction of a three-fingered toxin from snake venom with its target. *Structure* 1995; **3:** 1355–1366.
8. Kryger G, Silman I, Sussman JL. Three-dimensional structure of a complex of E2020 with acetylcholinesterase from *Torpedo californica. J Physiol (Paris)* 1998; **92:** 191–194.
9. Kryger G, Silman I, Sussman JL. Structure of acetylcholinesterase complexed with E2020 (Aricept): implications for the design of new anti-Alzheimer drugs. *Structure* 1999; **7:** 297–307.

10. Massoulié J, Sussman JL, Doctor BP *et al.* Recommendations for nomenclature in cholinesterases. In: Shafferman A, Velan B eds. *Multidisciplinary Approaches to Cholinesterase Functions.* New York: Plenum, 1992: 285–288.
11. Checler F, Grassi J, Vincent JP. Cholinesterases display genuine arylacylamidase activity but are totally devoid of intrinsic peptidase activities. *J Neurochem* 1994; **62:** 756–763.
12. Darboux I, Barthalay Y, Piovant M, Hipeau-Jacquotte R. The structure–function relationships in *Drosophila* neurotactin show that cholinesterasic domains may have adhesive properties. *EMBO J* 1996; **15:** 4835–4843.
13. Ichtchenko K, Hata Y, Nguyen T *et al.* Neuroligin 1: a splice site-specific ligand for β-neurexins. *Cell* 1995; **81:** 435–443.
14. Botti SA, Felder CE, Sussman JL, Silman I. Electrotactins: a class of adhesion proteins with conserved electrostatic and structural motifs. *Protein Engineering* 1998; **11:** 415–420.
15. Inestrosa NC, Alvarez A, Perez CA *et al.* Acetylcholinesterase accelerates assembly of amyloid-β-peptides into Alzheimer's fibrils: possible role of the peripheral site of the enzyme. *Neuron* 1996; **16:** 881–891.
16. Duval N, Massoulié J, Bon S. H and T subunits of acetylcholinesterase from *Torpedo*, expressed in COS cells, generate all types of globular forms. *J Cell Biol* 1992; **118:** 641–653.
17. Legay C, Bon S, Massoulié J. Expression of a cDNA encoding the glycolipid-anchored form of rat acetylcholinesterase. *FEBS Lett* 1993; **315:** 163–166.
18. Li Y, Camp S, Rachinsky TL, Getman D, Taylor P. Gene structure of mammalian acetylcholinesterase. Alternative exons dictate tissue-specific expression. *J Biol Chem* 1991; **266:** 23 083–23 090.
19. Bertrand C, Takke C, Cousin X, Toutant JP, Chatonnet A. *Zebrafish Development and Genetics.* Cold Spring Harbor, NY: Cold Spring Harbor Laboratory Press, 1996: 105.
20. Simon S, Massoulié J. Cloning and expression of acetylcholinesterase from *Electrophorus*: splicing pattern of the 3' exons in vivo and in transfected mammalian cells. *J Biol Chem* 1997; **273:** 33 045–33 055.
21. Cousin X, Bon S, Duval N, Massoulié J, Bon C. Cloning and expression of acetylcholinesterase from *Bungarus fasciatus* venom. A new type of COOH-terminal domain; involvement of a positively charged residue in the peripheral site. *J Biol Chem* 1996; **271:** 15 099–15 108.
22. Sikorav JL, Krejci E, Massoulié J. cDNA sequences of *Torpedo marmorata* acetylcholinesterase: primary structure of the precursor of a catalytic subunit; existence of multiple 5'-untranslated regions. *EMBO J* 1987; **6:** 1865–1873.
23. Legay C, Huchet M, Massoulié J, Changeux JP. Developmental regulation of acetylcholinesterase transcripts in the mouse diaphragm: alternative splicing and focalization. *Eur J Neurosci* 1995; **7:** 1803–1809.
24. Kaufer D, Friedman A, Seidman S, Soreq H. Acute stress facilitates long-lasting changes in cholinergic gene expression. *Nature* 1998; **393:** 373–377.
25. Chan RY, Adatia FA, Krupa AM, Jasmin BJ. Increased expression of acetylcholinesterase T and R transcripts during hematopoietic differentiation is accompanied by parallel elevations in the levels of their respective molecular forms. *J Biol Chem* 1998; **273:** 9727–9733.
26. Li Y, Camp S, Taylor P. Tissue-specific expression and alternative mRNA processing of the mammalian acetylcholinesterase gene. *J Biol Chem* 1993; **268:** 5790–5797.
27. Cousin X, Hotelier T, Giles K, Toutant JP, Chatonnet A. aCHEdb: the database system for ESTHER, the α/β fold family of proteins and the cholinesterase gene server. *Nucleic Acids Res* 1998; **26:** 226–228.
28. Fournier D, Karch F, Bride J-M, Hall LMC, Bergé J-B, Spierer P. *Drosophila melanogaster* acetylcholinesterase gene: structure, evolution and mutations. *J Mol Biol* 1990; **210:** 15–22.
29. Ferguson MAJ, Williams AF. Cell surface anchoring of proteins via glycosylphosphatidylinositol structures. *Annu Rev Biochem* 1988; **57:** 285–320.
30. Rosenberry TL, Toutant JP, Haas R, Roberts WL. Identification and analysis of glycoinosi-

tol phospholipid anchors in membrane proteins. *Methods Cell Biol* 1989; **32:** 231–255.
31. Haas R, Brandt PT, Knight J, Rosenberry TL. Identification of amine components in a glycolipid membrane-binding domain at the C-terminus of human erythrocyte acetylcholinesterase. *Biochemistry* 1986; **25:** 3098–3105.
32. Haas R, Jackson BC, Reinhold B, Foster JD, Rosenberry TL. Glycoinositol phospholipid anchor and protein C-terminus of bovine erythrocyte acetylcholinesterase: analysis by mass spectrometry and by protein and DNA sequencing. *Biochem J* 1996; **314:** 817–825.
33. Delahunty MD, Stafford FJ, Yuan LC, Shaz D, Bonifacino JS. Uncleaved signals for glycosylphosphatidylinositol anchoring cause retention of prescursor proteins in the endoplasmic reticulum. *J Biol Chem* 1993; **268:** 12 017–12 027.
34. Moran P, Caras IW. Proteins containing an uncleaved signal for glycophosphatidylinositol membrane anchor attachment are retained in a post-ER compartment. *J Cell Biol* 1992; **119:** 763–772.
35. Maxwell SE, Ramalingam S, Gerber LD, Udenfriend S. Cleavage without anchor addition accompanies the processing of a nascent protein to its glycosylphosphatidylinositol-anchored form. *Proc Natl Acad Sci USA* 1995; **92:** 1550–1554.
36. Bon S, Rosenberry TL, Massoulié J. Amphiphilic, glycophosphatidylinositol-specific phospholipase C (PI-PLC)-insensitive monomers and dimers of acetylcholinesterase. *Cell Mol Neurobiol* 1991; **11:** 157–172.
37. Bon S, Toutant JP, Méflah K, Massoulié J. Amphiphilic and nonamphiphilic forms of *Torpedo* cholinesterases. I. Solubility and aggregation properties. *J Neurochem* 1998; **51:** 776–785.
38. Bon S, Toutant JP, Méflah K, Massoulié J. Amphiphilic and nonamphiphilic forms of *Torpedo* cholinesterases. II. Electrophoretic variants and phosphatidylinositol phospholipase C-sensitive and -insensitive forms. *J Neurochem* 1998; **51:** 786–794.
39. Giles K. Interactions underlying subunit association in cholinesterases. *Protein Eng* 1997; **10:** 677–685.
40. Simon S, Krejci E, Massoulié J. A four-to-one association between peptide motifs: four C-terminal domains from cholinesterase assemble with one proline-rich attachment domain (PRAD) in the secretory pathway. *EMBO J* 1998; **17:** 6178–6187.
41. Bon S, Coussen F, Massoulié J. Quaternary associations of aceytlcholinesterase. II. The polyproline attachment domain of the collagen tail. *J Biol Chem* 1997; **272:** 3016–3021.
42. Gennari K, Brunner J, Brodbeck U. Tetrameric detergent-soluble acetylcholinesterase from human caudate nucleus: subunit composition and number of active sites. *J Neurochem* 1987; **49:** 12–18.
43. Inestrosa NC, Roberts WL, Marshall TL, Rosenberry TL. Acetylcholinesterase from bovine caudate nucleus is attached to membranes by a novel subunit distinct from those of acetylcholinesterases in other tissues. *J Biol Chem* 1987; **262:** 4441–4444.
44. Roberts WL, Doctor BP, Foster JD, Rosenberry TL. Bovine brain acetylcholinesterase primary sequence involved in intersubunit disulfide linkages. *J Biol Chem* 1991; **266:** 7481–7487.
45. Bon S, Massoulié J. Quaternary associations of acetylcholinesterase. I. Oligomeric associations of T subunits with and without the amino-terminal domain of the collagen tail. *J Biol Chem* 1997; **272:** 3007–3015.
46. Muller F, Dumez Y, Massoulié J. Molecular forms and solubility of acetylcholinesterase during the embryonic development of rat and human brain. *Brain Res* 1985; **331:** 295–302.
47. Feng G, Krejci E, Molgo J, Cunningham JM, Massoulié J, Sanes JR. Genetic analysis of collagen Q: roles in acetylcholinesterase and butyrylcholinesterase assembly and in synaptic structure and function. *J Cell Biol* 1999; **144:** 1349–1360.
48. Bon S, Massoulié J. Collagen-tailed and hydrophobic components of acetylcholinesterase in *Torpedo marmorata* electric organ. *Proc Natl Acad Sci USA* 1980; **77:** 4464–4468.
49. Schumacher M, Maulet Y, Camp S, Taylor P. Multiple messenger RNA species give rise to the structural diversity in acetylcholinesterase. *J Biol Chem* 1988; **263:** 18 979–18 987.

50. Sikorav JL, Duval N, Anselmet A *et al.* Complex alternative splicing of acetylcholinesterase transcripts in *Torpedo* electric organ: primary structure of the precursor of the glycolipid-anchored dimeric form. *EMBO J* 1988; **7:** 2983–2993.
51. Lev-Lehman E, Ginzberg D, Hornreich G *et al.* Antisense inhibition of acetylcholinesterase gene expression causes transient hematopoietic alterations in vivo. *Gene Ther* 1994; **1:** 127–135.
52. Soreq H, Patinkin D, Lev-Lehman E *et al.* Antisense oligoneucleotide inhibition of acetylcholinesterase gene expression induces progenitor cell expansion and suppresses hematopoietic apoptosis ex vivo. *Proc Natl Acad Sci USA* 1994; **91:** 7907–7911.
53. Patinkin D, Lev-Lehman E, Zakut H, Eckstein F, Soreq H. Antisense inhibition of butyrylcholinesterase gene expression predicts adverse hematopoietic consequences to cholinesterase inhibitors. *Cell Mol Neurobiol* 1994; **14:** 459–473.
54. Patinkin D, Seidman S, Eckstein F, Benseler F, Zakut H, Soreq H. Manipulations of cholinesterase gene expression modulate murine megakaryocytopoiesis in vitro. *Mol Cell Biol* 1990; **10:** 6046–6050.
55. Coussen F, Bonnerot C, Massoulié J. Stable expression of acetylcholinesterase and associated collagenic subunits in transfected RBL cell lines: production of GPI-anchored dimers and collagen-tailed forms. *Eur J Cell Biol* 1995; **67:** 254–260.
56. Seidman S, Sternfeld M, Ben Aziz-Aloya R, Timberg R, Kaufer-Nachum D, Soreq H. Synaptic and epidermal accumulations of human acetylcholinesterase are encoded by alternative 3'-terminal exons. *Mol Cell Biol* 1995; **1555:** 2993–3002.
57. Shapira M, Seidman S, Sternfeld M *et al.* Transgenic engineering of neuromuscular junctions in *Xenopus laevis* embryos transiently overexpressing key cholinergic proteins. *Proc Natl Acad Sci USA* 1994; **91:** 9072–9076.
58. Legay C, Mankal FA, Massoulié J, Jasmin BJ. Stability and secretion of acetylcholinesterase forms in skeletal muscle cells. *J Neurosci* 1999; **19:** 8252–8259.
59. Kerem A, Kronman C, Bar-Nun S, Shafferman A, Velan B. Interrelations between assembly and secretion of recombinant human acetylcholinesterase. *J Biol Chem* 1993; **268:** 180–184.
60. Velan B, Kronman C, Flashner Y, Shafferman A. Reversal of signal-mediated cellular retention by subunit assembly of human acetylcholinesterase. *J Biol Chem* 1994; **269:** 22 719–22 725.
61. Munro S, Pelham HR. A C-terminal signal prevents secretion of luminal ER proteins. *Cell* 1987; **48:** 899–907.
62. Sitia R, Neuberger M, Alberini C *et al.* Developmental regulation of IgM secretion: the role of the carboxy-terminal cysteine. *Cell* 1990; **60:** 781–790.
63. Rotundo RL, Rossi SG, Anglister L. Transplantation of quail collagen-tailed acetylcholinesterase molecules onto the frog neuromuscular synapse. *J Cell Biol* 1997; **136:** 367–374.
64. Krejci E, Coussen F, Duval N *et al.* Primary structure of a collagenic tail peptide of *Torpedo* acetylcholinesterase: co-expression with catalytic subunit induces the production of collagen-tailed forms in transfected cells. *EMBO J* 1991; **10:** 1285–1293.
65. Gisiger V, Stephens HR. Localization of the pool of G_4 acetylcholinesterase characterizing fast muscles and its alteration in murine muscular dystrophy. *J Neurosci Res* 1988; **19:** 62–78.
66. Sketelj J, Črne-Finderle N, Ribaric S, Brzin M. Interactions between intrinsic regulation and neural modulation of acetylcholinesterase in fast and slow skeletal muscles. *Cell Mol Neurobiol* 1991; **1:** 35–54.
67. Toutant JP, Massoulié J. Cholinesterases: tissue and cellular distribution of molecular forms and their physiological regulation. *Handbook Exp Pharmacol* 1988; **86:** 225–265.
68. Lømo T, Massoulié J, Vigny M. Stimulation of denervated rat soleus muscle with fast and slow activity patterns induces different expression of acetylcholinesterase molecular forms. *J Neurosci* 1985; **5:** 1180–1187.
69. Weinberg CB, Hall ZW. Junctional form of acetylcholinesterase restored at nerve-free

endplates. *Dev Biol* 1979; **68:** 631–635.
70. Krejci E, Legay C, Thomine S, Sketelj J, Massoulié J. Differences in expression of acetylcholinesterase and collagen Q control the distribution and oligomerization of the collagen-tailed forms in fast and slow muscles. *J Neurosci* 1999; in press.
71. Bon S, Cartaud J, Massoulié J. The dependence of acetylcholinesterase aggregation at low ionic strength upon a polyanionic component. *Eur J Biochem* 1978; **85:** 1–14.
72. Brandan E, Inestrosa NC. Binding of asymmetric forms of acetylcholinesterase to heparin. *Biochem J* 1984; **221:** 415–422.
73. Brandan E, Maldonado M, Garrido J, Inestrosa NC. Anchorage of collagen-tailed acetylcholinesterase to the extracellular matrix is mediated by heparan sulfate proteoglycans. *J Cell Biol* 1985; **101:** 985–992.
74. Peng HB, Xie H, Rossi SG, Rotundo RL. Acetylcholinesterase clustering at the neuromuscular junction involves perlecan and dystroglycan. *J Cell Biol* 1999; **145:** 911–921.
75. Deprez P, Inestrosa NC. Two heparin-binding domains are present on the collagenic tail of asymmetric acetylcholinesterase. *J Biol Chem* 1995; **270:** 11 043–11 046.
76. Deprez P, Signorelli J, Inestrosa NC. Effect of protamine on the solubilization of collagen-tailed acetylcholinesterase: potential heparin-binding consensus sequence in the tail of the enzyme. *Biochim Biophys Acta* 1995; **1252:** 53–58.
77. Donger C, Krejci E, Pou Serradell A *et al.* Mutation in the human acetylcholinesterase-associated collagen gene, ColQ, is responsible for congenital myasthenic syndrome with end-plate acetylcholinesterase deficiency (type Ic). *Am J Human Genet* 1998; **63:** 967–975.
78. Rossi SG, Rotundo RL. Localization of 'non-extractible' acetylcholinesterase to the vertebrate neuromuscular junctions. *J Biol Chem* 1993; **268:** 19 152–19 159.
79. Rossi SG, Rotundo RL. Transient interactions between collagen-tailed acetylcholinesterase and sulfated proteoglycans prior to immobilization in the extracellular matrix. *J Biol Chem* 1996; **271:** 1979–1987.
80. Ohno K, Brengman J, Tsujino A, Engel AG. Human endplate acetylcholinesterase deficiency caused by mutations in the collagen-like (ColQ) of the asymmetric enzyme. *Proc Natl Acad Sci USA* 1998; **95:** 9654–9659.
81. Hall ZW, Kelly RB. Enzymatic detachment of endplate acetylcholinesterase from muscle. *Nature New Biol* 1971; **232:** 62–64.
82. Gisiger V, Bélisle M, Gardiner PF. Acetylcholinesterase adaptation to voluntary wheel running is proportional to the volume of activity in fast, but not slow, rat hindlimb muscles. *Eur J Neurosci* 1994; **6:** 673–680.
83. Jasmin BJ, Gisiger V. Regulation by exercise of the pool of G_4 acetylcholinesterase characterizing fast muscles: opposite effects of running training in antagonist muscles. *J Neurosci* 1990; **10:** 1444–1454.
84. Massoulié J, Bon S. The molecular forms of cholinesterase and acetylcholinesterase in vertebrates. *Annu Rev Neurosci* 1982; **5:** 57–106.
85. Anselmet A, Fauquet M, Chatel JM, Maulet Y, Massoulié J, Vallette FM. Evolution of acetylcholinesterase transcripts and molecular forms during development in the central nervous system of the quail. *J Neurochem* 1994; **62:** 2158–2165.
86. Moreno RD, Campos FO, Dajas F, Inestrosa NC. Developmental regulation of mouse brain monomeric acetylcholinesterase. *Int J Dev Neurosci* 1998; **16:** 123–134.
87. Bon S, Méflah K, Musset F, Grassi J, Massoulié J. An immunoglobulin M monoclonal antibody, recognizing a subset of acetylcholinesterase molecules from electric organs of *Electrophorus* and *Torpedo*, belongs to the HNK-1 anti-carbohydrate family. *J Neurochem* 1987; **49:** 1720–1731.
88. Bacou F, Vigneron P. Neural influence on the expression of acetylcholinesterase molecular forms in fast and slow rabbit skeletal muscles. *Dev Biol* 1991; **145:** 356–366.
89. Scott LJC, Bacou F, Sanes JR. A synapse-specific carbohydrate at the neuromuscular junction: association with both acetylcholinesterase and a glycolipid. *J Neurosci* 1988; **8:** 932–944.
90. Méflah K, Bernard S, Massoulié J. Interactions with lectins indicate differences in the

carbohydrate composition of the membrane-bound enzymes acetylcholinesterase and 5'-nucleotidase in different cell types. *Biochimie* 1984; **66:** 59–69.

91. Vigny M, Bon S, Massoulié J, Leterrier F. Active-site catalytic efficiency of acetylcholinesterase molecular forms in *Electrophorus*, *Torpedo*, rat and chicken. *Eur J Biochem* 1978; **85:** 317–323.
92. Skau KA, Shipley MT. Phenylmethylsulfonyl fluoride inhibitory effects on acetylcholinesterase of brain and muscle. *Neuropharmacology* 1999; **38:** 691–698.
93. Skau K. Ethopropazine inhibition of AChE molecular forms. *Pharmacologist* 1981; **23:** 224.
94. Ogane N, Giacobini E, Messamore E. Preferential inhibition of acetylcholinesterase molecular forms in rat brain. *Neurochem Res* 1992; **17:** 489–495.
95. Enz A, Amstutz R, Boddeke H, Gmelin G, Malanowski J. Brain selective inhibition of acetylcholinesterase: a novel approach to therapy for Alzheimer's disease. *Prog Brain Res* 1993; **98:** 431–438.
96. Enz A, Chappuis A, Probst A. Different influence of inhibitors on acetylcholinesterase molecular forms G_1 and G_4 isolated from Alzheimer's disease and control brains. In: Shafferman A, Velan B, eds. *Multidisciplinary Approaches to Cholinesterase Functions*. New York: Plenum, 1992: 243–249.
97. Ogane N, Giacobini E, Struble R. Differential inhibition of acetylcholinesterase molecular forms in normal and Alzheimer disease brain. *Brain Res* 1992; **589:** 307–312.
98. Atack JR, Perry EK, Bonham JR, Candy JM, Perry RH. Molecular forms of acetylcholinesterase and butyrylcholinesterase in the aged human central nervous system. *J Neurochem* 1986; **47:** 263–277.
99. Arendt T, Brückner MK, Lange M, Bigl V. Changes in acetylcholinesterase and butyrylcholinesterase in Alzheimer's disease resemble embryonic development; a study of molecular forms. *Neurochem Int* 1992; **21:** 381–396.
100. Schegg KM, Harrington LS, Neilsen S, Zweig RM, Peacock JH. Soluble and membrane-bound forms of brain acetylcholinesterase in Alzheimer's disease. *Neurobiol Aging* 1992; **13:** 697–704.
101. Schegg KM, Nielsen S, Zweig RM, Peacock JH. Decrease in membrane-bound G_4 form of acetylcholinesterase in postmortem Alzheimer brain. *Prog Clin Biol Res* 1989; **317:** 437–452.
102. Younkin SG, Goodridge B, Katz J *et al.* Molecular forms of acetylcholinesterases in Alzheimer's disease. *Fed Proc* 1986; **45:** 2982–2988.
103. Morán MA, Mufson EJ, Gómez-Ramos P. Cholinesterases colocalize with sites of neurofibrillary degeneration in aged and Alzheimer's brain. *Acta Neuropathol* 1994; **87:** 284–292.
104. Wright CI, Geula C, Mesulam MM. Protease inhibitors and indoleamines selectively inhibit cholinesterases in the histopathologic structures of Alzheimer disease. *Proc Natl Acad Sci USA* 1993; **90:** 683–686.
105. Nakamura S, Kawashima S, Nakano S, Tsuji T, Araki W. Subcellular distribution of acetylcholinesterase in Alzheimer's disease: abnormal localization and solubilization. *J Neural Transm* 1990; **30(Suppl):** 13–23.
106. Castillo GM, Ngo C, Cummings J, Wight TN, Snow AD. Perlecan binds to the β-amyloid proteins (Aβ) of Alzheimer's disease, accelerates Aβ fibril formation, and maintains Aβ fibril stability. *J Neurochem* 1997; **69:** 2452–2465.
107. Castillo GM, Lukito W, Wight TN, Snow AD. The sulfate moieties of glycosaminoglycans are critical for the enhancement of β-amyloid protein fibril formation. *J Neurochem* 1999; **72:** 1681–1687.
108. Alvarez A, Opazo C, Alarcon R, Garrido J, Inestrosa NC. Acetylcholinesterase promotes the aggregation of amyloid–β-peptide fragments by forming a complex with the growing fibrils. *J Mol Biol* 1997; **272:** 348–361.
109. Campos EO, Alvarez A, Inestrosa NC. Brain acetylcholinesterase promotes amyloid–β-peptide aggregation but does not hydrolyze amyloid precursor protein peptides. *Neurochem Res* 1998; **23:** 135–140.
110. Calderon FH, von Bernhardi R, De Ferrari G *et al.* Toxic effects of acetylcholinesterase on neuronal and glial-like cells in vitro. *Mol Psy-*

chiatry 1998; **3:** 247–255.

111. Alvarez A, Alarcon R, Opazo C *et al.* Stable complexes involving acetylcholinesterase and amyloid-β peptide change the biochemical properties of the enzyme and increase the neurotoxicity of Alzheimer's fibrils. *J Neurosci* 1998; **18:** 3213–3223.
112. Inestrosa NC, Alarcon R. Molecular interactions of acetylcholinesterase with senile plaques. *J Physiol (Paris)* 1998; **92:** 341–344.
113. Karpel R, Sternfeld M, Ginzberg D, Guhl E, Graessmann A, Soreq H. Overexpression of alternative human acetylcholinesterase forms modulates process extensions in cultured glioma cells. *J Neurochem* 1996; **66:** 114–123.
114. Lockridge O, Bartels CF, Vaughan TA, Wong CK, Norton SE, Johnson FF. Complete amino acid sequence of human serum cholinesterase. *J Biol Chem* 1987; **262:** 549–557.
115. Legay C, Bon S, Vernier P, Coussen F, Massoulié J. Cloning and expression of a rat acetylcholinesterase subunit: generation of multiple molecular forms and complementarity with a *Torpedo* collagenic subunit. *J Neurochem* 1993; **60:** 337–346.

7

Mechanism of action of cholinesterase inhibitors

Elsa Reiner, Zoran Radić

Acylating and reversible cholinesterase inhibitors: interactions with catalytic and peripheral enzyme sites

The hydrolysis of substrates (S) of acetylcholinesterase (AChE; EC 3.1.1.7) and butyrylcholinesterase (BChE; EC 3.1.1.8) proceeds in three steps:

$$E + S \underset{k_{-1}}{\overset{k_1}{\rightleftarrows}} ES \xrightarrow{k_2} EA + P_1 \quad (1)$$

$$EA + H_2O \xrightarrow{k_3} E + P_2 \quad (2)$$

where ES is the enzyme–substrate Michaelis complex, EA is the acylated enzyme, and P_1 and P_2 are the products of substrate hydrolysis. When the substrate is acetylcholine, EA is the acetylated enzyme E–O–C(O)CH_3, and P_1 and P_2 are choline and the acetic acid. k_1 and k_{-1} are the rate constants for binding and dissociation of the substrate, and k_2 and k_3 are the rate constants for acylation and deacylation of the enzyme. Acylation and deacylation take place on the hydroxyl group of serine in the catalytic triad of the enzyme.

At steady-state, the enzyme activity v at a given substrate concentration s is:

$$v = \frac{V_m s}{K_m + s} = \frac{k_{cat} e_0 s}{K_m + s} \quad (3)$$

where V_m is the maximum activity, K_m is the Michaelis constant, k_{cat} is the catalytic constant, and e_0 is the total enzyme concentration.

It follows from reactions (1) and (2) that:

$$K_m = \frac{(k_{-1} + k_2)k_3}{k_1(k_2 + k_3)} \quad (4)$$

and

$$k_{cat} = \frac{k_2 k_3}{k_2 + k_3} \quad (5)$$

The Michaelis constant is a measure of the affinity of the substrate for the enzyme: the lower the value of K_m, the higher the affinity.

AChE has also a peripheral site which is catalytically inactive, but can bind substrates and other ligands. That site is located at the rim of the catalytic site gorge. Binding of substrates, like acetylcholine (ACh) and acetylthiocholine, to the peripheral site causes substrate inhibition, but it is not yet clear which step(s) in reactions (1) and (2) are most affected. BChE is also inhibited by some substrates, like benzoylcholine, and it is assumed that the mechanism of substrate inhibition is the same as that in AChE. For details concerning the structure and catalytic properties of cholinesterases (ChEs) the reader is referred to extensive reviews.[1–5]

Binding of some substrates to the peripheral site in BChE can cause an increase in activity termed substrate activation.[6–9] The same has recently been shown for the hydrolysis of some substrates by AChE (unpublished data).

Substrate inhibition and substrate activation are both defined by the dissociation constant K_{ss} of the enzyme–substrate complex in the peripheral site. Substrates can bind to the peripheral site either in the free enzyme (SE), or in the Michaelis complex (SES), or in the acylated enzyme (SEA). For the substrates studied so far, the K_{ss} constants are 100-fold or more higher than the corresponding K_m constants, which means that the affinity of the substrate for the peripheral site is considerably lower than for the catalytic site.

Substrate inhibition and substrate activation both cause the catalytic activity to deviate from equation (3) at substrate concentrations approaching K_{ss}. Under those conditions the enzyme activity is better described by:[6–10]

$$v = \frac{V_m[1 + b(s/K_{ss})]}{[1 + (K_m/s)][1 + (s/K_{ss})]} \tag{6}$$

where b represents the activity of ternary complexes (SES and SEA) relative to the activity of the ES complex. Values of b range from $b < 1$ in substrate inhibition to $b > 1$ in substrate activation. When $b = 1$, equation (6) reduces to equation (3).

For ACh and other ChE substrates K_m constants are usually in the micromolar range and K_{ss} constants in the millimolar range, well above typical physiological concentrations of ACh. Substrate inhibition of AChE could become physiologically relevant in vivo, during release of ACh from the presynaptic membrane, thus allowing more substrate to reach the postsynaptic membrane receptor. Activation of BChE activity in serum or liver could have a role in improving its efficacy in hydrolytic detoxification of dietary esters.

Acylating inhibitors (AB) react with ChEs in the same way as substrates.[1–5,11] The difference is only quantitative, as rates of deacylation (equation (2)) are very slow and with some inhibitors even approach zero. The enzyme therefore stays acylated by AB for a long time, and cannot hydrolyse substrates during that time. As deacylation is slow, it is usually possible to measure separately the acylation and deacylation steps, which is not the case with substrates.

Acylation, often termed progressive inhibition or irreversible inhibition,

$$\mathrm{E} + \mathrm{AB} \underset{k_{-1}}{\overset{k_1}{\rightleftarrows}} \mathrm{EAB} \xrightarrow{k_2} \mathrm{EA} + \mathrm{P_1} \quad (\text{overall: } \mathrm{E} + \mathrm{AB} \xrightarrow{k_a} \mathrm{EA}) \tag{7}$$

is defined by a second-order rate constant k_a, which describes the overall time course of acylation. EAB is a Michaelis-type complex between AB and the enzyme, and EA is the acylated enzyme. The constant:

$$K'_a = \frac{k_{-1} + k_2}{k_1} \tag{8}$$

is a measure of the affinity of AB for the enzyme; it is analogous to K_m for the substrate. When K'_a is much smaller than the AB concentration (ab), the time course of progressive inhibition follows the equation:

$$\ln\left(\frac{e_0}{e_t}\right) = \left(\frac{k_2}{K'_a}\right) ab\, t = k_a\, ab\, t \tag{9}$$

where e_0 and e_t are the concentrations of E at zero and time t of inhibition, and the ratio k_2/K'_a is equal to k_a.

Deacylation of the inhibited enzyme (analogous to equation (2)) is usually termed spontaneous reactivation, and is defined by a first-order rate constant k_r, which describes the time course of reactivation:

$$\ln\left(\frac{ea_0}{ea_t}\right) = k_r t \tag{10}$$

where ea_0 and ea_t are concentrations of the acylated enzyme at zero and time t of reactivation.

Reversible inhibitors form only an additive complex with the enzyme and this does not lead to any product formation. Their reaction is therefore defined by the rate constants of association to the enzyme (k_1) and dissociation from the enzyme (k_{-1}). Analogous to substrates, a reversible inhibitor (I) can bind either to the catalytic site (EI), to the peripheral site (IE) or to both sites (IEI) on the enzyme. The equilibrium dissociation constants (k_{-1}/k_1) for the enzyme–inhibitor complexes in the catalytic and peripheral sites are usually denoted by K_a and K_i, respectively. Low values of K_a or K_i correlate with high affinities of the reversible inhibitor for the enzyme.

Binding of reversible inhibitors to the catalytic site of the enzyme slows down the rate of substrate hydrolysis, due to competition between the substrate and the inhibitor. Equally, reversible inhibitors slow down progressive acylation of the catalytic site. The rate of substrate hydrolysis or progressive acylation is also affected when a reversible inhibitor binds to the peripheral site of the enzyme.[12–15]

Acylating inhibitors

Organophosphates

Esters of phosphoric, phosphonic and phosphinic acids, with no free hydroxyl group on the phosphorus atom, are acylating (phosphorylating) inhibitors of ChEs and other serine esterases. The substituents on the phosphorus atom can vary widely, and the cleavage occurs on the P–O, P–F, P–S, P–CN or other bond. The structures of some widely studied compounds, including the nerve agents sarin, soman, tabun and VX, are given in Table 7.1.

Organophosphate (OP) compounds with a sulphur atom double bonded to phosphorus (P=S) become ChE inhibitors only after enzymic oxidation to the corresponding P=O compound. Well-known examples are two pesticides: parathion (which oxidizes to paraoxon) and malathion (which oxidizes to malaoxon). Metrifonate is also not a direct ChE inhibitor. This organophosphate is used in the treatment of Alzheimer's disease (see Chapter 12) and as a drug against schistosomiasis.[20] In aqueous solutions, metrifonate rearranges spontaneously into DDVP (dichlorvos), which is a potent ChE inhibitor:

Metrifonate —(– HCl)→ DDVP

Inhibitor	*Formula*	k_a *(M^{-1} min^{-1})*	*Source**	*Ref.*
Ethoxylmethylphosphonyl toxogonine†	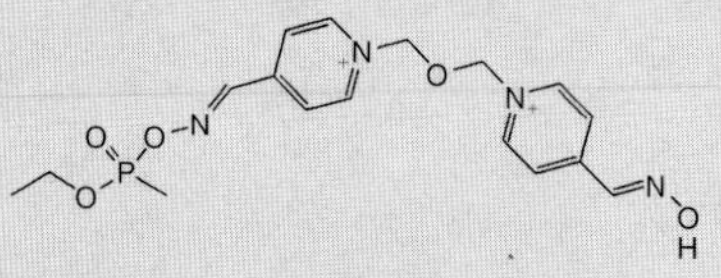	1.3×10^{10}	I	16
Soman	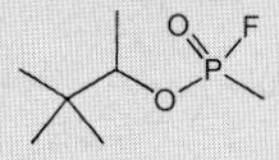	2.0×10^{8}‡	II	17
VX	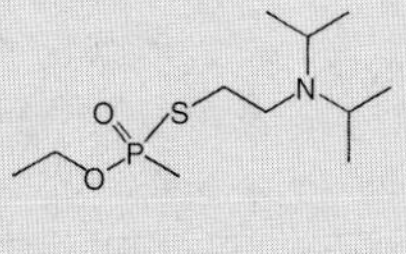	7.2×10^{7}‡	II	17
Tabun	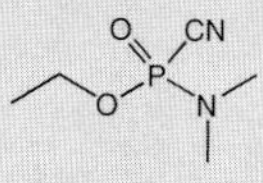	3.4×10^{7}‡	II	17
Sarin	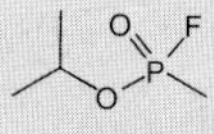	2.7×10^{7}‡	II	17
Paraoxon	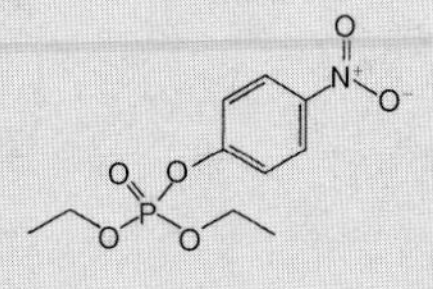	1.1×10^{6}	III	18
Malaoxon	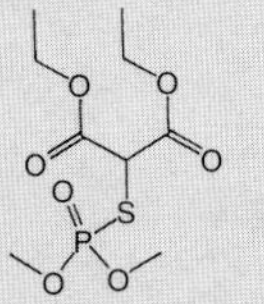	6.5×10^{4}	IV	19
DDVP	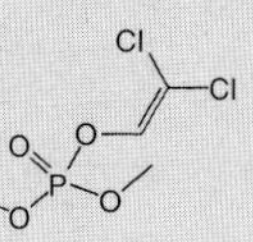	3.9×10^{4}	III§	
DFP	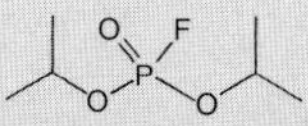	1.1×10^{4}	III	18

* Source: (I) mouse AChE in 20 mM HEPES buffer, pH 7.2, at 25°C; (II) human erythrocyte AChE in 100 mM phosphate buffer, pH 7.4, at 37°C; (III) mouse AChE in 100 mM phosphate buffer, pH 7.0, at 22°C; (IV) human erythrocyte AChE in 100 mM phosphate buffer, pH 7.6, at 25°C.
† This structure has not yet been completely confirmed.
‡ Constant for inhibition by a racemic mixture of a chiral organophosphate.
§ Z Radić, unpublished results.

Table 7.1
Organophosphate inhibitors of AChE: second-order rate constants (k_a) of inhibition.

ChEs are stereospecific, and the inhibitory potency of enantiomers can differ considerably.[19]

Phosphorylated ChEs undergo two reactions: one is spontaneous reactivation (analogous to equation (2)) and the other is dealkylation, termed ageing. The aged enzyme can no longer undergo spontaneous reactivation. An example of ageing is dealkylation of the diisopropylphosphorofluoridate (DFP) inhibited enzyme:

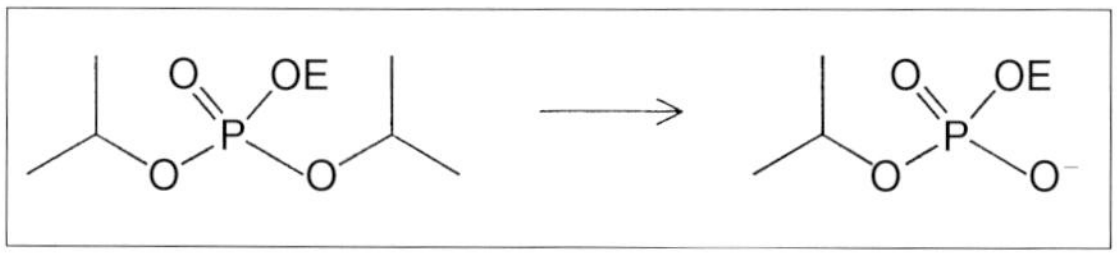

Crystal structures of soman-, sarin- and DFP-aged AChE have recently been determined, confirming the above reaction[21] (see Chapter 2).

Rate constants of phosphorylation (k_a; equations (7) and (9)), spontaneous reactivation (k_r; equation (10)) and ageing span a broad range of values depending on the OP compound and on the enzyme source.

Oximes (Fig. 7.1) are more efficient reactivators than water, because they are stronger nucleophiles. The oxime-catalysed reactivation proceeds via a Michaelis-type complex between phosphorylated enzyme and reactivator, resulting in the free enzyme and phosphorylated oxime. Phosphorylated oximes are potent cholinesterase inhibitors.[16] As they are very unstable, re-inhibition of the reactivated enzyme has not been consistently reported. However, recently it has been found that some phosphonylated derivatives of TMB-4 and toxogonin have life times of 30–60 min, unlike the lifetimes of 2PAM and HI-6 derivatives, which appear to be undetectably short.[16]

Oximes are not only reactivators of phosphorylated ChEs, but they are also reversible inhibitors of these enzymes. Pyridinium and imidazolium oximes bind to the catalytic or peripheral, or both, sites on the enzyme, causing a decrease in the rate of substrate hydrolysis and also a decrease in the rate of acylation of the catalytic site by OP compounds.[22–24]

The toxicity of OP compounds is primarily due to phosphorylation of AChE. Concerning the reactions described above, a given OP

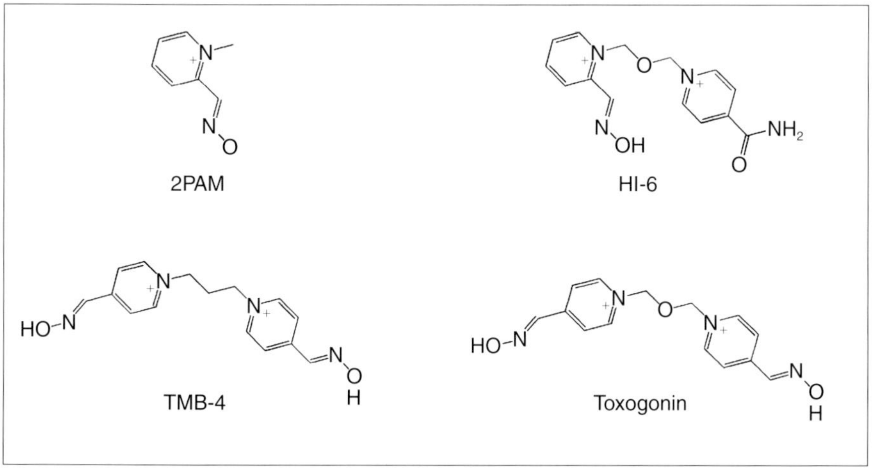

Figure 7.1 *Common oxime reactivators of phosphylated AChE.*

compound is potentially more toxic the higher the rate constants of phosphorylation and ageing, and the lower the rate constant of spontaneous reactivation. Oximes are applied in the therapy of poisoning.

Carbamates

Esters of carbamic acids, of the type shown in Table 7.2, are acylating (carbamylating) ChE inhibitors. They are used as therapeutics (e.g. physostigmine and pyridostigmine) and pesticides (e.g. sevin and carbofuran). Their toxicity is primarily due to inhibition of AChE.

Inhibitor	*Formula*	k_a *(M^{-1} min^{-1})*	*Source**	*Ref.*
Neostigmine	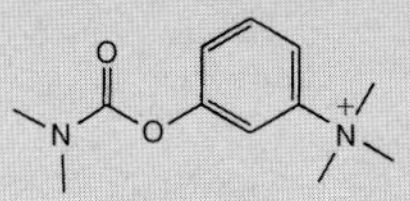	5.7×10^6	I	18
Physostigmine (Eserine)	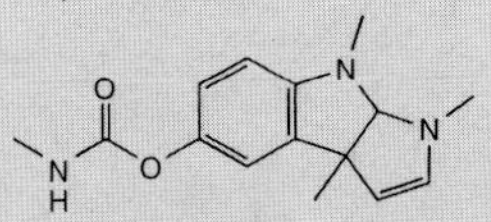	1.8×10^6	I	18
Carbofuran	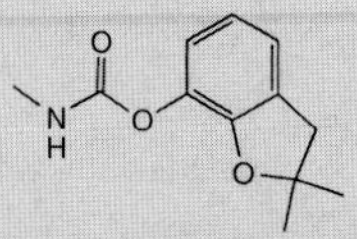	1.1×10^6	I†	
Ro-02-0683	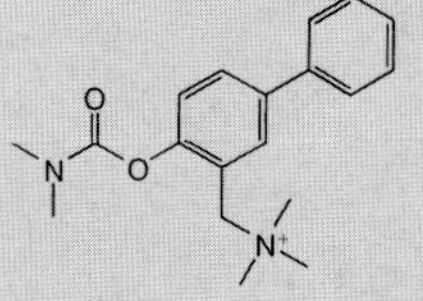	7.4×10^5	II	25
Carbaryl (Sevin)	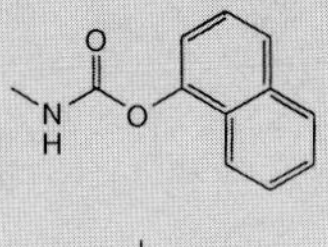	2.9×10^4	I†	
Bambuterol	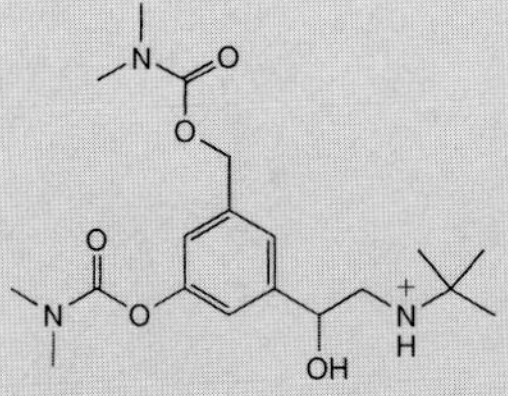	3.6×10^2	II	25

* Source: (I) mouse AChE in 100 mM phosphate buffer, pH 7.0, at 22°C; (II) mouse AChE in 100 mM phosphate buffer, pH 7.4, at 25°C.
† Z Radić, unpublished results.

Table 7.2
Carbamate inhibitors of AChE: second-order rate constants (k_a) of inhibition.

The mechanism of reaction of carbamates with ChEs is analogous to the reaction of OP compounds (equation (7)), and rate constants of carbamylation (k_a) also span a broad range, like rate constants of phosphorylation. Rates of spontaneous reactivation (equation (10)) depend on the substituents on the nitrogen atom. *N*-Unsubstituted carbamates reactivate most quickly, with half-lives of the order of minutes. *N*-Substituted carbamates reactivate more slowly. Rates of spontaneous reactivation decrease with increasing length and branching of the substituents on the nitrogen atom, becoming as slow as rates of spontaneous dephosphorylation.[11] Because some carbamylated enzymes reactivate rapidly, carbamates have for many years been incorrectly considered as reversible inhibitors.

Nucleophilic compounds such as oximes and hydroxamic acids do not increase decarbamylation very much, which is different to the case of dephosphorylation.

Reversible inhibitors

Reversible inhibitors typically interact with ChEs very rapidly. *N*-Methyl acridinium, one of micromolar inhibitors of both AChE and BChE, associates with AChE at diffusion-limited rates and dissociates in few hundred microseconds (Table 7.3). The full equilibration of the complex is reached within a few milliseconds. The more potent inhibitor tacrine dissociates from AChE 100 times slower, but still equilibrates in about one second. The most potent AChE reversible inhibitors, however, dissociate from enzyme very slowly, forming tight-binding complexes that may require days to equilibrate.

Fast equilibrating inhibitors

The large amount of inhibition data available in the literature shows that reversible inhibitors bind to AChE at more than one site. Two frequently identified sites are the active centre and the peripheral site. Five out of the six crystal structures of AChEs complexed with reversible inhibitors (tacrine, huperzine A, edrophonium, aricept and decamethonium) described to date reveal inhibitor bound in the choline binding site of the AChE active centre. Fasciculin 2 (Fas2) binds only to the peripheral site and aricept and decamethonium span the two sites. Since kinetic and binding studies indicate that AChE substrates in the course of their enzymatic hydrolysis bind both to the choline binding site and to the peripheral site, the two ligands, substrate and inhibitor can compete for binding, both at the choline binding site of the active centre and at the peripheral site on AChE. Competition at the choline binding site directly prevents substrate turnover, by mutually exclusive binding. Inhibitor binding at the peripheral site, however, blocks substrate turnover indirectly. This can occur through conformational change in the enzyme, through electrostatic interaction of cationic inhibitors with cationic ACh in the course of the catalytic reaction, or through steric and/or electrostatic restriction of substrate entry en route to the active-centre gorge while confronting ligand bound at the peripheral site.

When both the substrate and inhibitor bind to two sites on AChE, their associations result in formation of eight possible enzyme–substrate–inhibitor complexes: ES, SE, SES, EI, IE, IEI, IES and SEI.

If it is assumed that binding of a ligand to one site has no effect on the dissociation constant for binding to another site, the following equation can be derived:[11]

$$\frac{v\,\boldsymbol{i}}{v_0 - v} = \frac{(1 + K_s/s)(1 + s/K_{ss})}{(1/K_a + 1/K_i + \boldsymbol{i}/K_aK_i)K_s/\boldsymbol{s} + 1/K_i + K_s/K_aK_{ss}} \qquad (11)$$

Inhibitor	*Structure*	K_i *or* K_a *(nM)*	k_1 *(M^{-1} min^{-1})*	k_{-1} *(min^{-1})*	*Source†*	*Ref.*
TFK⁺	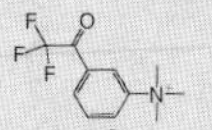	000 0053	2.1×10^{11}	0.0011	I	26
Fas2		0.0023	2.3×10^{8}	0.0045	I	26, 27
TFK^0	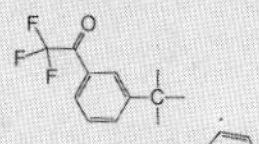	0.0037	3.0×10^{9}	0.011	I	26
Ambenonium	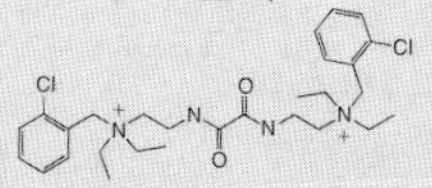	0.12	3.1×10^{9}	0.78	II	28
BW286C51	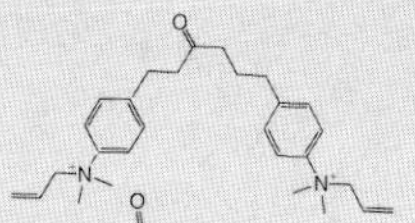	2.8	2.8×10^{10}	36	I	8‡
E2020	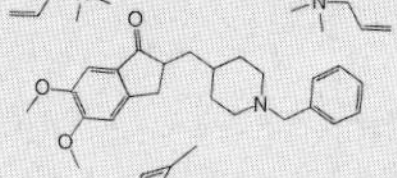	4.3	–	–	III	29
Huperzine A	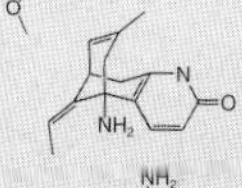	24	9.3×10^{5}	0.022	IV	30
Tacrine	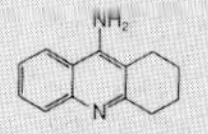	40	1.0×10^{10}	400	I	8‡
Edrophonium	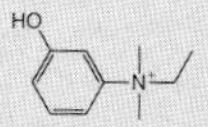	130	1.0×10^{10}	1300	I	8‡
N-Methylacridinium	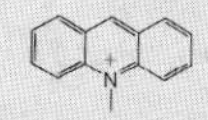	1000	4.8×10^{10}	45 000	V	31
Propidium	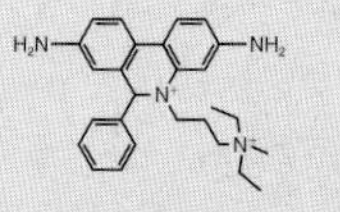	1000	2.0×10^{10}	20 000	I‡	–
Coumarin	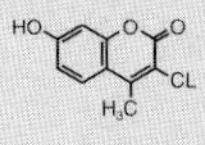	1300	2.4×10^{10}	30 000	I‡	–
Decamethonium	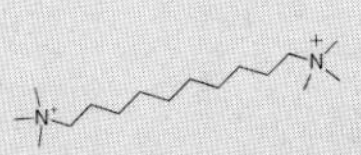	3500	–	–	I	8

* For all listed inhibitors, except for Fas2, propidium and coumarin, the given constants reflect interaction at the catalytic site of the enzyme (K_a), while for Fas2, propidium and coumarin the K_i constant reflects interaction at the peripheral site of AChE. Structures are not drawn to scale.

† Source: (I) mouse AChE in 100 mM phosphate buffer, pH 7.0, at 22°C; (II) human erythrocyte AChE in 20 mM phosphate buffer, pH 7.0, at 25°C; (III) electric eel AChE in 100 mM phosphate buffer, pH 8.0, at 25°C; (IV) bovine AChE in 50 mM phosphate buffer, pH 8.0, at 23°C; (V) human erythrocyte AChE in 100 mM phosphate buffer, pH 7.0, at 25°C.

‡ Z Radić, unpublished results.

Table 7.3

Reversible inhibitors of AChE: equilibrium dissociation constants (K_i or K_a) and rate (k_1, k_{-1}) constants of inhibition.

where v_0 and v are the enzyme activities in the absence and presence of inhibitor at concentration i, respectively, measured with substrate at concentration s. The constants K_a and K_i are the enzyme/inhibitor dissociation constants for the catalytic and peripheral enzyme sites, and K_s and K_{ss} are the corresponding enzyme/substrate dissociation constants. The expression on the left-hand side of equation (11), frequently termed the 'apparent inhibition constant', is a non-linear function of the substrate concentration. When the inhibitor binds to only one of the two sites, equation (11) simplifies to linear functions. These relationships hold when all the equilibria are established rapidly.

An alternative approach has recently been introduced for AChE inhibition reactions where binding of the inhibitory ligand significantly slows down the equilibration of the system.[32] In such an event analytical solution of the reaction is not possible, and individual constants of inhibition can be derived only through numeric simulation.

Analysis of the reversible inhibition of BChE over the wide range of substrate concentrations is complicated by substrate activation, and is not addressed in this chapter.

Tight binding inhibitors

Fast association and slow dissociation of an inhibitor with AChE results in a tight inhibitory complex. Tight-binding inhibitors with equilibrium dissociation constants in the picomolar range or below (e.g. Fas2, ambenonium, trifluoroacetophenones (TFK^+ and TFK^0) and, to some extent, huperzine A) effectively inhibit AChE at concentrations that approach the minimal enzyme concentrations required for reliable activity assays. Under these experimental conditions equation (11) cannot be used to evaluate inhibition constants since the concentration of unbound inhibitor is changing during inhibition. The equilibrium dissociation constants of inhibition can then be derived as a ratio of the first-order dissociation (k_{-1}) and second-order association (k_1) rate constants, determined in separate experiments. Alternatively, equations corrected for ligand depletion such as the one proposed by Ackermann and Potter, cf. [26] can be used.

Structure–activity relationships in the mechanism of cholinesterase inhibition and reactivation

The approach of the inhibitory ligand to the target molecule and its entry into the binding site is the first step in any inhibition reaction. The common structural features of all good ChE inhibitors include the presence of positive charge and/or aromatic or hydrophobic substituent groups that facilitate the first reaction step. Trifluoroacetophenones, fasciculins (Fas1, Fas2 and Fas3) and ambenonium, the most potent AChE inhibitors, with dissociation constants in the femtomolar and picomolar range, bear both aromatic and positively charged groups (see Table 7.3 and Chapter 2). Every catalytic subunit of AChE is an electrostatic dipole, having an electric field collinear with the axis of its active-centre gorge.[33,34] The attraction of positively charged ligands by this field, into the active centre, is particularly notable with the binding of large multivalent cations, such as fasciculins, propidium and gallamine. The large size prevents these ligands from entering the approximately 5 × 9 Å wide opening of the active centre (measured by the accessibility to water) and they remain trapped at the enzyme peripheral site. This site will also trap smaller aromatic neutral ligands (e.g. 3-chloro-7-hydroxy-4-methylcoumarin is an effective peripheral site AChE inhibitor) that are not electrostatically attracted into the active-centre gorge.[35] Bound at the peripheral site, the ligands inhibit catalytic activity both by restricting access to the catalytic centre[31] and by compromising chemical reaction in the centre.[36] Thus only smaller, or more elongated, and preferentially cationic, ligands proceed through the active-centre gorge guided by anionic residues, Asp74(72), and to some extent by Glu202(Glu199) and Glu450(Glu443),[37] and must pass through another constriction 4 × 8 Å wide, further narrowed by 2 Å indentation, formed by Tyr124(121) and Tyr337(330). Once bound to the peripheral site or active centre of the enzymes, the ligands are primarily stabilized through interaction with multiple aromatic amino acid residues. The aromatic rings with branched alkyl substituents, found in ethopropazine and bambuterol, can fit into the choline binding site only by trimming substituent side-chains or by substituting the Tyr337(330) of AChE with the smaller Ala.[8,25] This residue, found in BuChE, renders bambuterol and ethopropazine very potent BuChE and poor AChE inhibitors. Long and slender molecules having two aromatic or positively charged groups 10–12 Å apart possess an advantageous ability to interact simultaneously with the choline binding site and the peripheral site. Ambenonium, A2020 (Aricept, Donepezil) and BW284C51 (see Table 7.3) are subnanomolar and nanomolar AChE inhibitors. The complementarity of the AChE and inhibitor surfaces is, however, best exemplified in the perfect fit of fasciculins (see Chapter 2).

Covalent inhibitors of AChE, similar to substrates, have two additional structural requirements for the efficient inhibition related to their mechanism of action (see equation (7)). First, upon approach to the enzyme active centre they must form a reversible complex close to the active serine residue. Second, in the course of their chemical reaction, they appear fixed at two points: by covalent bonding to Ser203(200) and by accommodating their phosphonyl or carbonyl oxygen atom in the oxyanion hole of the enzyme.[21,38] This determines the orientations of the leaving group and the remaining substituents in the corresponding transition state. During the course of the reaction the substrates and carbamates are thought to change their geometry, such as trifluoroacetophenones (see Chapter 2), from trigonal–planar to transitional tetrahedral while organophosphates change from tetrahedral to transitional pentacoordinate, trigonal bipyramidal (Fig. 7.2). The potencies of acylating inhibitors also depend on the

Figure 7.2

The proposed catalytic mechanism of AChE: (a–c) ACh hydrolysis; (d–f) phosphorylation of enzyme by DDVP. A stereoview from the oxyanion hole within the AChE active-centre gorge. The mouse AChE active-centre gorge is represented by vertical cross-section through the solvent-accessible surface and by selected amino acid side-chains. The opening of the gorge is at the top of each stereo panel, in the plane of the paper. Peripheral-site residues are displayed in white, acyl pocket (F295 and F297) and choline binding site residues (W86, Y337 and D74) in blue. Residues of the catalytic triad (E334, H447 and S203) involved in the chemical reaction are colour coded by atom type (C, yellow; O, red; N, blue; H, white; P, pink; Cl, green) and by substrate (ACh) and inhibitor (DDVP). (a) A model of the ACh Michaelis complex (denoted by ES in equation (1)) stabilized by a three-point interaction: the $N^+(CH_3)_3$ group by choline binding site residues, the C=O group approaching the oxyanion hole, and the $-CH_3$ group by the acyl pocket. (b) The active Ser203 rendered nucleophylic by E334 and His447 forms a covalent bond with the carbonyl carbon atom of ACh; the transition state is tetrahedral with sp^3 geometry. (c) Departure of the choline leaves Ser203 briefly acetylated (EA in equation (2)) before the attack by a water molecule deacetylates Ser203, rendering it ready to hydrolyse another molecule of

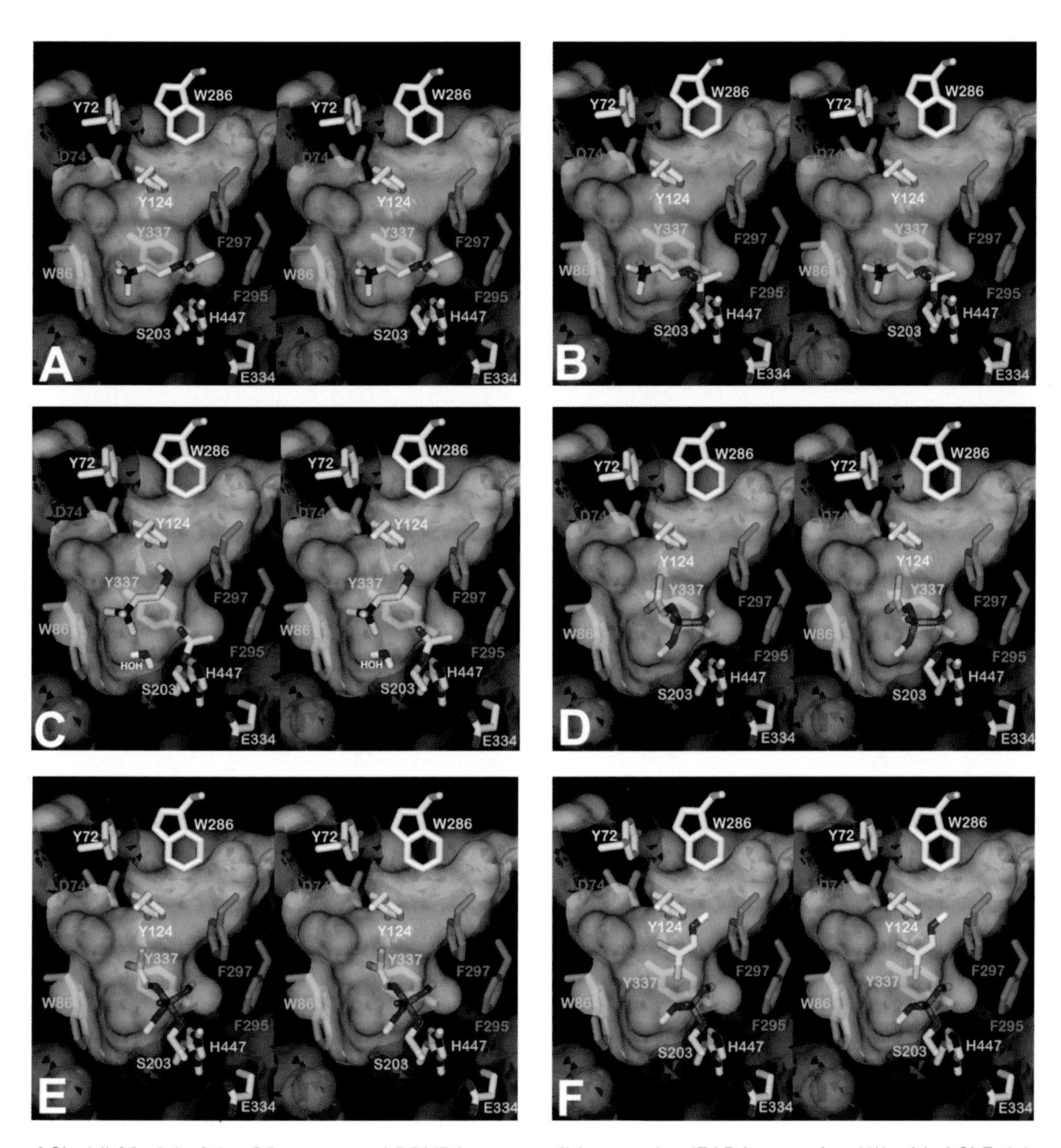

ACh. (d) Model of the OP compound DDVP in a reversible complex (EAB in equation (6)) with AChE. (e) Model of pentacoordinate, trigonal bipyramidal transition state of DDVP formed by covalent binding of Ser203 to the phosphorus atom of DDVP. One alkoxy group of DDVP is stabilized in the acyl pocket and the other in the choline binding site; the phosphonyl oxygen atom is in the oxyanion hole and the leaving group is pointing towards the opening of the AChE gorge. (f) Model of tetrahedral phosphorylated AChE (EA in equation (6)) formed by the breakdown of the highly unstable pentacoordinate transition state. The 1-hydroxy-2,2-dichlorovinyl leaving group of DDVP is departing from the active-centre gorge.

properties of their leaving groups. The acyl pocket of AChE, which is formed up by two phenylalanines located 5 and 6 Å from the active serine residue, efficiently stabilizes substituents of up to propyl group in size. Elongated and flexible substituents, such as dimethylbutyl, more easily escape the steric limitations of the acyl pocket. In an extreme example,[38] the dimethylmorpholinooctyl substituent of carbamate MF268 is stabilized by Trp279 of the peripheral site instead. The size of the acyl pocket in BChE is larger due to the absence of two phenylalanines, and this enzyme, unlike AChE, is efficiently inhibited by carbamates and organophosphates with branched, bulky ester groups such as iso-OMPA (tetraisopropyl pyrophosphoramide). Aliphatic, cationic leaving groups of organophosphates interact primarily with the 13 Å distant Asp74(72),[39] while aromatic cationic groups interact with both Asp74(72) and Tyr337(330). Carbamates with charged aromatic leaving groups show a similar dependence.[25]

The asymmetry of the AChE active-centre gorge renders stereoisomers of chiral inhibitors differently potent. Organophosphates with an asymmetrically substituted phosphorus atom usually favour the S_p enantiomers, and this preference is determined by the size of the AChE acyl pocket.[39] Preference for stereoisomers of reversible inhibitors is usually determined by the residues of the choline binding site.

The structure–activity relationships for oxime reactivation of phosphorylated AChE are further complicated by the diversity of organophosphates covalently attached to the active serine residue. All effective AChE reactivators are charged oxime compounds, typically pyridinium derivatives, containing either one or two linked cationic pyridinium groups with oxime substituents. The angle of attack of phosphorus for monopyridinium/monoxime compounds (2PAM), and bisquaternary monoximes (HI-6) or dioximes (toxogonin, TMB-4) varies, resulting in differences in their reactivation efficiencies. Monoquaternary oximes are typically stabilized by the choline binding site, while bis-quaternary oximes have two points of interaction.[40] For HI-6 the second point is shown to be the peripheral site.[40]

Enzymes hydrolysing acylating cholinesterase inhibitors

Organophosphate (OP) compounds are substrates of phosphoric triester hydrolases (EC 3.1.8), which are subdivided into paraoxonase (EC 3.1.8.1) and DFPase (EC 3.1.8.2).[41] They are named after their characteristic substrates, paraoxon and DFP.

The enzymes were first described by Mazur,[42] and were named A-esterases by Aldridge.[43] Other names have also been suggested (e.g. OP compound hydrolases and phosphorotriestrases).[41,44] The substrate specificity of these enzymes is broad and includes phenylacetate, the characteristic substrate of arylesterase (EC 3.1.1.2), and also esters of carbamic acids. For details concerning the specificity, distribution and catalytic properties of these enzymes, the reader is referred to extensive reviews.[2,5,44]

Phosphoric triester hydrolases are widely distributed in mammalian tissues, particularly serum and liver, and in lower species, including micro-organisms. DFPase has also been found in plants.[45]

The physiological role of phosphoric triester hydrolases is not known. As OP compounds are their substrates, the enzymes are important for the detoxication of organophosphates. Mammalian paraoxonases are associated with high-density lipoprotein (HDL) and

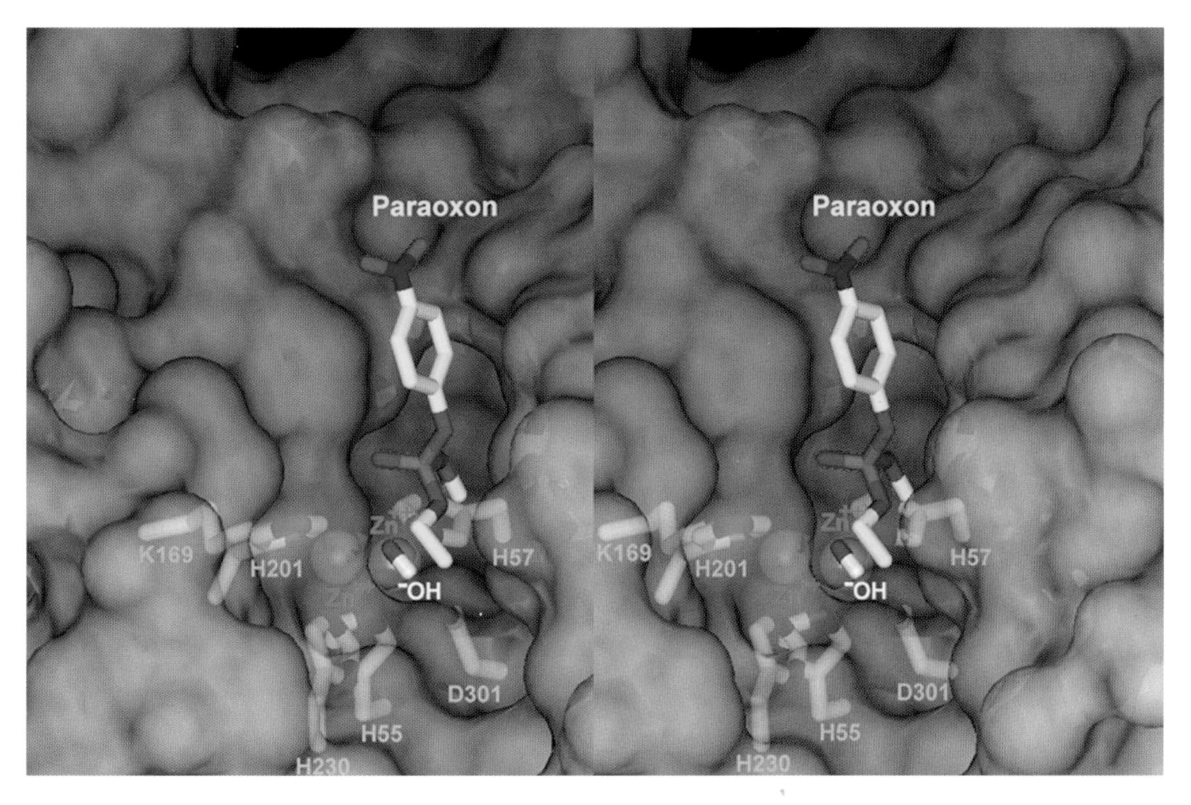

Fig. 7.3
The active-centre cleft of the phosphotriesterase from Pseudomonas diminuta[48] with the substrate paraoxon manually docked. The enzyme is represented with the semitransparent water-accessible surface and with the side-chains of four histidines, an aspartate and a lysine involved in the coordination of two cationic zinc atoms. A nucleophilic hydroxyl group was also found in the coordination of zinc atoms, representing the reactive species in the hydrolysis of substrate. Paraoxon was positioned similarly to the inhibitory substrate analogue diethyl-4-methylbenzylphosphonate, with the phosphonyl oxygen atom in the oxyanion hole, but with the p-nitrophenyl leaving group facing the enzyme exterior (as suggested by Vanhooke et al.[48]). Atoms: C, yellow; N, blue; O, red; P, pink; Zn, light blue.

they might play a role in the protection of low-density lipoprotein (LDL) from oxidative modification.[46,47]

Divalent cations, such as Ca^{2+}, Zn^{2+}, Mn^{2+}, Co^{2+} and Mg^{2+}, are co-factors of the phosphoric triester hydrolases. Therefore, chelating agents, such as EDTA, are inhibitors. The enzyme from *Pseudomonas diminuta* has been crystallized; it has a binuclear metal centre embedded in a cluster of histidine residues. Fig. 7.3 shows paraoxon docked in the catalytic site of this enzyme.[48] Histidine residues have also an important role in the catalytic mechanism of human serum paraoxonase.[49]

Phosphoric triester hydrolases cleave the same bond in OP compounds that is cleaved in the reaction of ChEs with OPs (the phosphorus–4-nitrophenyl bond in paraoxon, and the phosphorus–fluoride bond in DFP). The mechanism of hydrolysis is, however, different. Activated H_2O bound to the catalytic site of phosphoric triester hydrolases directly attacks the phosphorus centre, resulting in product formation. The catalytic specificities (k_{cat}/K_m) of these enzymes range from rates that are diffusion limited to rates which are up to eight orders of magnitude lower, depending on the substrate and on the enzyme source.[50]

Phosphoric triester hydrolases are stereospecific, like ChEs. However, the enantiomer which is the better substrate for phosphoric triester hydrolases is usually the less efficient inhibitor of ChEs.[51]

The activity of paraoxonase in human sera is polymodally distributed due to two genetic polymorphisms.[52,53] The distribution pattern is ethnically dependent. In Caucasian populations the low activity mode comprises about 50% of the population. Comparative studies of paraoxonase activities in healthy and diseased groups have shown that individuals with hyperlipoproteinaemia or diabetes, or in elderly demented people (including those with Alzheimer's disease) have, on average, lower paraoxonase activities than non-diseased individuals.[47,54–57]

The hydrolysis of paraoxon in human sera, and other mammalian sera, is not completely inhibited by EDTA; about 5–10% activity is EDTA insensitive.[52,54,57,58] The distribution of this activity is unimodal, but asymmetric. There is no correlation between EDTA-sensitive and EDTA-insensitive activities, indicating that different enzymes are involved in these two reactions. There is a greater difference between healthy and diseased individuals in the EDTA-insensitive than the EDTA-sensitive paraoxonase activity.

Phosphoric triester hydrolases play a role in the detoxification of OP compounds, and studies on animals have shown that they are promising antidotes in OP poisoning.[50,59–61] It has also been suggested that the enzymes can be used as decontaminating agents against OP compounds.

References

1. Massoulié J, Bacou F, Barnard E, Chatonnet A, Doctor BP, Quinn DM, eds. *Cholinesterases: Structure, Function, Mechanism, Genetics and Cell Biology*. Washington: American Chemical Society, 1991.
2. Reiner E, Lotti M, eds. *Enzymes Interacting with Organophosphorus Compounds. Chem-Biol Interact* 1993: **87**.
3. Quinn DM, Balasubramanian AS, Doctor BP, Taylor P, eds. *Enzymes of the Cholinesterase Family*. New York: Plenum, 1995.
4. Doctor BP, Taylor P, Quinn DM, Rotundo RL, Gentry MK, eds. *Structure and Function of Cholinesterases and Related Proteins*. New York: Plenum, 1998.
5. Reiner E, Simeon-Rudolf V, Doctor BP *et al.*, eds. *Esterases Reacting with Organophosphorus Compounds. Chem-Biol Interact* 1999; **119–120.**
6. Cauet G, Friboulet A, Thomas D. Substrate activation and thermal denaturation kinetics of the tetrameric and the trypsin-generated monomeric forms of the horse serum butyrylcholinesterase. *Biochim Biophys Acta* 1987; **912:** 338–342.
7. Cauet G, Friboulet A, Thomas D. Horse serum butyrylcholinesterase kinetics: a molecular mechanism based on inhibition studies with dansyl aminoethyltrimethylammonium. *Biochem Cell Biol* 1987; **65:** 529–535.
8. Radić Z, Pickering NA, Vellom DC, Camp S, Taylor P. Three distinct domains in the cholinesterase molecule confer selectivity for acetyl- and butyrylcholinesterase inhibitors. *Biochemistry* 1993; **32:** 12 074–12 084.
9. Masson P, Froment MT, Bartels CF, Lockridge O. Asp70 in the peripheral anionic site of human butyrylcholinesterase. *Eur J Biochem* 1996; **235:** 36–48.
10. Webb JL. *Enzyme and Metabolic Inhibitors*, Vol I. New York: Academic Press, 1963: 32–48.
11. Aldridge WN, Reiner E. *Enzyme Inhibitors as*

Substrates: Interaction of Esterases with Esters of Organophosphorus and Carbamic Acids. Amsterdam: North-Holland, 1972.

12. Radić Z, Reiner E, Simeon V. Binding sites on acetylcholinesterase for reversible ligands and phosphorylating agents: a theoretical model tested on haloxon and phosphostigmine. *Biochem Pharmacol* 1984; **33:** 671–677.
13. Radić Z, Reiner E, Taylor P. Role of the peripheral anionic site on acetylcholinesterase: inhibition by substrates and coumarin derivatives. *Mol Pharmacol* 1991; **39:** 98–104.
14. Radić Z, Taylor P. The effect of peripheral site ligands on the reaction kinetics of phosphyl and carboxyl esters with acetylcholinesterase. In: Doctor BP, Taylor P, Quinn DM, Rotundo RL, Gentry MK, eds. *Structure and Function of Cholinesterases and Related Proteins.* New York: Plenum, 1998: 211–214.
15. Mallender WD, Szegletes T, Rosenberry TL. Organophosphorylation of acetylcholinesterase in the presence of peripheral site ligands — distinct effects of propidium and fasciculin. *J Biol Chem* 1999; **274:** 8491–8499.
16. Luo C, Saxena A, Smith M *et al.* Phosphoryl oxime inhibition of acetylcholinesterase during oxime reactivation is prevented by edrophonium. *Biochemistry* 1999; **38:** 9937–9947.
17. Reiner E. Inhibition of acetylcholinesterase by 4,4'-bipyridine and its effect upon phosphylation of the enzyme. *Croat Chim Acta* 1986; **59:** 925–931.
18. Radić Z, Gibney G, Kawamoto S, MacPhee-Quigley K, Bongiorno C, Taylor P. Expression of recombinant acetylcholinesterase in a baculovirus system: kinetic properties of glutamate 199 mutants. *Biochemistry* 1992; **31:** 9760–9767.
19. Rodriguez OP, Muth GW, Berkman CE, Kim K, Thompson CM. Inhibition of various cholinesterases with the enantiomers of malaoxon. *Bull Environ Contam Toxicol* 1997; **58:** 171–176.
20. *Metrifonate and Dichlorvos: Theoretical and Practical Aspects, Symposium Proceedings. Acta Pharmacol Toxicol* 1981; **49(Suppl V)**.
21. Millard CB, Kryger G, Ordentlich A *et al.* Crystal structures of aged phosphonylated acetylcholinesterase: nerve agent reaction products at the atomic level. *Biochemistry* 1999; **38:** 7032–7039.
22. Reiner E, Škrinjarić-Špoljar M, Simeon-Rudolf V. Binding sites on acetylcholinesterase and butyrylcholinesterase for pyridinium and imidazolium oximes, and other reversible ligands. *Period Biol* 1996; **98:** 325–329.
23. Reiner E, Simeon-Rudolf V, Škrinjarić-Špoljar M. Kinetics and mechanisms of reversible and progressive inhibition of acetylcholinesterase. In: Becker R, Giacobini E, eds. *Cholinergic Basis for Alzheimer Therapy.* Boston: Birkhauser, 1991: 63–67.
24. Škrinjarić-Špoljar M, Simeon-Rudolf V. Reaction of usual and atypical human serum cholinesterase phenotypes with progressive and reversible ligands. *J Enzyme Inhib* 1993; **7:** 169–174.
25. Kovarik Z, Radić Z, Grgas B, Škrinjarić-Špoljar M, Reiner E, Simeon-Rudolf V. Amino acid residues involved in the interaction of acetylcholinesterase and butyrylcholinesterase with the carbamates Ro-02-0683 and Bambuterol, and Terbutaline. *Biochim Biophys Acta* 1999; **1433:** 261–271.
26. Radić Z, Quinn DM, Vellom DC, Camp S, Taylor P. Allosteric control of acetylcholinesterase catalysis by fasciculin. *J Biol Chem* 1995; **270:** 20 391–20 399.
27. Radić Z, Duran R, Vellom DC, Li Y, Cervenansky C, Taylor P. Site of fasciculin interaction with acetylcholinesterase. *J Biol Chem* 1994; **269:** 11 233–11 239.
28. Hodge AS, Humphrey DR, Rosenberry TL. Ambenonium is a rapidly reversible noncovalent inhibitor of acetylcholinesterase, with one of the highest known affinities. *Mol Pharmacol* 1992; **41:** 937–942.
29. Nochi S, Asakawa N, Sato T. Kinetic study on the inhibition of acetylcholinesterase by 1-benzyl-4-[(5,6-dimethoxy-1-indanon)-2-yl]methylpiperidine hydrochloride (E2020). *Biol Pharm Bull* 1995; **18:** 1145–1147.
30. Ashani Y, Peggins JO, Doctor BP. Mechanism of inhibition of cholinesterases by huperzine A. *Biochem Biophys Res Commun* 1992; **184:** 719–726.
31. Rosenberry TL, Rabl CR, Neumann E. Binding of the neurotoxin fasciculin 2 to the acetylcholinesterase peripheral site drastically reduces the association and dissociation rate

constants for *N*-methylacridinium binding to the active site. *Biochemistry* 1996; **35:** 685–690.

32. Szegletes T, Mallender WD, Rosenberry TL. Nonequilibrium analysis alters the mechanistic interpretation of inhibition of acetylcholinesterase by peripheral site ligands. *Biochemistry* 1998; **37:** 4206–4216.
33. Gilson MK, Straatsma TP, McCammon JA *et al.* Open back door in a molecular dynamics simulation of acetylcholinesterase. *Science* 1994; **263:** 1276–1278.
34. Porschke D, Creminon D, Cousin X, Bon C, Sussman J, Silman I. Electrooptical measurements demonstrate a large permanent dipole moment associated with acetylcholinesterase. *Biophys J* 1996; **70:** 1603–1608.
35. Simeon-Rudolf V, Kovarik Z, Radić Z, Reiner E. Reversible inhibition of acetylcholinesterase and butyrylcholinesterase by 4,4'-bipyridine and by a coumarin derivative. *Chem-Biol Interact* 1999; **119:** 119–127.
36. Eastman E, Wilson J, Cerveñansky C, Rosenberry TL. Fasciculin 2 binds to the peripheral site on acetylcholinesterase and inhibits substrate hydrolysis by slowing a step involving proton transfer during enzyme acylation. *J Biol Chem* 1995; **270:** 19 694–19 701.
37. Radić Z, Kirchhoff PD, Quinn DM, McCammon JA, Taylor P. Electrostatic influence on the kinetics of ligand binding to acetylcholinesterase — distinctions between active centre ligands and fasciculin. *J Biol Chem* 1997; **272:** 23 265–23 277.
38. Bartolucci C, Perola E, Cellai L, Brufani M, Lamba D. 'Back door' opening implied by the crystal structure of a carbamoylated acetylcholinesterase. *Biochemistry* 1999; **38:** 5714–5719.
39. Hosea NA, Radić Z, Tsigelny I, Berman HA, Quinn DM, Taylor P. Aspartate 74 as a primary determinant in acetylcholinesterase governing specificity to cationic organophosphonates. *Biochemistry* 1996; **35:** 10 995–11 004.
40. Ashani Y, Radić Z, Tsigelny I *et al.* Amino acid residues controlling reactivation of organophosphonyl conjugates of acetylcholinesterase by mono- and bisquaternary oximes. *J Biol Chem* 1995; **270:** 6370–6380.
41. *Enzyme Nomenclature. Recommendations (1992) of the Nomenclature Committee of the International Union of Pure and Applied Chemistry.* San Diego: Academic Press, 1992.
42. Mazur A. An enzyme in animal tissues capable of hydrolysing the phosphorus–fluorine bond of alkylfluorophosphates. *J Biol Chem* 1946; **164:** 271–289.
43. Aldridge WN. Serum esterases. 2. An enzyme hydrolysing diethyl-*p*-nitrophenyl phosphate (E600) and its identity with the A-esterase of mammalian sera. *Biochem J* 1953; **53:** 117–124.
44. Reiner E, Aldridge WN, Hoskin FCG, eds. *Enzymes Hydrolysing Organophosphorus Compounds.* Chichester: Ellis Horwood, 1989.
45. Hoskin FCG, Walker JE, Mello CM. Organophosphorus acid anhydrolase (OPAA) in slime mold, duck weed and mung bean: a continuous search for a physiological role and a natural substrate. *Chem Biol Interact* 1999; **119–120:** 399–404.
46. La Du BN, Aviram M, Billecke S *et al.* On the physiological role(s) of paraoxonases. *Chem Biol Interact* 1999; **119–120:** 379–388.
47. Mackness IM, Durrington PN, Ayub A, Mackness B. Low serum paraoxonase: a risk factor for atherosclerotic disease? *Chem Biol Interact* 1999; **119–120:** 389–398.
48. Vanhooke JL, Benning MM, Raushel HM. Three dimensional structure of the zinc-containing phosphotriesterase with the bound substrate analog diethyl-4-methylbenzylphosphonate. *Biochemistry* 1996; **35:** 6020–6025.
49. Doorn JA, Sorenson RC, Billecke SS, Hsu C, La Du BN. Evidence that several conserved histidine residues are required for hydrolytic activity of human paraoxonase/arylesterase. *Chem Biol Interact* 1999; **119–120:** 235–242.
50. Disioudi BD, Miller CE, Lai K, Grimsley JK, Wild JR. Rational design of organophosphorus hydrolase for altered substrate specificities. *Chem Biol Interact* 1999; **119–120:** 211–224.
51. Hong S-B, Rauschel FM. Stereochemical preference for chiral substrates by the bacterial phosphotriesterase. *Chem Biol Interact* 1999; **119–120:** 225–234.
52. Geldmacher von Mallinckrodt M, Diepgen M. The human serum paraoxonase polymorphism and specificity. *Toxicol Environ Chem*

1988; **18:** 79–196.

53. Adkins S, Gan KN, Mody M, La Du BN. Molecular basis for the polymorphic forms of human serum paraoxonase/arylesterase: glutamine or arginine at position 191 for the respective A or B allozymes. *Am J Hum Genet* 1993; **52:** 598–608.
54. Reiner E, Simeon V. Catalytic properties and classification of phosphoric triester hydrolases: EDTA-sensitive and EDTA-insensitive paraoxonases in sera of non-diseased and diseased population groups. In: Mackness MI, Clerc M, eds. *Esterases, Lipases and Phospholipases: From Structure to Clinical Significance.* New York: Plenum, 1994: 57–64.
55. Mackness MI. The human serum paraoxonase polymorphism and atherosclerosis. In: Mackness MI, Clerc M, eds. *Esterases, Lipases and Phospolipases: From Structure to Clinical Significance.* New York: Plenum, 1994: 65–73.
56. Pavković E, Simeon V, Reiner E, Sučić M, Lipovac V. Serum paroxonase and cholinesterase activities in individuals with lipid and glucose metabolism disorders. *Chem Biol Interact* 1993; **87:** 179–182.
57. Reiner E, Simeon-Rudolf V, Škrinjarić-Špoljar M. Catalytic properties and distribution profiles of paraoxonase and cholinesterase phenotypes in human sera. *Toxicol Lett* 1995; **82/83:** 447–452.
58. Reiner E, Pavković E, Radić Z, Simeon V. Differentiation of esterases reacting with organophosphorus compounds. *Chem Biol Interact* 1993; **87:** 77–83.
59. Hoskin FCG. An organophosphorus detoxifying enzyme unique to squid. In: Gilbert DL, Adelman WJ, Arnold JM, eds. *Squid as Experimental Animals.* New York: Plenum, 1990: 469–480.
60. Broomfield CA. A purified recombinant organophosphorus acid anhydrase protects mice against soman. *Chem Biol Interact* 1993; **87:** 279–284.
61. Costa LG, Li WF, Richter RJ, Shih DM, Lusis A, Furlong CE. The role of paraoxonase (PON1) in the detoxication of organophosphates and its human polymorphism. *Chem Biol Interact* 1999; **119–120:** 429–438.

8

Neuroanatomy of cholinesterases in the normal human brain and in Alzheimer's disease

Marsel Mesulam

Introduction

The human brain contains two cholinesterases (ChEs): acetylcholinesterase (AChE), encoded by a gene on chromosome 7; and butyrylcholinesterase (BuChE), encoded by a gene on chromosome 3. AChE (EC 3.1.1.7) is by far the more prominent of the two, and the only one consistently associated with cholinergic pathways. Its critical role in cholinergic neurotransmission is based on its ability to terminate the action of acetylcholine (ACh) through rapid catalytic hydrolysis. BuChE (EC 3.1.1.8) is present in much lower concentrations and displays a much more restricted distribution. Its physiological role remains unknown since it does not seem to have a natural substrate in the human central nervous system (CNS). The molecular biology, structure, post-translational processing, catalytic properties and functions of the ChEs are described in great detail in other chapters in this book. The purpose of this chapter is to review the distribution of AChE and BuChE in the human brain from the vantage point of systems neuroanatomy.

Although AChE and BuChE are products of two different genes, they both belong to the Type B carboxylesterase gene family. Through differential splicing and additional post-translational modifications, ChEs display a bewildering variety of quaternary structures, solubilities, ionic interactions and glycosylation states. This diversity allows ChEs to exist in intracellular compartments, to be inserted within membranes, anchored to extracellular basal laminae, transported intra-axonally and secreted as soluble protein.[1–3] The basic building block of ChEs is a catalytic unit in the form of a globular monomer. Six major molecular forms (designated according to the number of catalytic units they contain) are recognized, three globular (G1, G2 and G4) and three asymmetric (A4, A8 and A12), the latter being associated with a triple-strand collagen-like tail. The differential splicing of AChE yields three mRNA subtypes (known as the synaptic, read through, erythrocytic forms). The synaptic subtype (AChE-S) is the principal species in the mature brain and appears to exist mostly in the globular tetrameric form (G4), the readthrough subtype (AChE-R) appears to exist mostly in the form of soluble monomers (G1), and the erythrocytic subtype (AChE-E) exists predominantly in red blood cells in the form of glycophosphoinositol (GPI) anchored G2 dimers.[4]

In all molecular forms of ChEs hydrolysis occurs through a catalytic triad involving the hydroxyl group of an active serine, the imidazole group of a histidine and a negatively charged glutamate residue. The molecule contains a 20 Å relatively hydrophobic gorge which leads to the catalytic site and which influences substrate specificities for ACh versus butyrylcholine (BCh).[5,6] In addition to the catalytic site and hydrophobic gorge, the ChE molecule contains a peripheral anionic site which plays an important role in inhibitor specificity.

ChEs also appear to have non-catalytic properties related to cell differentiation, process extension, and dendritic modeling. These extensively documented but poorly understood plasticity-related functions appear to be mediated through the peripheral anionic rather than the catalytic site of the ChE molecule.[4,7–9] The expression of ChEs by non-neural cells and their homologies to the adhesion proteins gliotactin, glutactin and neurotactin provide additional circumstantial evidence that ChEs may serve non-catalytic signaling functions related to proliferation, morphogenesis, and neuroplasticity.[4]

Methods for the anatomical demonstration of ChEs

Three major approaches are available for the neuroanatomical visualization of AChE and BuChE in human brains obtained at autopsy: enzyme histochemistry, immunocytochemistry, and in situ hybridization.

Enzyme histochemistry

Enzyme histochemistry, based on the precipitation of an insoluble reaction product at sites of enzyme activity, is by far the most flexible method and provides the best microscopic detail. It can be based on the two-step Koelle or direct-coloring Karnovsky–Roots method.[10–12] Death-to-fixation intervals of up to 30 h and subsequent time- and temperature-controlled aldehyde fixation have allowed excellent staining. A reaction product obtained with acetylthiocholine (but not butyrylthiocholine) as the substrate, and inhibited by BW284C51 (but not iso-octamethyl pyrophosphoramide (OMPA)) indicates AChE activity, whereas a reaction product obtained with butyrylthiocholine as the substrate and inhibited by iso-OMPA indicates BuChE activity.

Immunocytochemistry

Immunohistochemistry, based on antibodies that recognize specific AChE or BuChE epitopes, can also be used in the human brain.[13–15] This method helps to determine the location of the enzyme proteins. Specificity is demonstrated by showing that immunologically irrelevant immunoglobulins fail to yield similar patterns of immunoreactivity. In general, the distribution of immunoreactivity overlaps that of enzyme activity and helps to confirm that the catalytic activity is linked to the appropriate protein. The immunohistochemical method is slightly less flexible than enzyme histochemistry and requires shorter post-mortem intervals, usually under 8 h.

In situ hybridization

This method for the anatomical characterization of ChEs involves the localization of ChE mRNA by in situ hybridization.[16] The identification of mRNA at enzymatically ChE-positive perikarya helps to confirm that the enzyme is synthesized locally rather than derived from presynaptic inputs. The in situ method does not label axons, even in neurons that have ChE-positive axons. In general, perikarya which contain ChE enzyme activity or immunopositivity also express the corresponding ChE mRNA.

AChE in cholinergic neurons

A neuron is said to be *cholinergic* when it synthesizes ACh for the purpose of neurotransmission, and *cholinoceptive* when it contains nicotinic and/or muscarinic receptors which respond to the transmitter action of ACh. Most, if not all, cholinergic neurons are probably also cholinoceptive.[17] Their cholinoceptive receptors are located at traditional somatodendritic postsynaptic sites and at presynaptic axonal sites. The cholinergic activation of the axonal autoreceptors tends to decrease the release of ACh through feedback inhibition.

The most specific neuroanatomical identification of cholinergic neurons occurs through the immunohistochemical demonstration of choline acetyltransferase (ChAT), the synthetic enzyme for ACh. In the human CNS such cholinergic neurons occur in the basal forebrain (the Ch1–Ch4 cell groups), the striatum, the pedunculopontine nucleus (Ch5), the laterodorsal tegmental nucleus (Ch6), the medial nucleus of the habenula (Ch7), the parabigeminal nucleus (Ch8), the motor nuclei of cranial nerves, the parasympathetic neurons of the brainstem and the motor neurons of the spinal cord. All of these cholinergic neurons are also characterized by an intense AChE enzyme activity and immunoreactivity which can be designated 'AChE-rich' and which extends into the perikaryon, proximal dendritic tree, and axon. The AChE-rich axons of these cholinergic neurons almost always display varicose swellings which appear to make synaptic contacts with cholinoceptive neurons. These cholinergic AChE-rich neurons are also characterized by positive in situ hybridization for AChE mRNA.[16]

The AChE of cholinergic neurons is synthesized in the perikaryon and then transported to dendrites and axons where it becomes inserted into membranes as an ectoenzyme. The major role of the AChE is to terminate the action of ACh. Its presence in such high quantities in the somatodendritic as well as the axonal domains of cholinergic neurons probably reflects the additional cholinoceptive properties of these neurons and the presence of autoreceptors along their axons. In the human brain, all known cholinergic neurons are also AChE-rich. In the rodent, the cerebral cortex contains ChAT-positive (presumably cholinergic) neurons that do not express detectable AChE.[18] Analogous cholinergic interneurons are not present in the cerebral cortex of the adult human brain.

AChE in cholinoceptive neurons

Non-cholinergic (i.e. ChAT-negative) but AChE-rich neurons are found in almost all parts of the brain (Figs 8.1 and 8.2). There is considerable evidence to suggest that these neurons are cholinoceptive. Examples include the glutamatergic pyramidal neurons of the cerebral cortex, the nitric oxide producing juxtacortical polymorphic neurons, the gabaergic neurons of the reticular thalamic nucleus, the dopaminergic neurons of the substantia innominata, the histaminergic neurons of the hypothalamus, the serotonergic neurons of the raphe nuclei, and the noradrenegic neurons of the nucleus locus coeruleus.[19–24] In the cerebral cortex, AChE-rich intracortical pyramidal neurons in layers 3 and 5 respond to ACh through the mediation predominantly of m1 and nicotinic receptors, whereas the AChE rich polymorphic neurons of the juxtacortical U-fiber zone respond to ACh through the mediation predominantly of m2 receptors.[19]

In contrast to cholinergic neurons where high levels of AChE extend into the axon, the AChE of non-cholinergic cholinoceptive neurons remains confined to the dendrites and perikarya. Thus, AChE-rich neurons can be divided into two groups: a cholinergic group characterized by AChE-rich axons, and a non-cholinergic group with AChE-negative axons.[25] Perhaps this distinction reflects the absence of axonal cholinoceptive receptors in non-cholinergic neurons, thus obviating the need for catalytic AChE activity along the axon. As in the case of neurons that are both

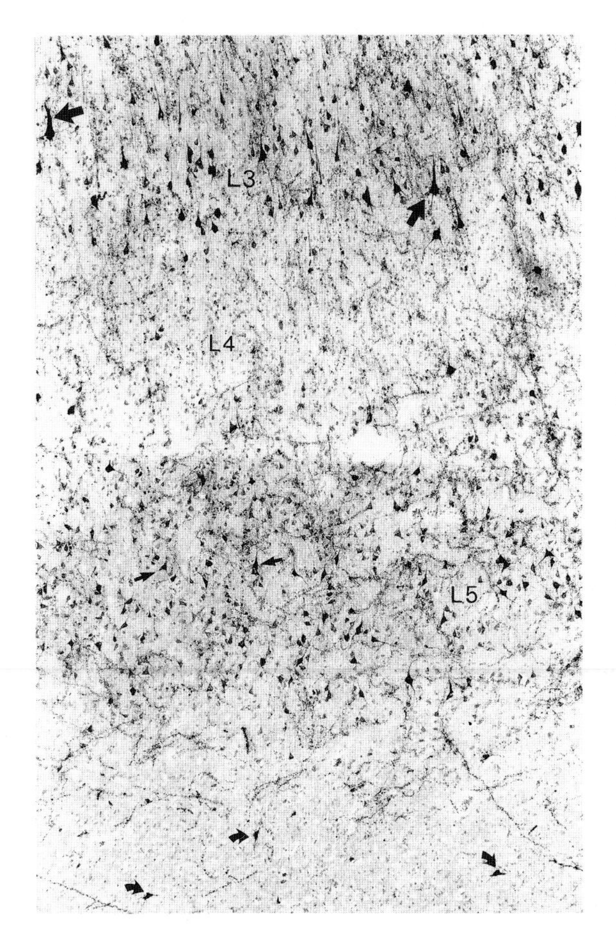

Figure 8.1
Laminar distribution of AChE-rich perikarya in peristriate cortex (area 19) of a 42-year-old woman with no known neurological disease, stained with a modified Koelle method for AChE histochemistry. Layers 3 and 5 (L3 and L5) contain many AChE-rich neurons (straight arrows), the vast majority of which are pyramidal in shape. Layer 4 (L4) contains virtually no AChE-rich neurons except for a few displaced L3 or L5 perikarya. All of these cortical layers also contain less intensely reactive AChE-positive perikarya. A second group of AChE-rich perikarya is located in layer 6 and the subjacent juxtacortical U-fiber zone of the white matter. Many of the neurons (curved arrows) are non-pyramidal in shape. The juxtacortical AChE-rich neurons contain nitric oxide synthase and respond to cholinergic innervation through the mediation of m2 receptors, whereas the intracortical AChE-rich neurons are glutamatergic and respond to cholinergic stimulation through the mediation of m1 and nicotinic receptors.[19] Magnification: ×132. (From Mesulam and Geula.[10])

cholinergic and cholinoceptive, the primary purpose of the AChE in non-cholinergic cholinoceptive neurons would also seem to be the catalytic hydrolysis of ACh.

In the rodent brain, dopaminergic neurons of the substantia nigra can release AChE in response to glutamatergic inputs. This AChE seems to bind neuronal membranes through its peripheral anionic site to mediate poorly characterized but non-catalytic signaling functions.[9] Whether such a release of AChE also occurs in the human is unknown, although the human CSF contains readily detectable AChE enzyme activity. Cholinergic deafferentation does not decrease the number of cortical AChE-rich cholinoceptive neurons. This has been demonstrated in rats where damage to the cholinergic neurons of the nucleus basalis did not influence the distribution of AChE-rich cortical neurons, and in patients with Alzheimer's disease where the nearly complete loss of cortical cholinergic innervation does not necessarily cause a comparable loss of AChE-rich cortical neurons.[26,27] Thus, the expression of AChE in cholinoceptive neurons does not seem to be under the strong regulatory influence of presynaptic cholinergic afferents.

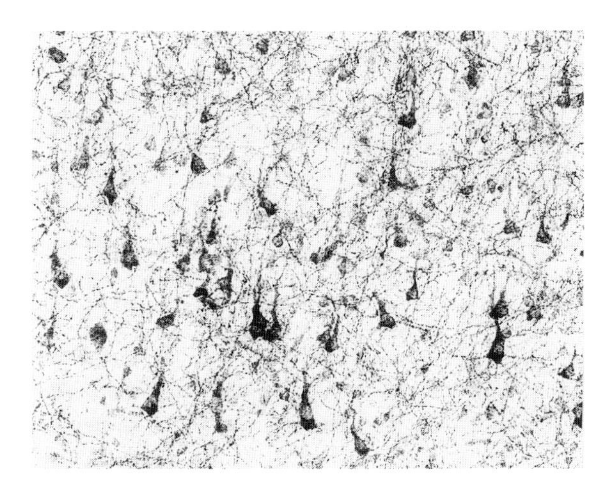

Figure 8.2
AChE immunostaining with a monoclonal antibody to human AChE shows numerous positive pyramidal neurons in layer 3 of area 6 in the brain of a 20-year-old woman with no known neurological disease. Immunopositive perikarya have AChE immunopositivity in the dendrites but not in the axons. Immunopositive axons are also seen but come from the cholinergic and AChE-rich neurons of the nucleus basalis. Magnification: ×207. (From Mesulam and Geula.[10])

All AChE-rich neurons are probably cholinoceptive. However, the differential distribution of AChE-rich cortical neurons and their developmental patterns suggest that there may be another group of cholinoceptive neurons with much lower perikaryal AChE levels. For example, the AChE-rich pattern of neocortical pyramidal neurons in layers 3 and 5 does not emerge until approximately 10 years of age despite the presence of a heavy cholinergic innervation in the cerebral cortex since the time of birth (Fig. 8.3). This developmental pattern is quite specific to intracortical pyramidal neurons since the juxtacortical AChE-rich polymorphic neurons are much more prominent at birth than during adulthood.[10,28,29] Furthermore, in the adult brain the limbic–paralimbic parts of the cerebral cortex contain a lower density of AChE neurons, although these are the parts of the cerebral cortex with the highest density of cholinergic afferents (Fig. 8.4).[10] It is therefore reasonable to assume that the AChE-rich perikarya of layers 3 and 5 represent only one subset of

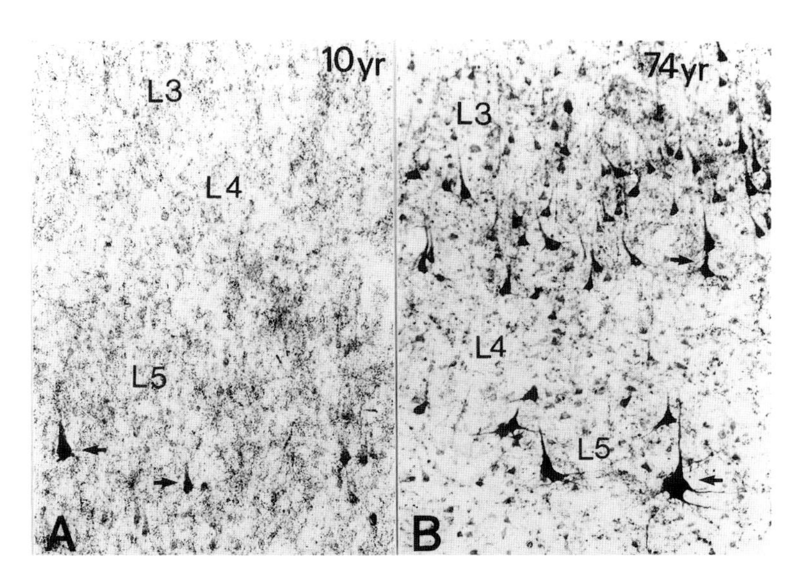

Figure 8.3
Ontogenetic patterns of AChE-rich perikarya demonstrated with a modified Koelle method for enzyme histochemistry. (A) Primary motor cortex (area 4) from a 10 year old. A few Betz cells in layer 5 (L5) yield an AChE-rich reaction (arrows). Magnification ×132. (B) Same cortical region (area 4) at age 74 years. There is a much greater number of AChE-rich neurons (arrows) in layers 3 and 5 (L3 and L5). Magnification ×126. (From Mesulam and Geula.[10])

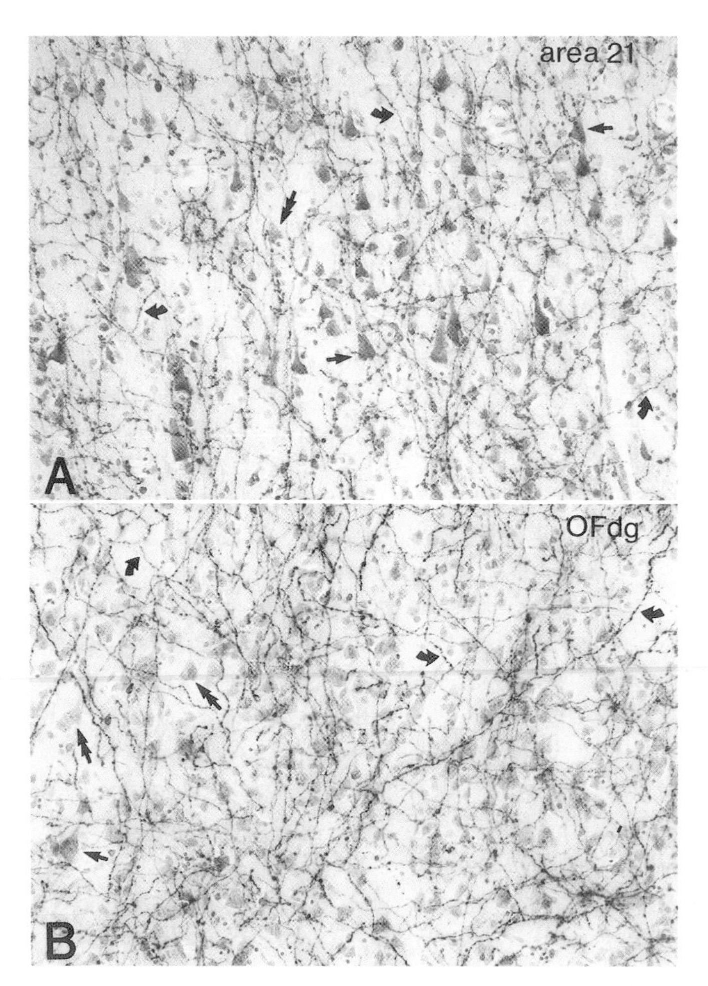

Figure 8.4
Relationship between AChE-rich axons and perikarya as demonstrated with a modified Karnovsky–Roots method. (A) Layer 3 of visual association cortex (area 21) in a 22-year-old woman with no known neurological disease. Counterstaining was done with neutral red. There are many AChE-rich perikarya (straight arrows) and AChE-rich axons (curved arrows). AChE-negative perikarya (double arrow) are relatively few. Magnification ×233. (B) Layer 3 of non-isocortical paralimbic cortex (dysgranular orbitofrontal cortex (OFdg) in a 74-year-old man with no known neurological disease. AChE-rich axons (curved arrows) are denser than in (A) and come in frequent contact with perikarya. However, AChE-rich perikarya (straight arrow) are rare, whereas AChE-negative perikarya (double arrows) are widespread. Magnification ×235. These two photomicrographs indicate that the AChE-rich perikaryal density has no positive correlation with the density of AChE-rich (cholinergic) axons. (From Mesulam and Geula.[10])

cortical cholinoceptive neurons and that their AChE-rich staining pattern reflects affiliations that transcend the requirements of standard cholinergic transmission. The high level of AChE in these neurons is either needed for purposes other than the hydrolysis of ACh, or these neurons sustain a special type of cholinergic transmission which necessitates a particularly rapid and complete hydrolysis of ACh.

Is AChE always associated with cholinergic transmission, or can it be expressed by neurons which are neither cholinergic nor cholinoceptive? Observations in the cerebellum and retina offer information that is potentially relevant to this question. In the cerebellum, for example, intense AChE activity occurs despite a relatively sparse cholinergic innervation. In the adult retina, where there are no cholinergic cells and where only amacrine cells are cholinoceptive, AChE expression can be detected not only in amacrine cells but also in photoreceptors, the one component of the retina that is subject to continuous remodeling.[4] Presumably the

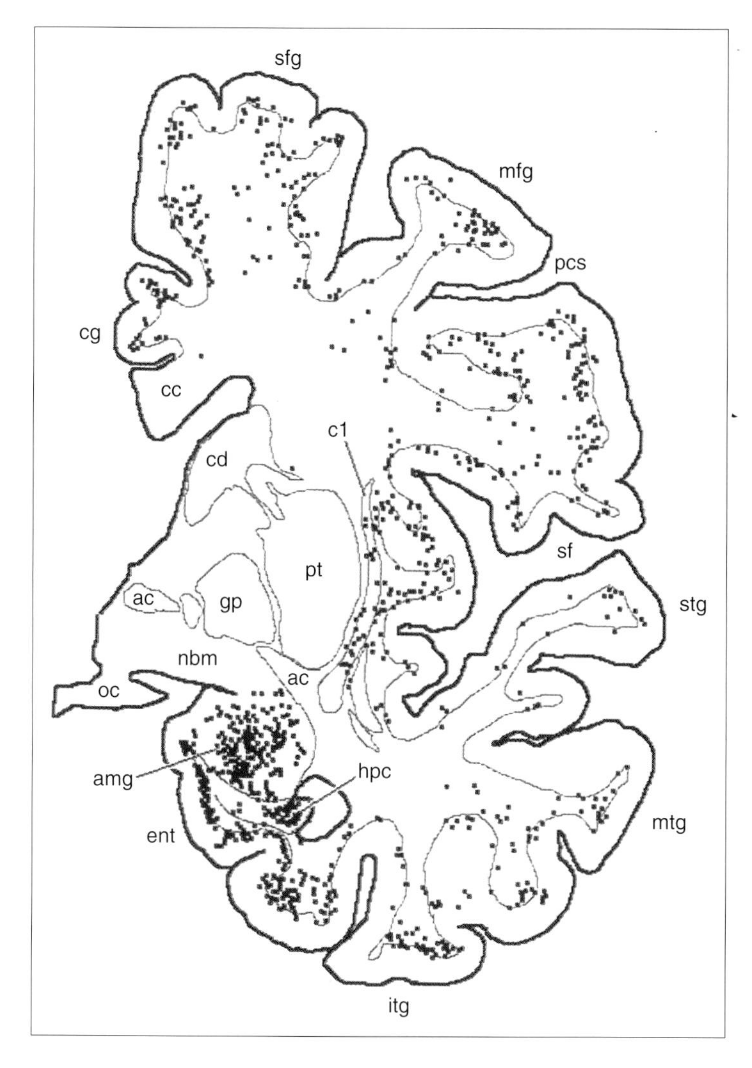

Figure 8.5
The distribution of BuChE immunopositive neurons in a coronal section of the human brain. Each square represents one immunopositive neuron. These neurons are found in all cortical areas and their density is particularly high in limbic structures such as the amygdala and the hippocampus. In neocortex, most of the BuChE-positive neurons are located in layers 5 and 6 and in the juxtacortical white matter. These neurons are at least two orders of magnitude less numerous than AChE-rich cortical neurons. ac, Anterior commissure; amg, amygdala; cc, corpus callosum; cd, caudate; cg, cingulate gyrus; cl, claustrum; ent, entorhinal cortex; gp, globus pallidus; hpc, hippocampus; itg, inferior temporal gyrus; lv, lateral ventricle; mfg, middle frontal gyrus; mtg, middle temporal gyrus; nbm, nucleus basalis of Meynert; oc, optic chiasm; pcs, pre-central sulcus; pt, putamen; sf, Sylvian fissure; sfg, superior frontal gyrus; stg, superior temporal gyrus.

AChE at these sites mediates non-catalytic roles which may be related to plasticity.

BuChE in neurons

The human cerebral neocortex also contains BuChE-rich neurons.[14,30] The number of these neurons is approximately two orders of magnitude less than the number of AChE-rich neurons. Enzyme histochemistry and immunohistochemistry with monoclonal antibodies indicate that neocortical BuChE-rich neurons are found mostly in layer 6 and in the immediately adjacent juxtacortical region, usually embedded within the U-fibers.[14] Limbic structures such as the amygdala, hippocampus, and entorhinal cortex appear to contain a slightly higher density of BuChE-rich neurons (Fig. 8.5). In the hippocampal complex, BuChE-neurons are located mostly within CA4 and in the stratum oriens, a white-matter layer which is analogous to the U-fiber band

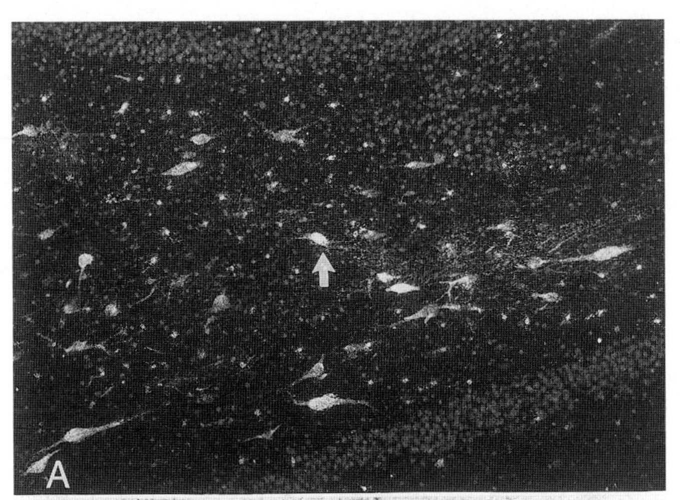

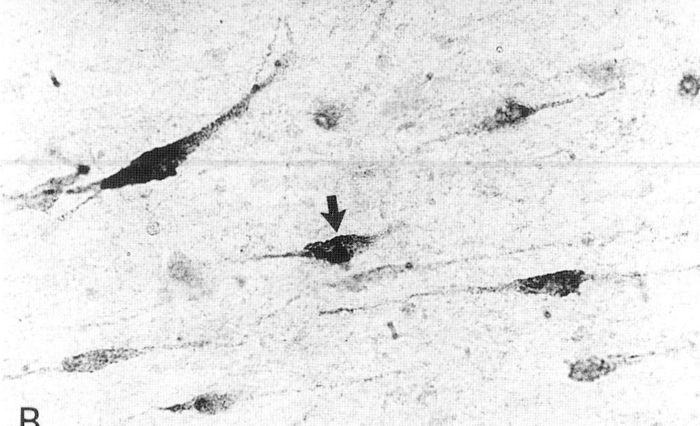

Figure 8.6
(A) Dark field photomicrograph of BuChE-rich neurons in the CA4 region of the hippocampus shown with a modified Koelle method for enzyme histochemistry. Magnification: ×25. (B) BuChE immunoreactivity in the stratum oriens of the hippocampus. Magnification: ×100. Arrows point to examples of BuChE-rich neurons.

of the neocortex (Fig. 8.6). Many of the BuChE-rich cortical and hippocampal neurons are also AChE-rich and appear to be cholinoceptive. The BuChE reactivity of cholinergic perikarya is quite variable. There are no BuChE-rich axons in the cerebral cortex.

ChEs in neuroglia

The astroglia and oligodendroglia of the human cerebral cortex contain immunohistochemically and enzymatically identified AChE and BuChE.[14,31] The AChE-rich glia are found in all cortical layers, whereas those that are BuChE-rich are found predominantly in the deeper cortical layers which also contain the majority of the BuChE-rich neurons. Glial AChE displays optimal catalytic pH and inhibitor sensitivities which are distinctly different from those of neuronal AChE.[32] Since all AChE molecular forms have identical catalytic units, these differences may reflect peculiarities of the three-dimensional conformation or molecular environment. There are no obvious regional variations in the density of AChE-rich neuroglia. However, BuChE-rich glia display a higher density in the entorhinal and inferotemporal areas compared to primary sensory cortices.[31]

The role of glial ChEs remains mysterious. They may contribute to the catalytic inactivation of the ACh released from neurons. They may also participate in non-cholinergic phenomena related to neuroplasticity and neuronal–glial interactions. In cultured glia, for example, AChE levels increase in response to differentiating agents such as dibutyryl cyclic-AMP. Furthermore, in transfected glia, the nature of the AChE isoform that is expressed influences cell size and process extension.[33] ChEs also have high sequence homologies with gliotactin, a transmembrane protein expressed by peripheral glia in *Drosophila*, and may play analogous roles in regulating glial–axonal interactions and the blood–brain barrier.[34]

ChEs in Alzheimer's disease

The changes in ChEs in Alzheimer's disease (AD) can be summarized under two headings:

- the loss of AChE in cholinergic pathways that are vulnerable to AD, and
- the appearance of AChE and BuChE in pathological markers such as amyloid deposits and neurofibrillary tangles.

Loss of AChE in the cholinergic axons and cholinoceptive neurons of the cerebral cortex

The AChE-rich cholinergic axons of the cerebral cortex are severely depleted and mirror the loss of cortical cholinergic axons in AD.[35] The cortical cholinergic depletion, and the concomitant loss of axonal AChE, are among the earliest and most severe components of the neuropathoogy in AD (see Fig. 8.7(A) and 8.7(D)).

The finding that biochemically determined cortical ChAT and AChE levels are significantly decreased in advanced but not early AD may appear to challenge this statement.[36] However, the biochemical determination of AChE is a very poor indicator of cortical cholinergic innervation because the enzyme is found not only in cholinergic axons but also in cholinoceptive perikarya, neuroglia, and the plaques and tangles of AD. The interpretation of ChAT activity is equally problematic. In the monkey, for example, nearly total destruction of the basal forebrain cholinergic neurons gives rise to a cortical ChAT depletion of 48–74%, whereas substantial but partial lesions induce depletions that do not exceed 12–62%.[37] Cortical ChAT levels are therefore not very sensitive indicators of cortical cholinergic depletion. One possibility for the discrepancy is a hypothetical up-regulation of ChAT in the remaining cortical axons. A more accurate assessment of cortical cholinergic denervation requires the microscopic quantitation of AChE-rich or ChAT-rich cortical axons.[38] Such methods show that cortical cholinergic depletion starts even before the onset of clinical dementia and that it becomes exceedingly severe in advanced AD.[38]

The loss of axonal AChE is most severe in the temporal lobe and least pronounced in primary sensory-motor areas. Cholinoceptive pyramidal neurons also show a decrease in AChE enzyme reactivity, although this is less consistent and somewhat unrelated to the severity of cholinergic deafferentation.[39] Some patients, for example, may continue to display a nearly normal density of AChE-rich cholinoceptive perikarya in regions of the cortex which sustain a severe loss of AChE-rich axons. Subcortical AChE-rich axons and

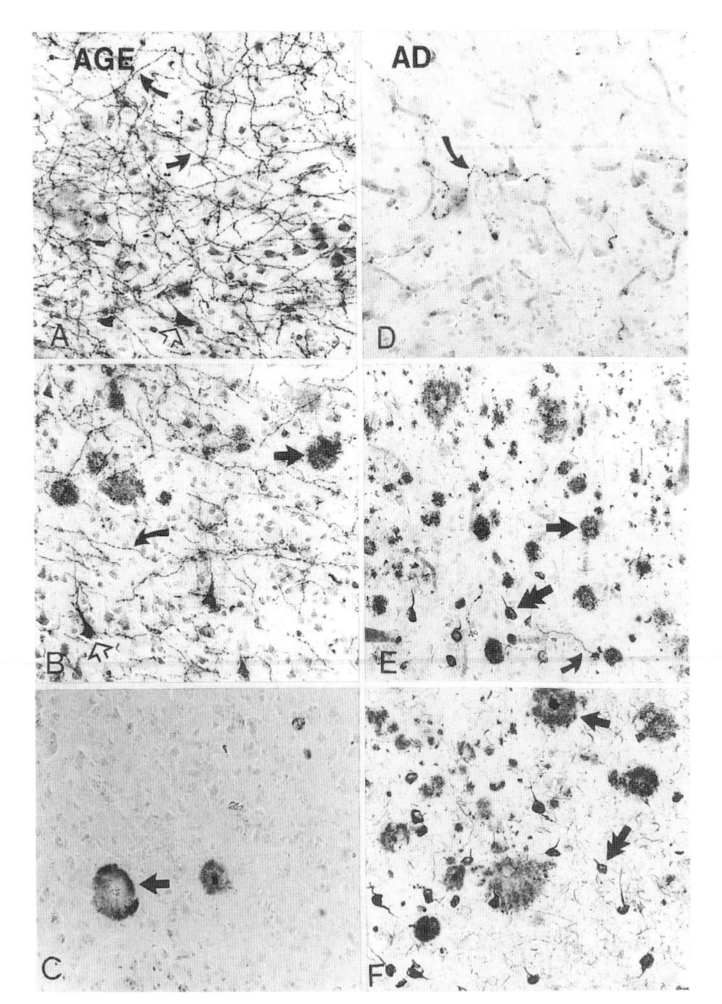

Figure 8.7

(A–C) From a 91-year-old non-demented man; (D–F) from a 78-year-old man with AD. ChE histochemistry with a modification of the Karnovsky–Roots method. (A vs D) AChE incubation at pH 8, a pH which is optimal for demonstrating enzyme activity in AChE-rich perikarya and cholinergic axons. Ethopropazine (7.2 mg/l) was used to inhibit BuChE activity. The cerebral cortex of the normal aged brain (A) contains a dense net of AChE-rich axons (curved arrows) and AChE-rich cell bodies (open arrow). Synaptic swellings can be seen along the axons. In contrast, the AD brain (D) contains extremely few AChE-rich fibers (curved arrow) and almost no AChE-rich cell bodies. (B vs E) AChE incubation at pH 7.0, a pH which is more effective for the demonstration of enzyme activity in plaques and tangles. In the 91-year-old non-demented brain (B), the AChE-rich axons (curved arrow) are less intensely stained than at pH 8.0 (A). AChE-rich cell bodies (open arrow) are visible. At this pH, some fibrillar plaques (solid straight arrow) are also AChE positive. In the AD brain stained for AChE at pH 7 (E), intense enzyme activity is present, but almost all of it is located within plaques (straight arrow) and tangles (double arrow). A few solitary AChE-rich axons (curved arrow) are still visible. In comparison to the incubation at pH 8 (D), the AChE activity in plaques and tangles is much more intense at pH 7 (E). The morphology of normal AChE-rich neurons (open arrow) (B) is easily differentiated from that of the AChE-rich tangles in AD (double arrow) (E). (C vs F) BuChE activity at pH 7.0. In the cerebral cortex of the nondemented 91-year-old man (C), few BuChE-rich plaques are present (arrow). BuChE-rich axons and neurons are not present. In the brain from the patient with AD (F), large numbers of plaques (single arrow) and tangles (double arrow) are BuChE-rich. Virtually no BuChE activity was present at pH 8 in either of the brains. All photomicrographs are from layers 4 to 6 in midtemporal visual association cortex (area 21). Magnification: ×160. (From Mesulam et al.[48])

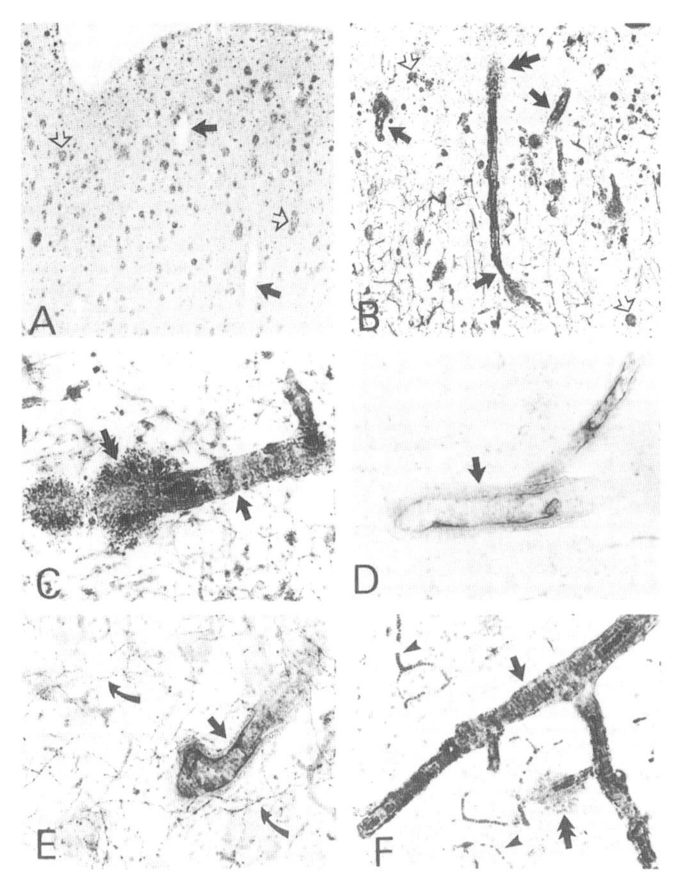

Figure 8.8
Occipital cortex. (A) Patient without amyloid angiopathy. Tissue is stained for AChE histochemistry at pH 6.8 with a modified Karnovsky–Roots method. Plaques display the AChE reaction product (open arrows). Vessel walls (solid arrows) do not give a prominent AChE reaction. (B) Patient with amyloid angiopathy stained for AChE as in (A). In addition to plaques (open arrows), vessel walls (solid arrows) also give an intense AChE reaction. Some of the AChE reaction product infiltrates the adjacent neuropil (double arrow). Magnification: ×49. (C) AChE reaction at pH 6.8. Vessel walls (solid arrows) are intensely AChE positive. AChE-positive plaque-like material extends into the neuropil (double arrows). (D) Same incubation as in (c), but in the presence of the specific AChE inhibitor BW284C51 (10^{-4} M). The AChE reaction of the vessel wall is almost completely gone. The vessel wall (solid arrow) yields only a non-specific stain. (E) Same reaction as in (A), but at pH 8.0. The vessel wall staining (black arrow) is much reduced. However, the few remaining AChE-rich cholinergic axons (curved arrows) give a more intense enzymatic reaction than at pH 6.8. (F) The BuChE reaction at pH 6.8. Vessel walls (solid arrow) are intensely BuChE positive. Some of the BuChE-positive material extends from the vessel wall into the neuropil (double arrow). A variable degree of BuChE reaction product deposition was also seen in red blood cells (arrowhead in (F)). Magnification (C–F): ×200. (From Mesulam et al.[42])

the AChE-rich perikarya of the brainstem and striatum remain intact, showing that the cortical cholinergic denervation in AD is selective and not part of a general cholinergic deficiency.

Emergence of AChE and BuChE in plaques, tangles and amyloid angiopathy

Light- and electron-microscopic investigations show that the vast majority of amyloid plaques, neurofibrillary tangles and vessels with amyloid angiopathy display intense AChE and BuChE activity (Figs 8.7 and 8.8).[40–44] ChE activity is one of the very few markers seen in all three major neuropathological markers of AD. The plaque-, tangle-, and angiopathy-bound AChE has enzymatic properties which are distinctly different from those of neuronal AChE (Fig. 8.9). Thus, the AChE enzyme activity associated with the histopathological markers of AD is more

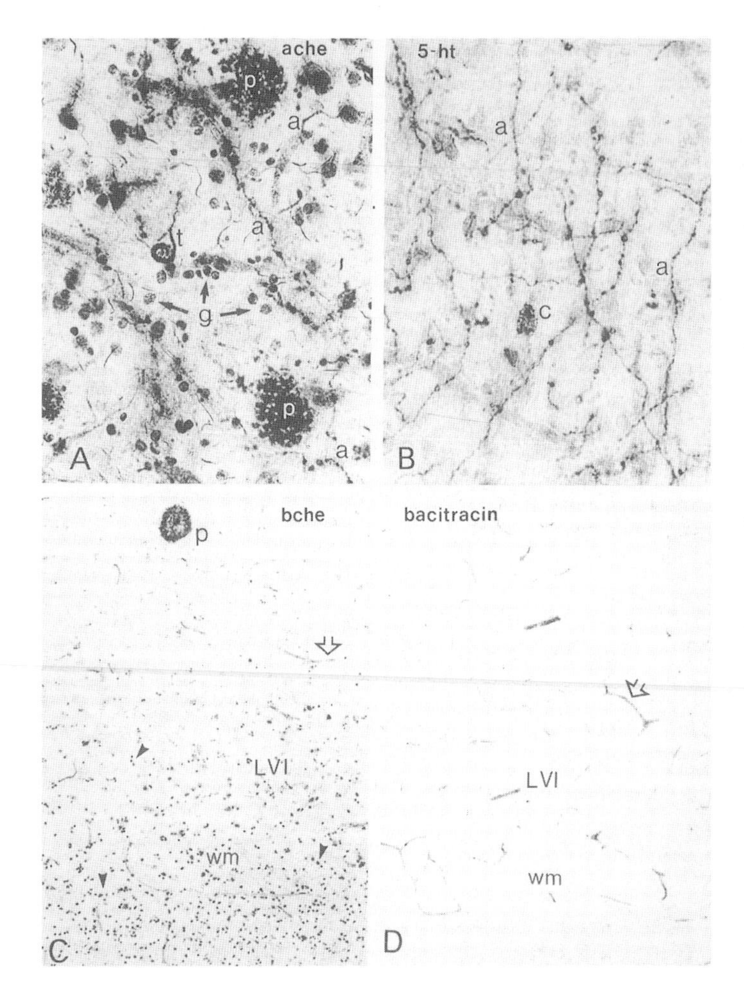

Figure 8.9
Inhibitor sensitivity of ChE-positive neuroglia demonstrated with a modified Karnovsky–Roots method. (A) Medial frontal area, layer 5 in the brain of an 82-year-old patient with AD. At a pH of 6.8 AChE activity is seen in plaques (p) and tangles (t) and in the few remaining cholinergic axons (a). Many neuroglia (g) are also AChE positive. (B) Same brain region processed using the same AChE reaction but in the presence of 0.05 mM 5-HT (serotonin). The AChE in plaques, tangles, and glia is almost completely inhibited, but the neuronal AChE in axons (a) and neuronal cells (c) remains relatively unchanged. Magnification (A, B): ×578. (C) Inferotemporal area from the brain of a 71-year-old control subject. There is intense BuChE activity in rare plaques (p) and numerous neuroglia (arrowheads). Erythrocyte-containing vessels yield a faint but non-specific reaction (open arrow). Note that BuChE-positive glia are much denser in layer 6 (LVI) and in the subcortical white matter (wm). (D) Same part of the brain processed with the identical BuChE reaction in the presence of 0.5 mM bacitracin. The non-specific vessel staining persists (open arrow), but the plaque- and tangle-associated and glial BuChE is completely inhibited. (From Wright et al.[31])

intense at a pH of 6.8–7.0 than at a pH of 8 which is optimal for AChE in normal axons and perikarya. Furthermore, the AD-related AChE is more sensitive to inhibition by indoleamines (such as 5-HT) and protease inhibitors (such as bacitracin) and less sensitive to BW284C51. These enzymatic features are similar to those of glia, leading to the inference that the AD-related AChE may represent an actively produced glial substance rather than the remnants of degenerated cholinergic axons.[31,32]

The AD brain contains far more BuChE enzyme activity than the normal brain. The potential usefulness of BuChE as a biomarker for the diagnosis of AD has not yet been fully

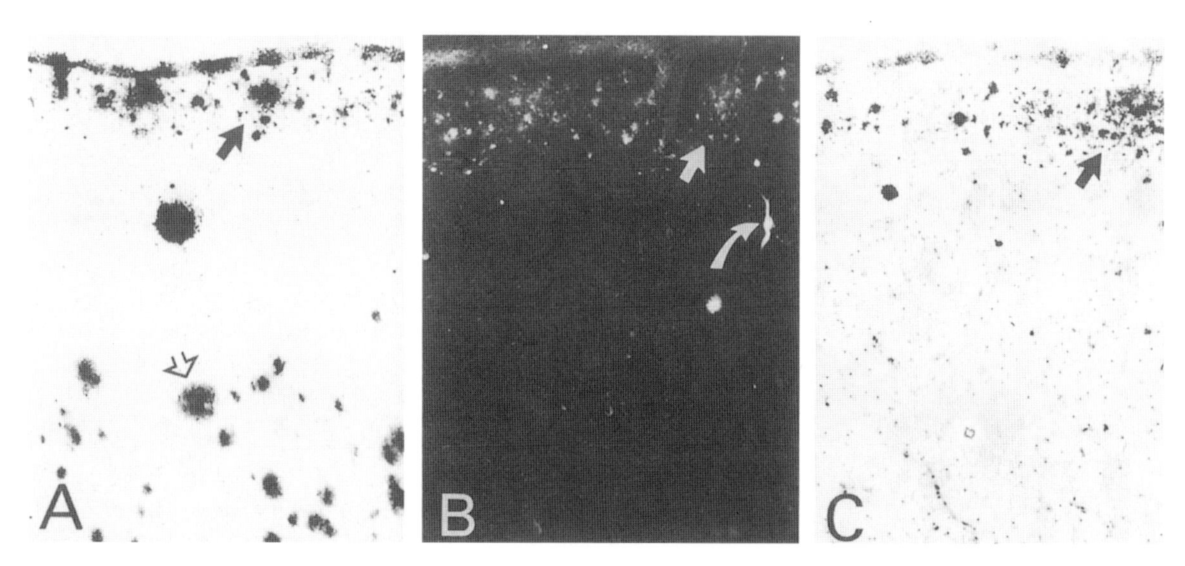

Figure 8.10
Regional comparison of thioflavin s histofluorescence and BuChE histochemistry (obtained with a modified Karnovsky–Roots method) in adjacent sections from the perirhinal cortex of a 90-year-old non-demented individual. (A) β-Amyloid (Aβ) immunohistochemistry reveals Aβ deposits in superficial and deeper layers (solid and open arrows, respectively). (B) Thioflavin S (indicating the presence of a fibrillar β-pleated structure) stains the plaques found in the superficial (straight arrow) but not deeper layers. The curved arrow points to a tangle. (C) BuChE enzyme histochemistry in an adjacent section. Only the thioflavin S-positive plaques contain BuChE activity (arrow).

explored. In AD, BuChE activity is found almost exclusively in the amyloid plaques, neurofibrillary tangles, and vessels with amyloid angiopathy (see Figs 8.7 to 8.9). There is a tendency for this AD-related BuChE activity to be more widespread in the deeper layers of the cerebral cortex where BuChE-rich glia are also more numerous, leading to the suggestion that neuroglia may provide the source of BuChE as well. The intensity of plaque-bound BuChE seems to vary according to the maturation stage of the plaque (Fig. 8.10). Amyloid plaques in AD are deposited in an initially diffuse form which does not induce local tissue injury. Eventually, these plaques assume a β-pleated fibrillar structure associated with neuritic degeneration and dementia. Diffuse amyloid deposits tend not to display BuChE enzyme activity, whereas the vast majority of compact plaques do, providing circumstantial support for the conjecture that BuChE may participate in pathological maturation of the amyloid plaque into a toxic fibrillar form.[45] The pathophysiological role of ChEs in AD remains quite mysterious. Both AChE and BuChE are said to influence the in vitro aggregation state of amyloid, perhaps through non-catalytic mechanisms related the peripheral anionic site.[46,47]

Overview and conclusions

In the human brain, AChE activity is 10–100 times higher in subcortical structures, such as the striatum, thalamus, and brainstem, than in the cerebral cortex. A very large number of CNS neurons contain AChE, but only a subgroup displays the particularly intense AChE activity that can be designated AChE-rich. All known cholinergic neurons of the human brain are AChE-rich. These are also the only neurons with AChE-rich axons. All AChE-rich axons are therefore cholinergic. The CNS contains additional AChE-rich neurons which lack an AChE-rich axon and which are almost universally cholinoceptive but not cholinergic.

In the cerebral cortex, most of the AChE-rich, non-cholinergic, cholinoceptive neurons have a pyramidal shape, lie in layers 3 and 5, and are more numerous in association than limbic cortex. These neurons have a remarkable developmental pattern and assume their AChE-rich pattern only after the age of 10 years. The distribution of BuChE displays an entirely different pattern. Most cortical BuChE-rich neurons are non-pyramidal, lie predominantly in deeper cortical layers, are more numerous in limbic cortices and show no comparable developmental pattern. In addition to neurons, many neuroglia in the CNS express both AChE and BuChE.

In AD, the cerebral cortex displays a profound loss of cholinergic AChE-rich axons and a variable decrease in AChE-rich cholinoceptive neurons. At the same time, histopathological markers of AD, such as the amyloid plaques, neurofibrillary tangles and, vessels with amyloid angiopathy, express intense AChE and BuChE activity. The AChE in these lesions displays enzymatic properties that differ from that of normal neurons and that are similar to those of glial AChE. Furthermore, the AD-related histopathological lesions with BuChE activity tend to be more numerous in cortical layers which display higher densities of BuChE-positive glia. The AD-related ChEs therefore seem to have a glial origin. These ChEs may actively participate in the pathophysiology of AD, although the mechanisms for this relationship remain to be elucidated.

The human CNS contains dopaminergic, serotonergic, glutamatergic, gabaergic, noradrenergic, histaminergic and peptidergic neurons that are AChE-rich and cholinoceptive. ChE inhibitors would be expected to exert a widespread postsynaptic influence on all of these non-cholinergic pathways. In AD, where progressively fewer cortical axons are left to release ACh, ChE inhibitors may become progressively less effective for promoting cholinergic transmission in the cerebral cortex. In fact, in advanced disease, the main outcome of ChE inhibition in the cerebral cortex may be to inhibit the plaque- and tangle-bound ChEs. The therapeutic desirability of inhibiting these AD-related ChEs, especially the BuChE, remains to be explored.

Acknowledgments

This work was supported in part by NS 20285 and an NIA Alzheimer's disease center AG 13854.

References

1. Toutant J-P, Massoulie J. Acetylcholinesterase. In: Kenny AJ, Turner AH, eds. *Mammalian Ectoenzymes.* New York: Elsevier, 1987: 289–328.
2. Massoulié J, Bon S. The molecular forms of cholinesterase and acetylcholinesterase in vertebrates. *Annu Rev Neurosci* 1982; **5:** 57–106.
3. Li Y, Camp S, Rachinsky TL, Getman D, Taylor P. Gene structure of mammalian acetylcholinesterase. *J Biol Chem* 1991; **266:** 23 083–23 090.
4. Grisaru D, Sternfeld M, Eldor A, Glick D, Soreq H. Structural roles in acetylcholinesterase variants in biology and pathology. *Eur J Biochem* 1999; in press.
5. Hovel M, Sussmaan JL, Krejci E *et al.* Conver-

sion of acetylcholinesterase to butyrylcholinesterase: modeling and mutagenesis. *Proc Natl Acad Sci USA* 1992; **82:** 10 827–10 831.
6. Sussman J, Silman I. Acetylcholinesterase: structure and use as a model for specific cation–protein interactions. *Curr Opin Struct Biol* 1992; **2:** 721–729.
7. Layer PG, Weikert T, Alber R. Cholinesterases regulate neurite growth of chick nerve cells in vitro by means of a non-enzymatic mechanism. *Cell Tissue Res* 1993; **273:** 219–226.
8. Beeri R, Le Novere N, Mervis R *et al.* Enhanced hemicholinium binding and attenuated dendrite branching in cognitively impaired acetylcholinesterase-transgenic mice. *J Neurochem* 1997; **69:** 2441–2451.
9. Greenfield SA. A non-cholinergic function for acetylcholinesterase. In: Quinn DM, Balasubramanian AS, Doctor BP, eds. *Enzymes of the Cholinesterase Family.* New York: Plenum, 1995: 415–421.
10. Mesulam M-M, Geula C. Acetylcholinesterase-rich neurons of the human cerebral cortex: cytoarchitectonic and ontogenetic patterns of distribution. *J Comp Neurol* 1991; **306:** 193–220.
11. Karnovsky MJ, Roots L. A 'direct coloring' thiocholine method for cholinesterases. *J Histochem Cytochem* 1964; **12:** 219–221.
12. Koelle GB, Friedenwald JS. A histochemical method for localizing cholinesterase activity. *Proc Soc Exp Biol Med* 1949; **70:** 617–622.
13. Fambrough DM, Engel AG, Rosenberry TL. Acetylcholinesterase of human erythrocytes and neuromuscular junctions: Homologies revealed by monoclonal antibodies. *Proc Natl Acad Sci USA* 1982; **79:** 1078–1082.
14. Mesulam M-M, Geula C, Brimijoin S, Smiley JF. Butyrylcholinesterase immunochemistry in human cerebral cortex. *Soc Neurosci Abstr* 1995; **21:** 1976.
15. Mesulam M-M, Geula C, Cosgrove R, Mash D, Brimijoin S. Immunocytochemical demonstration of axonal and perikaryal acetylcholinesterase in human cerebral cortex. *Brain Res* 1991; **539:** 233–238.
16. Landwehrmeyer B, Probst A, Palacios JM, Mengod G. Expression of acetylcholinesterase messenger RNA in human brain: an in situ hybridization study. *Neuroscience* 1993; **57:** 615–634.
17. Smiley JF, Mesulam M-M. Cholinergic neurons of the nucleus basalis of Meynert (Ch4) receive cholinergic, catecholaminergic, and GABAergic synapses: an electron microscopic investigation in the monkey. *Neuroscience* 1998; **00:** 000–000.
18. Levey AI, Wainer BH, Rye DB, Mufson EJ, Mesulam MM. Choline acetyltransferase-immunoreactive neurons intrinsic to rodent cortex and distinction from acetylcholinesterase-positive neurons. *Neuroscience* 1984; **13:** 341–353.
19. Smiley JF, Levey AI, Mesulam M-M. Infracortical interstitial cells concurrently expressing m2-muscarinic receptors, AChE, and NADPH-d in the human and monkey cerebral cortex. *Neuroscience* 1998; **84:** 755–769.
20. Heckers S, Geula C, Mesulam MM. Cholinergic innervation of the human thalamus: dual origin and differential nuclear distribution. *J Comp Neurol* 1992; **325:** 68–82.
21. Mesulam M-M, Geula C. Nucleus basalis (Ch4) and cortical cholinergic innervation in the human brain: observations based on the distribution of acetylcholinesterase and choline acetyltransferase. *J Comp Neurol* 1988; **275:** 216–240.
22. Mesulam M-M, Geula C, Bothwell MA, Hersh LB. Human reticular formation: cholinergic neurons of the pedunculopontine and laterodorsal tegmental nuclei and some cytochemical comparisons to forebrain cholinergic neurons. *J Comp Neurol* 1989; **283:** 611–633.
23. Mesulam MM, Dichter M. Concurrent acetylcholinesterase staining and γ-aminobutyric acid uptake of cortical neurons in culture. *J Histochem Cytochem* 1981; **29:** 306–308.
24. Okinaka S, Yoshikawa M, Uono M *et al.* Distribution of cholinesterase activity in the human cerebral cortex. *Am J Phys Med* 1961; **40:** 135–146.
25. Mesulam M-M, Geula C. Overlap between acetylcholinesterase-rich and choline acetyltransferase-positive (cholinergic) axons in human cerebral cortex. *Brain Res* 1992; **577:** 112–120.
26. Mufson EJ, Kehr AD, Wainer BH, Mesulam MM. Cortical effects of neurotoxic damage to

the nucleus basalis in rats: persistent loss of extrinsic cholinergic input and lack of transsynaptic effect upon the number of somatostatin-containing, cholinesterase-positive, and cholinergic cortical neurons. *Brain Res* 1987; **417:** 385–388.

27. Heckers S, Ohtake T, Wiley RG, Lappi DA, Geula C, Mesulam M-M. Complete and selective cholinergic denervation of rat neocortex and hippocampus but not amygdala by an immunotoxin aganst the p75 NGF receptor. *J Neurosci* 1994; **14:** 1271–1289.
28. Mesulam M-M, Geula C. Acetylcholinesterase-rich pyramidal neurons in the human neocortex and hippocampus: absence at birth, development during the life span, and dissolution in Alzheimer's disease. *Ann Neurol* 1988; **24:** 765–773.
29. Kostovic I, Skavic J, Strinovic D. Acetylcholinesterase in the human frontal associative cortex during the period of cognitive development: early laminar shifts and late innervation of pyramidal neurons. *Neurosci Lett* 1988, **90.** 107–112.
30. Darvesh S, Grantham DL, Hopkins DA. Distribution of butyrylcholinesterase in the human amygdala. *J Comp Neurol* 1998; **393:** 374–390.
31. Wright CI, Geula C, Mesulam MM. Neuroglial cholinesterases in the normal brain and in Alzheimer's disease: relationship to plaques, tangles, and patterns of selective vulnerability. *Ann Neurol* 1993; **34:** 373–384.
32. Wright CI, Geula C, Mesulam M-M. Protease inhibitors and indoleamines selectively inhibit cholinesterases in the histopathologic structures of Alzheimer disease. *Proc Natl Acad Sci USA* 1993; **90:** 683–686.
33. Karpel R, Sternfeld M, Ginzberg D, Guhl E, Graessman A, Soreq H. Overexpression of alternative human acetylcholinesterase forms modulates process extension in cultured glioma cells. *J Neurochem* 1996; **66:** 114–123.
34. Auld VJ, Fetter RD, Broadie K, Goodman CS. Gliotactin, a novel transmembrane protein on peripheral glia, is required to form the blood–brain barrier in *Drosophila*. *Cell* 1995; **81:** 757–767.
35. Geula C, Mesulam M-M. Cholinergic systems and related neuropathological predilection patterns in Alzheimer disease. In: Terry RD, Katzman R, Bick KL, eds. *Alzheimer Disease*. New York: Raven, 1994: 263–294.
36. Davis KL, Mohs RC, Marin D *et al.* Cholinergic markers in elderly patients with early signs of Alzheimer's disease. *J Am Med Assoc* 1999; **281:** 1401–1406.
37. Voytko ML, Olton DS, Richardson RT, Gorman LK, Tobin JR, Price DL. Basal forebrain lesions in monkeys disrupt attention but not learning and memory. *J Neurosci* 1994; **14:** 167–186.
38. Geula C, Mesulam MM. Cortical cholinergic fibers in aging and Alzheimer's disease: a morphometric study. *Neuroscience* 1989; **33;** 469–481.
39. Heckers S, Geula C, Mesulam MM. Acetylcholinesterase-rich pyramidal neurons in Alzheimer's disease. *Neurobiol Age* 1992; **13:** 455–460.
40. Mesulam MM, Morán A. Cholinesterases within neurofibrillary tangles related to age and Alzheimer's disease. *Ann Neurol* 1987; **22:** 223–228.
41. Mesulam M-M. Cortical cholinesterases in Alzheimer's disease: anatomical and enzymatic shifts from the normal pattern. In: Becker R, Giacobini E, eds. *Cholinergic Basis for Alzheimer Therapy*. Boston: Birkhäuser, 1991: 25–30.
42. Mesulam M, Carson K, Price B, Geula C. Cholinesterases in the amyloid angiopathy of Alzheimer's disease. *Ann Neurol* 1992; **31:** 565–569.
43. Carson KA, Geula C, Mesulam MM. Electron microscopic localization of cholinesterase activity in Alzheimer brain tissue. *Brain Res* 1991; **540:** 204–208.
44. Friede RL. Enzyme histochemical studies of senile plaques. *J Neuropathol Exp Neurol* 1965; **24:** 477–491.
45. Guillozet AL, Smiley JF, Mash DC, Mesulam M-M. Butyrylcholinesterase in the life cycle of amyloid plaques. *Ann Neurol* 1997; **42:** 909–918.
46. Inestrosa NC, Alvarez A, Pérez CA *et al.* Acetylcholinesterase accelerates assembly of amyloid-β-peptides into Alzheimer's fibrils: possible role of the peripheral site of the enzyme. *Neuron* 1996; **16:** 881–891.

47. Barber KL, Mesulam M-M, Krafft GA, Klein WL. Butyrylcholinesterase (BChE) alters the aggregation state of Aβ amyloid. *Soc Neurosci Abstr* 1996; **22:** 1172.

48. Mesulam M-M, Geula C, Morán A. Anatomy of cholinesterase inhibition in Alzheimer's disease: effect of physostigmine and tetrahydroaminoacridine on plaques and tangles. *Ann Neurol* 1987; **22:** 683–691.

9

Measurement of cholinesterase activity

Israel Hanin, Bertalan Dudas

Quantitative procedures for the measurement of cholinesterase (ChE) activity have undergone a variety of developments over the years. These procedures have been based on the specific characteristics of the enzyme, the investigator's ability to measure the chemical substrates employed, and the products obtained from the enzymatic reaction between these selected substrates and the enzyme. A number of comprehensive reviews on the measurement of ChE can be found in the literature.[1–8] These reviews should serve as an excellent source of general information to the interested reader. For the sake of simplicity, the generic term 'ChE' used in this chapter represents either acetylcholinesterase (AChE), butyrylcholinesterase (BuChE, also known as pseudo-cholinesterase), or both. Both enzyme types might be present in a tissue extract. Hence, identifying the action of one in the presence of the other can be achieved by including, in the reaction medium, a selective inhibitor of the enzyme that is not of concern to the investigator; thus only the enzyme of interest would be active and its action could be quantitated.

The general principle employed for measuring ChE in each case is quite straightforward: a choline ester (e.g. acetylcholine (ACh)) is exposed to ChE which, in an aqueous medium, cleaves this substrate to form choline and the corresponding acid (Fig. 9.1). Hence, in the case of ACh the products of hydrolysis by AChE would be choline and acetic acid. If butyrylcholine (BuCh) were the substrate, the products of hydrolysis by BuChE would be choline and butyric acid. Similarly, if a thiocholine ester were used as a substrate, the products would be thiocholine and the corresponding acid; and so on. These products of enzymatic cleavage can be quantitated using a variety of methodological approaches.

Tracing the progression of the methods developed for the quantitative estimation of ChE is most interesting from a historical perspective, because it illustrates the gradual improvement in, and sophistication of, available analytical chemical techniques and the ingenuity of investigators in developing these methods.

Some of the earliest approaches to be used employed basic analytical chemical techniques to assess ChE activity (e.g. changes in pH or release of CO_2 as a result of the enzyme activity). Subsequently, electrometric, photometric, gasometric, fluorometric, polarographic, and spectrophotometric methods were developed. As radiochemical analyses became more common and it was feasible to synthesize reliable and selective radiochemical substances, radiometric approaches became popular, particularly because of their ability to measure, in a reproducible manner, very small amounts of enzyme activity in tissue extracts. Figure 9.2 illustrates the variety of approaches that have been developed over the past five

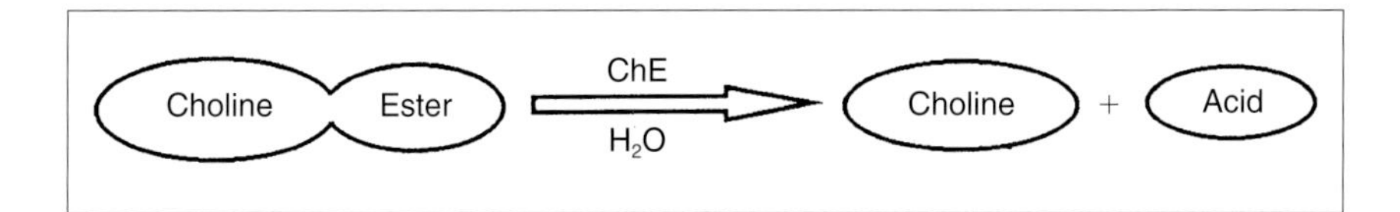

Fig. 9.1
Schematic general representation of the mode of action of ChEs.

decades in an attempt to measure ChE. Some key examples are surveyed briefly below. Comprehensive critiques of these various methodologies can be found in the literature[1–8] (see, in particular, Holmstedt[1] and Silver[4]).

What, then, have been the approaches used over time for the measurement of ChE activity? The following examples, for purposes of illustration, focus on approaches which have been employed for the measurement of AChE activity. In every case, substituting BuCh for ACh, and BuChE for AChE, would provide a similar scenario for BuChE activity.

The earliest measurements were made of the substrate remaining in the medium after a prescribed exposure to AChE (see Fig. 9.2(A)). Hestrin's colorimetric approach[9] was employed in such studies. Hydroxylamine was added to the medium, which reacted with the substrate, ACh, to form acethydroxamic acid. Presence of ferric chloride and HCl in the medium resulted in a red-purple complex of the ACh–acethydroxamic product, which could be readily quantitated using a spectrophotometer. Bioassay was also employed by some investigators to measure the amount of ACh remaining in the medium. Nowadays there are several very sensitive techniques to measure ACh and BuCh, involving gas chromatography, high performance liquid chromatography (HPLC), and combined gas chromatography–mass spectrometry.[10] They are not, however used for the measurement of ChE activity, since more efficient and less complex techniques have since been developed (see below).

Other investigators approached the question by measuring the amount of acetic acid formed following the hydrolysis of ACh by AChE. The Michel method, first employed in 1949,[11] measured the rate of change in pH during hydrolysis of ACh by AChE. A pH meter was used for such measurements (electrometric method). Various indicators (e.g. phenol red, bromthymol blue, and litmus) were also used to measure AChE activity, employing either visual or spectrophotometric analyses (photometric recording). A variation on the theme was employed, in which the pH was maintained constant in the reaction medium by means of careful titration of the medium (titrimetric method) with standard alkali solution. This procedure could be automated. Yet another approach is the so-called gasometric method, which measures CO_2 liberation following an interaction between the acetic acid formed during the hydrolysis of ACh by AChE, and bicarbonate ions in the medium. A Cartesian diver technique, originally described by Linderstrom-Lang and Glick in 1938,[12] was used for this purpose. Giacobini and his colleagues refined and miniaturized the technique to the extent that it could

Fig. 9.2
(A) Methodology developed for the measurement of AChE between 1949 and 1961. (B) Measures of AChE activity developed between 1961 and 1983, based on the original histochemical technique of Koelle and Friedenwald (1949). (C) Radiometric approach to measure AChE activity. (For details, see text.)

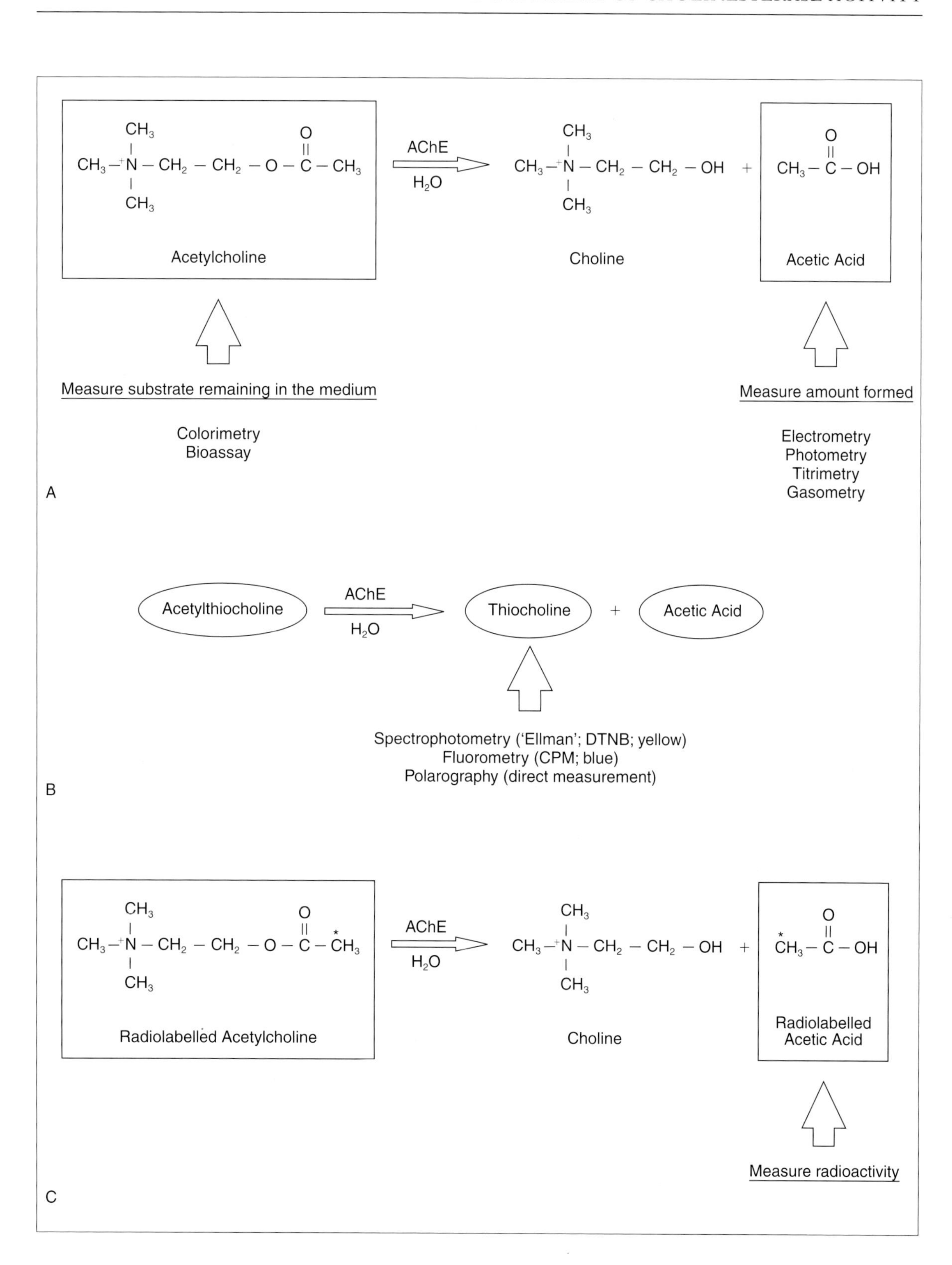

CH3
CH3—+N – CH2 – CH2 – O – C – CH3
CH3
O
Acetylcholine
AChE
H2O
CH3—+N – CH2 – CH2 – OH
Choline
+
CH3 – C – OH
Acetic Acid
Measure substrate remaining in the medium
Colorimetry
Bioassay
Measure amount formed
Electrometry
Photometry
Titrimetry
Gasometry
A
Acetylthiocholine
AChE
H2O
Thiocholine
Acetic Acid
Spectrophotometry ('Ellman'; DTNB; yellow)
Fluorometry (CPM; blue)
Polarography (direct measurement)
B
Radiolabelled Acetylcholine
Choline
Radiolabelled
Acetic Acid
Measure radioactivity
C

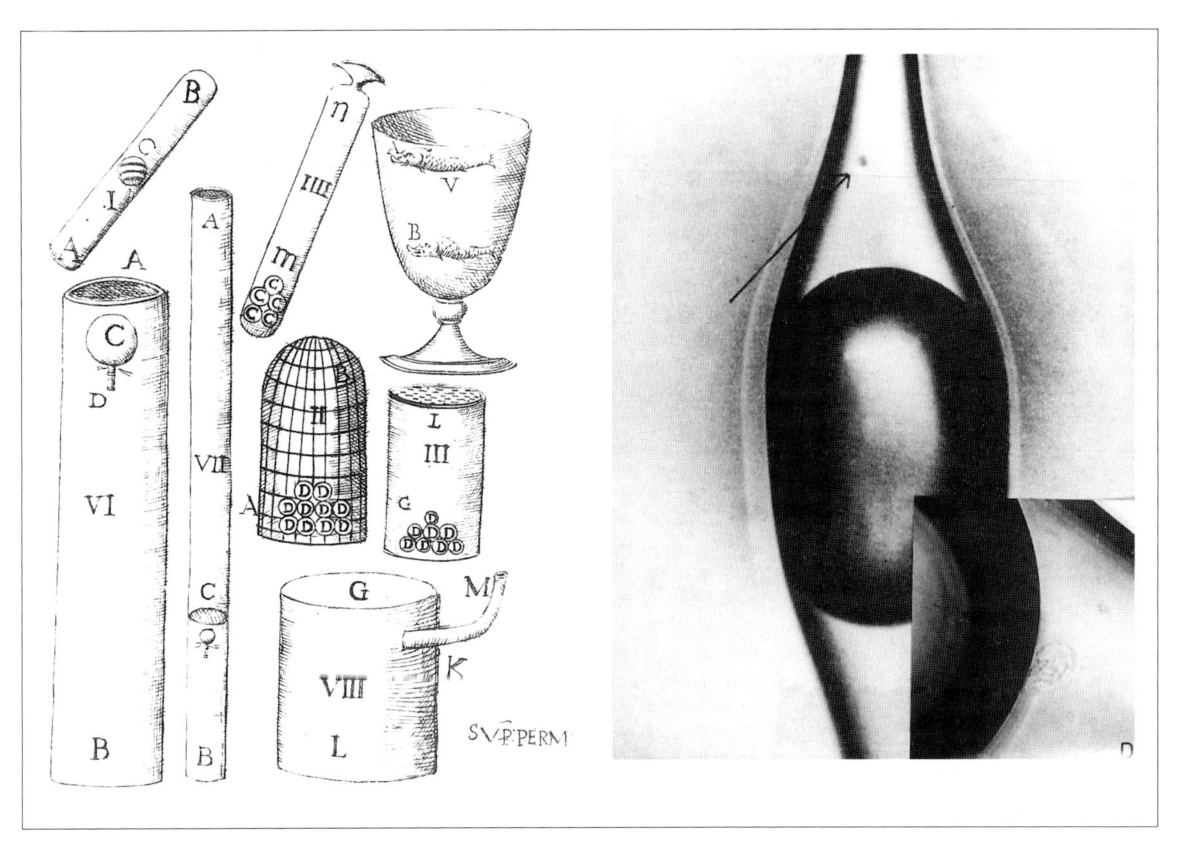

Fig. 9.3
Left: Original drawings of Cartesian divers of different sizes and shapes described in 1648 by R Magiotti, a pupil of Galileo Galilei in Pisa. The divers were suspended inside a vessel containing water. Right: A nerve cell body isolated from a sympathetic ganglion (arrow) of the rat placed inside the Cartesian-diver. Insert: a cell body (diameter 30 μm) shown inside the diver close to the CO_2 bubble. (From Giacobini.[13])

be used to determine AChE activity in single nerve cells dissected from spinal and sympathetic ganglia[13] (Fig. 9.3) and from single anterior horn cells.[14]

Histochemical techniques for in situ localization of ChE have been used since the late 1940s. Koelle and Friedenwald[15] were the first to show that thioesters of choline were excellent substrates for the histochemical analysis of ChE. An extensive array of histochemical analyses has since been developed based on this primary finding; these analyses are elaborated on in detail in Silver's review.[4] They are not discussed in this chapter, since the focus of the present chapter pertains to quantitative measurement approaches for AChE. What is important is that the chemical principle initiated by Koelle and Friedenwald was incorporated by Ellman *et al.*[16] into their spectrophotometric approach to measuring

ChE activity, which is now used in many laboratories. In this modification of the original Koelle and Friedenwald chemical reaction, acetylthiocholine is used as a substrate for AChE (see Fig. 9.2(B)). The product of hydrolysis, thiocholine, is allowed to react with 5,5'-dithio-bis(2-nitrobenzoic acid) (DTNB) in the medium, generating, in the process, 5-thio-2-nitrobenzoic acid anion, which is bright yellow. The rate of production of the yellow product is measured at 405 nm by spectrophotometry. The reaction is linear and allows calculation of the activity of the enzyme. While this technique is not as sensitive as, for example, the radiometric method, it is reproducible and simple to set up in the laboratory. It can also be automated (see, for example, Hanin *et al.*[17]), and hence adapted to routine and efficient laboratory application. A fluorometric variation on the above theme has been developed by Parvari *et al.*[18] based on the principle that thiocholine, produced by the same chemical reaction as shown above, can react with *N*-[4-(7-diethylamino-4-methylcoumarin-3-yl)phenyl]maleimide (CPM), to form an intensely blue fluorescent compound. The product is measured at 473 nm (excitation 390 nm) using a spectrofluorimeter. This technique is also linear, relatively straightforward, and considerably more sensitive than the Ellman method (picomolar, vs nmolar for the latter). Another approach involves the direct polarographic measurement of thiocholine produced from the enzymatic hydrolysis, by AChE, of acetylthiocholine.[19]

The radiometric approach (see Fig. 9.2(C)) relies on utilizing a radiolabeled analog of the substrate. Thus, ACh, labeled in the acetate moiety, is subjected to AChE hydrolysis. The products of this reaction are choline and radiolabeled acetate. After terminating the reaction, the labeled acetate is selectively isolated and measured by liquid scintillation spectrometry.[20–23] This method is most sensitive (picomolar) and reproducible. In fact, Koslow and Giacobini[24] have employed this approach to determine the AChE activity in single cells. This technique allows the detection of about 7 pmol ACh/h, and allows measurement of up to 150 samples in a single experiment.

In summary, the past 50 years have witnessed a progression of methods employed to measure ChE activity. Some have weathered the test of time better than others. Factors that have influenced the choice of one method over another have been its selectivity, sensitivity, reliability, ease of adaptation in the laboratory, and cost. Modifications of some of these methods have enabled their application for the analysis of ChE activity in infinitely low quantities (e.g. in single cells). By including a selective inhibitor of one of the enzymes in the medium, it is possible selectively to measure the activity of the other enzyme, in cases where it is anticipated that both AChE and BuChE are present in the tissue. The two most commonly used techniques today for the measurement of ChE activity are a modification of the original 'Ellman' method, and the radiometric approach.

References

1. Holmstedt B. Distribution and determination of cholinesterases in mammals. *Bull WHO* 1971; **44:** 99–107.
2. Augustinsson K-B. Determination of activity of cholinesterases. In: Glick D, ed. *Analysis of Biogenic Amines and Their Related Enzymes.* New York: Interscience, 1971: 217–269.
3. Augustinsson K-B. Assay methods for cholinesterases. In: Glick D, ed. *Methods of Biochemical Analysis*, Vol 5. New York: Interscience, 1957: 1–63.
4. Silver A, ed. *The Biology of Cholinesterases.* New York: Elsevier, 1974: 56–98.
5. Hanin I, Goldberg AM. Appendix I: Quantitative assay methodology for choline, acetyl-

choline, choline acetyltransferase, and acetylcholinesterase. In: Goldberg AM, Hanin I, eds. *Biology of Cholinergic Function.* New York: Raven, 1976: 647–654.

6. Brzin M, Sketelj J, Klinar B. Cholinesterases. In: Lajtha A, ed. *Handbook of Neurochemistry*, Vol 4, 2nd edn. New York: Plenum, 1983: 251–259.
7. Rakonczay Z. Mammalian brain acetylcholinesteraase. *Neuromethods, Neurotransmitter Enzymes.* 1984; **5:** 323–331.
8. Whittaker M. *Cholinesterase.* New York: Karger, 1986: 86–97.
9. Hestrin S. The reaction of acetylcholine and other carboxylic acid derivatives with hydroxylamine, and its analytical application. *J Biol Chem* 1949; **180:** 249–261.
10. Potter PE, Hanin I. Estimation of choline and acetylcholine and analysis of acetylcholine turnover rates in vivo. In: Whittaker VP, ed. *Handbook of Experimental Pharmacology*, Vol 86, *The Cholinergic Synapse.* Berlin: Springer-Verlag, 1988: 103–124.
11. Michel HO. An electrometric method for the determination of red blood cell and plasma cholinesterase activity. *J Lab Clin Med* 1949; **34:** 1564–1568.
12. Linderstrom-Lang K, Glick D. Micromethod for determination of cholinesterase activity. *CR Trav Lab Carlsberg, Ser Chim* 1938; **22:** 300–306.
13. Giacobini E. Quantitative determination of cholinesterase in individual sympathetic cells. *Acta Physiol Scand* 1959; **45:** 238–254.
14. Giacobini E, Holmstedt B. Cholinesterase content of certain regions of the spinal cord as judged by histochemical and Cartesian diver techniques. *Acta Physiol Scand* 1958; **42:** 12–27.
15. Koelle GB, Friedenwald JS. A histochemical method for localizing cholinesterase activity. *Proc Soc Exp Biol Med* 1949; **70:** 617–622.
16. Ellman GL, Courtney DK, Andres V, Featherstone RM. A new and rapid colorimetric determination of acetylcholinesterase activity. *Biochem Pharmacol* 1961; **7:** 88–95.
17. Hanin I, Yaron A, Ginzburg D, Soreq H. The cholinotoxin AF64A differentially attenuates in vitro transcription of the human cholinesterase genes. In: Hanin I, Yoshida M, Fisher A, eds. *Alzheimer's and Parkinson's Diseases.* New York: Plenum, 1995: 339–345.
18. Parvari R, Pecht I, Soreq H. A microfluorometric assay for cholinesterases, suitable for multiple kinetic determinations of picomoles of released thiocholine. *Anal Biochem* 1983; **133(2):** 450–456.
19. Kramer DN, Cannon PL Jr, Guilbault GG. Electrochemical determination of cholinesterase and thiocholine esters. *Anal Chem* 1962; **34:** 842–845.
20. Reed DJ, Goto K, Want CH. A direct radioisotopic assay for acetylcholinesterase. *Anal Biochem* 1966; **16:** 59–64.
21. Potter LT. A radiometric microassay of acetylcholinesterase. *J Pharmacol Exp Ther* 1967; **156:** 500–506.
22. McCaman MW, Tomey LR, McCaman RE. Radiometric assay of acetylcholinesterase activity in submicrogram amounts of tissue. *Life Sci* 1968; **7:** 233–244.
23. Fonnum F. Radiochemical microassays for the determination of choline acetyltransferase and acetylcholinesterase activities. *Biochem J* 1969; **115:** 465–472.
24. Koslow SH, Giacobini E. An isotopic micromethod for the measurement of cholinesterase activity in individual cells. *J Neurochem* 1969; **16:** 1523–1528.

10

Preclinical pharmacology of cholinesterase inhibitors

Giancarlo Pepeu

Introduction

The pharmacology of cholinesterase (ChE) inhibitors began more than 70 years ago when Loewi and Navratil[1] suggested that physostigmine might inhibit an enzyme inactivating acetylcholine (ACh). The enzyme was discovered shortly after, first in plasma and then in several tissues, as described in Chapter 1. In the two following decades, the synthesis of organophosphorus esters, developed first as insecticides and then as weapons for chemical warfare, prompted many toxicological studies aimed at defining the effects and potencies of these agents and the risks connected with their use. This subject, together with the description of the symptoms of acute and chronic ChE inhibitor intoxication, is dealt with in Chapter 11. From these studies, stemmed the interest in the relationship between the cholinergic system of the brain and behaviour, and the ChE inhibitors became important tools for investigating the role of the cholinergic neuronal network of the brain. Reviews of the pioneering work in this field have been written by Bignami *et al.*[2] and Warburton and Wesnes.[3]

The discovery in 1976 that in the brain of patients affected by Alzheimer's disease (AD) there is a degeneration of the forebrain cholinergic system[4] and the ensuing cholinergic hypothesis of geriatric memory dysfunction[5] prompted the search for agents able to correct the cholinergic hypofunction. ChE inhibitors were proposed as potential therapeutic agents and became the object of many investigations. The aims of the studies were:

- to define the efficacy of the existing agents and of new ChE inhibitors in correcting cognitive deficits in animal models of AD;
- to identify the cognitive functions responding to cholinergic activation; and
- to correlate the degree of ChE inhibition with changes in brain ACh levels and behavioural effects.

Part of these investigations has been reviewed by Giacobini and Quadra.[6]

In this chapter an attempt is made to answer a series of questions which have both heuristic and practical meaning, since a better knowledge of the cholinergic system and ChE inhibitor action may help in improving the therapy for AD.

The general term 'ChE inhibitor' is used throughout this chapter, although in fact these can be divided into selective and non-selective types, according to their affinity for acetylcholinesterase (AChE) and butyrylcholinesterase (BuChE).[7] The questions to be answered are the following:

- Is acute and chronic ChE inhibitor administration followed by an increase in brain ACh levels, and what is the relationship between the degree of ChE inhibition and increased ACh levels?
- How does the cholinergic system adapt to

prolonged ChE inhibition: do high extracellular ACh levels persist?
- What are the consequences of the increase in cholinergic activity, induced by ChE inhibition, on other neurotransmitter systems?
- Do ChE inhibitors improve cognitive behaviour in normal animals?
- Which cognitive deficits can be improved by ChE inhibitors in animal models of AD?

Brain ACh levels after acute and chronic ChE inhibitor administration

Since the first paper by Dubois *et al.*,[8] it has been repeatedly demonstrated that inhibition of ChE by organophosphate (OP) is associated with an increase in brain ACh levels 1–2 h after treatment.[2] Giarman and Pepeu[9] demonstrated that the toxic signs induced by the administration of 1 mg/kg of tetraethylpyrophosphate (TEPP) were associated with a doubling of the ACh level in the whole brain in the rat. Physostigmine 1 mg/kg caused a 50% increase in rat-brain ACh content.[10] In rats treated with a single injection of paraoxon (0.75 mg/kg i.p.) severe cholinergic symptoms were associated with a 100% increase in brain ACh levels and 83% ChE inhibition.[11] Conversely, after three daily doses of 0.3 mg/kg i.p., the symptoms appeared with a 50% brain ACh increase and 55% ChE inhibition. More recently, by measuring tetrahydroaminoacridine (THA) levels in the brain and its IC_{50} for AChE inhibition, Nielsen *et al.*[12] concluded that an 80–90% AChE inhibition was necessary in order to obtain an increase of about 30% in ACh levels in the rat cerebral cortex and striatum, and a 22% increase in the hippocampus.

A correlation can be demonstrated between the increase in extracellular ACh levels and the degree of ChE inhibition after acute administration. Giacobini *et al.*[13] demonstrated that a short-lasting 15% ChE inhibition in the whole brain was paralleled by a 200% increase in ACh level, while a 35% inhibition was accompanied by a peak increase of 2100%. However, 4 h after a single administration of diisopropylphosphorofluoridate (DFP) a slow gradual increase in extracellular ACh levels with a ChE inhibition up to 70% followed by a sharp six-fold increase in ACh level with 80–90% inhibition was observed.[14] Scali *et al.*[15] demonstrated that a 16% inhibition of cortical ChE induced by metrifonate administration (30 mg/kg os) was followed by only a small, not statistically significant, increase in extracellular ACh level. They also observed that 45 min after the administration of 80 mg/kg of metrifonate, there was a 77% ChE inhibition and a four fold increase in ACh extracelular levels in the cerebral cortex of ageing rats and a two-fold increase in young rats. However, a 73% ChE inhibition in the hippocampus was accompanied by a two-fold increase in ACh extracellular levels in young rats and only a 30% irregular increase in ageing rats.

A comparison between these experiments is difficult, since different ChE inhibitors (i.e. those selective for AChE, e.g. rivastigmine, or non-selective inhibitors such as DFP and metrifonate) were used. In addition, in some cases ChE inhibition was measured in the whole brain,[13] while in others[15] ChE inhibition and ACh extracellular levels were measured in the same area. Taken together, all these findings demonstrate that, although in general a large ChE inhibition is necessary in order to induce a significant increase in ACh levels, differences in the drugs used, the region investigated and the age of the rats may confound the relationship between ChE inhibition and ACh levels.

Even more uncertain are the relationships

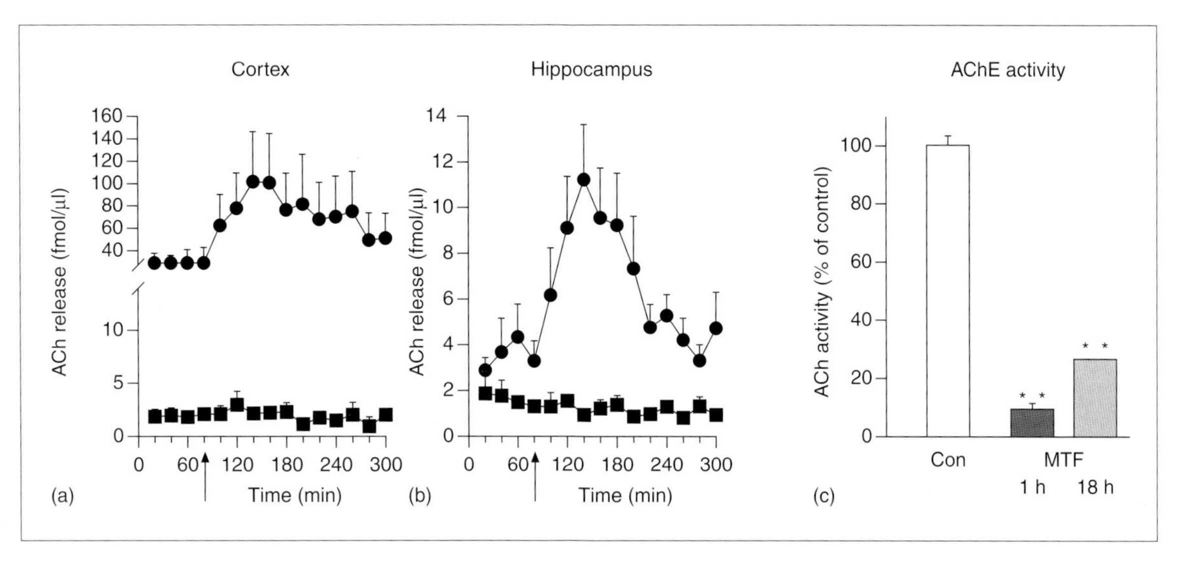

Fig. 10.1
*(a, b) The effect of metrifonate on extracellular ACh levels in the cerebral cortex and hippocampus. Metrifonate (80 mg/kg, p.o., ●) or vehicle (1 ml/kg, p.o., ■) was given twice daily for 21 days to 22 to 24-month-old Fisher 344 rats. The arrow indicates a challenge with metrifonate. The ACh level was measured 18 h after the last administration. (a) Note the large increase in the basal ACh levels in the parietal cortex of treated rats and the further increase in ACh levels following challenge with metrifonate. (b) Note the small increase in the basal levels of ACh in the hippocampus of treated rats. Significant differences between the two experimental groups were calculated using one-way ANOVA. (c) The effect of metrifonate treatment on ChE activity in whole-brain homogenates. ChE activity was detected 1 and 18 h after the last treatment. Con, vehicle; MTF, metrifonate. Significant differences between the experimental groups were calculated using one-way ANOVA followed by Duncan's test (** $p < 0.01$ vs controls). (Adapted from Giovannini et al.[17])*

between ChE inhibition and ACh levels in chronic treatment. In rats treated for 15 days with heptastigmine, AChE activity measured 24 h after discontinuation showed a 16% inhibition associated with extracellular ACh levels twice that observed in the saline-treated controls.[16] Giovannini *et al.*[17] demonstrated that after 21 days of metrifonate administration, 18 h after discontinuation there was a 75% ChE inhibition in the whole brain with a 15-fold increase in the cortical ACh extracellular level in comparison with saline-treated rats (Fig. 10.1). Under the same experimental conditions, 18 h after the last administration of tacrine there was a 15% ChE inhibition and no difference in basal extracellular ACh levels between the treated and the control rats.

In conclusion, the correlation between the degree of ChE inhibition and the increase in brain ACh levels is approximate, and the information available does not allow one to define which is the minimal ChE inhibition resulting in a significant increase in ACh extracellular levels in all brain regions.

Adaptation of the brain cholinergic system to prolonged ChE inhibition associated with increased ACh levels

After 2 or 3 weeks of daily ChE inhibitor administration resulting in a persistent ChE inhibition, extracelular ACh levels in the cerebral cortex remain higher than in untreated rats. A challenge dose of ChE inhibitor brings about an increase in ACh levels as large as that in rats treated for the first time[16,17] (see Fig. 10.1). No change in choline acetyltransferase (ChAT) activity was detected after 21 days of metrifonate treatment,[17] confirming a previous observation.[18] Indeed, no changes were seen in ACh synthesis and choline uptake in rat brain after 22 days of DFP treatment, resulting in a persistent increase in total brain ACh.[18] These findings indicate that no relevant end-product inhibition of ACh synthesis occurs in the presence of high ACh levels. However, the levels of brain choline were significantly decreased as also reported by Testylier and Dykes,[19] a finding demonstrating that de novo synthesis of ACh occurred utilizing free choline stores.

There is extensive evidence reported in the literature demonstrating that the main adaptive mechanism is a down-regulation of muscarinic receptors. It has been demonstrated[20] that after 10 weeks of 75–90% inhibition of blood ChE induced by chlorpyrifos injections in rats there was a 40% decrease in muscarinic receptor density in the cortex and hippocampus, as measured by binding of the non-selective ligand [^{3}H]QNB. Raiteri *et al.*[21] demonstrated a significant decrease in the negative-feedback mechanism mediated by muscarinic presynaptic autoreceptors in response to chronic treatment with paraoxon. It appears[22] that the cholinergic system is modulated by homeostatic, dynamic mechanisms that make relatively rapid adjustments to perturbations in the concentration of extracellular ACh. During continuous exposure to cholinergic agonists, muscarinic receptors undergo various forms of desensitization, from uncoupling of the receptor from its G-protein to sequestration in intracellular compartments. A rapid and dramatic change in subcellular compartmentalization of presynaptic m2 muscarinic receptors in the striatum has been observed[23] after a single in vivo administration of the muscarinic agonist oxotremorine.

The final result of the desensitization of the pre- and postsynaptic receptors should be a balance between an enhanced ACh release due to the loss of the inhibitory feedback mechanisms and a reduced postsynaptic cholinergic response. It is difficult to predict the extent to which this balance may affect the long-term therapeutic use of ChE inhibitors by reducing their efficacy.

Changes in other neurotransmitter systems brought about by the increase in cholinergic activity, induced by ChE inhibition

Although it is well known that the cholinergic system influences other neurotransmitter systems, few investigations of the effects of the activation of the cholinergic system following ChE inhibition have been reported. According to Nielsen *et al.*[12] THA administered, at a dose which did not increase ACh levels, to rats caused a modest increase in DOPAC levels in the whole brain. Conversely, a clear-cut increase in extracellular noradrenaline levels and a smaller increase in dopamine was shown after systemic administration of heptylphysostigmine.[24] An increase in noradrenaline after systemic administratin of a large dose of metrifonate and an increase in dopamine only after small doses of metrifonate were found in rat cortex.[25] Serotonin levels were not affected. The systemic administration of donepezil at the dose of 2 mg/kg i.p., which in rat cortex caused a 35% ChE inhibition associated with a 2100% increase in ACh extracellular level, was accompanied by increases of 100% and 80% in extracellular levels in the cortex of noradrenaline and dopamine, respectively.[13] It has been shown that noradrenaline inhibits ACh release in the cerebral cortex by acting on presynaptic α_2-heteroreceptors.[26] The possibility of enhancing the effects of ChE inhibition either by blocking the α_2-receptors or by inhibiting noradrenaline release has been investigated.[27,28] It has been shown that only the combination of ChE inhibitors with a selective α_2-antagonist such as idazoxan is able to potentiate the effect of ChE inhibition on extracellular ACh levels.

There is no information available about the consequences of a diffuse cholinergic activation on the function of the GABAergic and glutamatergic systems, both of which are affected by AD.[29,30] It has been demontrated[31] that concomitant with a local application of physostigmine is a marked increase in the evoked adenosine release from the cerebral cortex of freely moving rats. Whether this adenosine increase has a behavioural consequence is unknown.

There is no information available on the effect of long-term ChE inhibitor treatment on brain monoamines and neurotransmitter amino acids, and the meaning of the increase in monoamine observed after acute treatment has not been clarified.

Do ChE inhibitors improve cognition in normal animals?

Most of the initial experimental work on the effect of ChE inhibitors on learning and memory was aimed at defining the impairment caused by toxic doses. However, more than 20 years ago, when discussing the possible role of the cholinergic system in memory, Deutsch[32] proposed that: 'as physostigmine and DFP prolong the effects of ACh by inhibiting AChE, then these drugs should strengthen weak memories but weaken strong memories because there will be excessive ACh which will result in a depolarization block'. In the same year, McGaugh[33] reviewed the papers demonstrating that in normal rats learning is enhanced by post-training administration of low doses of physostigmine, and suggested that the cholinergic mechanisms were involved in memory storage. These findings have been confirmed by others using both physostigmine and metrifonate (Table 10.1). The active doses are reported, but in most experiments a bell-shaped dose–effect curve was obtained.

It has been reported[37] that both tacrine and donepezyl improve the acquisition of a passive avoidance response in 16-day-old rats in which there is an age-dependent cognitive impairment presumably related to the not yet fully developed brain cholinergic system.[38] It should be noted that there are no reports available on changes in learning and memory in normal animals chronically treated with ChE inhibitors.

An improvement in long-term memory processes was seen in normal humans during an 80-min infusion of low doses of physostigmine,[39] thus confirming the animal studies. Wetherell,[40] by using a similar drug schedule, which induces a 30% whole-blood ChE inhibition, suggested that physostigmine might improve performance on memory tests by

Drug	Dose	Animals	Time of treatment	Task	Ref.
Physostigmine	0.025–0.05 mg/kg	Mice	Pre/post-training	Passive avoidance	34
Physostigmine	0.32–32 μg/kg	Monkeys	Before sessions	Visual recognition	35
Metrifonate	10–30 mg/kg	Rats	Before sessions	Morris water maze	14
Metrifonate	12.5 mg/kg	Rats	Before sessions	Active avoidance	36

Table 10.1
Improvement in cognition in normal animals after ChE inhibitor administration

affecting nicotinically mediated perceptual processes in addition to, or perhaps even instead of, memory.

Which cognitive deficits can be improved by ChE inhibitors in animal models of AD?

While the number of papers on investigations into the effects of ChE inhibitors on cognitive processes in normal animals is small, much larger is the number of papers which demonstrate the effectiveness of ChE inhibitors in attenuating the cognitive deficit in putative animal models of AD. The most commonly used models are scopolamine-treated rats, rats with a lesion of the forebrain cholinergic pathways, and ageing animals, mostly rats.[41]

Scopolamine

It is certainly not unexpected that drugs which increase ACh levels in the synaptic cleft remove or prevent the effects of scopolamine, a reversible non-selective muscarinic receptor antagonist. Physostigmine has frequently been used to reverse the cognitive deficits induced by scopolamine and in investigations of the role of the cholinergic system in learning and memory. Pepeu and Pazzagli[10] demonstrated in rats that physostigmine restored the maze performance impaired by scopolamine and the decrease in brain ACh levels. Sarter *et al.*[42] have summarized several papers describing the antagonism of scopolamine amnesia by physostigmine in active avoidance and delayed non-matching to sample tasks.

In an extensive investigation on the ability of heptylphysostigmine to reverse scopolamine impairment of long-term and working memory in rodents, Dawson *et al.*[43] obtained a dose-dependent reversion in all cognitive tests, but not in motor coordination or motivation. Similarly, metrifonate 30–100 mg/kg os restored the working memory performance of scopolamine-treated rats tested in water maze and passive avoidance tasks.[44] It has also been shown[45] that the impairment of the passive avoidance response can be prevented by metrifonate at doses of 10–15 mg/kg os. At these doses, neither ChE inhibition nor an increase in cortical ACh levels could be detected. We may therefore assume that small undetectable increases in ACh levels at critical synapses are sufficient to ameliorate the cognitive impairment induced by scopolamine.

Wang and Tang[46] compared in rats the effectiveness of (−)-huperzine, donepezil and tacrine in antagonizing the disruptive effects of scopolamine on radial maze performance, and found that all three drugs showed a bell-shaped dose–effect curve and oral huperzine was more effective than the other drugs.

Finally, it should be mentioned that both physostigmine and tacrine completely restored the impairment in water maze and passive avoidance performance induced by the administration of mecamilamine, a nicotinic antagonist.[47]

Forebrain cholinergic lesions

In 1992, Sarter *et al.*[42] reviewed 23 papers reporting on the effects of physostigmine, tacrine and galantamine on the deficit in spatial and working memory, as assessed by maze performance and avoidance learning, respectively, in rats with lesions of the forebrain cholinergic nuclei. In most of these papers, ChE inhibitors were active after both acute and chronic treatment, irrespective of the test used, but with a very narrow range of active doses. However, single administrations of physostigmine did not improve a persistent impairment in sustained attention induced by extensive lesions of cortical cholinergic afferents by intracortical injection of 192-Ig–saporin.[48]

In rats with a unilateral lesion of the forebrain cholinergic nuclei, metrifonate (7.5 and 12.5 mg/kg os) improved the performance in the passive avoidance and Morris' water maze tasks.[45] Similar results had been obtained with metrifonate in rats with a lesion of the medial septum.[44] However, according to Beninger *et al.*,[49] physostigmine was more effective in restoring working memory than reference memory impaired by a unilateral lesion of the nucleus basalis. It should be mentioned that THA (0.05 and 5 mg/kg), did not facilitate performance recovery in a T-maze in rats with a lesion of the entorhinal cortex, which does not specifically damage the cholinergic pathways.[50]

Ageing animals

Ageing animals show a brain cholinergic hypofunction[51] associated with cognitive impairment which makes them a useful model for investigating the potential therapeutic activity of ChE inhibitors.[52] Bartus *et al.*[53] were among the first to demonstrate that appropriate doses of physostigmine and tacrine were able to correct in old monkeys the age-related memory impairment, as evaluated by using a delayed response task measuring recent memory. It has recently been shown that tacrine (1–3 mg/kg p.o.) increases ACh extracellular levels in the cortex and hippocampus and concomitantly restores object recognition and improves passive avoidance behaviour in 22 to 24-month-old rats.[54] Metrifonate at doses of 10–30 mg/kg p.o. has been shown to improve reference and working memory, as tested by water-maze performance, in 23-month-old rats but not in 27-month-old rats. However, metrifonate improved passive avoidance performance in both old and very old groups of rats.[44] The dose of 30 mg/kg p.o. metrifonate, which caused only 14% ChE inhibition in the cerebral cortex of 22 to 24-month-old rats, and no increase in ACh extracellular levels, was unable to restore object recognition. The recovery was observed after the administration of 80 mg/kg p.o. of metrifonate, resulting in an 85% ChE inhibition and a large increase in extracellular levels.[15] The reasons for this discrepancy are difficult to understand, unless we admit that object recognition requires a more diffuse cholinergic activation than passive avoidance and maze navigation. In 28-month-old rats the daily administration for 15 days of heptastigmine (0.6 mg/kg s.c.) prevented the deterioration of spatial memory but had no effect on non-spatial memory, as evaluated by an object-recognition test in which the pretest and test phases were separated by 45 days.[16] Metrifonate has been also tested on ageing rabbits in which doses of 12.5 and 24 mg/kg os for 7 weeks induced a 40% red cell ChE inhibition and enhanced both acquisition and retention of an eyeblink conditioned task.[55] The facilitating effect of chronic metrifonate treatment on the acquisition of a hippocampus-dependent task such as eyeblink conditioning may be caused by an increased excitability of CA1 pyramidal neurons.[56] It may be assumed that this cellular mechanism could be responsible for the facilitatory effect of ChE inhibitors on the acquisition of other behavioural tasks.

Conclusions

Answers, albeit sometimes incomplete, have been given to all the questions asked. The literature on preclinical pharmacology of ChE inhibitors is extensive, but few long-term studies have been reported. From the information available it may be concluded that the administration of ChE inhibitors is followed by ChE inhibition and an increase in extracellular ACh levels in all brain areas. In turn, this is accompanied by an improvement in learning and memory deficits, involving working, spatial and reference memories, associated with cholinergic hypofunction. However, no straightforward correlation can be established between the intensity of ChE inhibition, the increase in ACh and the behavioural changes. In some cases, improvements have been observed with doses of ChE inhibitors the effects of which on ChE and ACh levels were undetectable. Nevertheless, this does not justify the hypothesis of other mechanisms of action for which we have no evidence. Finally, the experiments in animals, particularly those in ageing rats and monkeys, and in putative models of AD support the clinical effects demonstrated in patients treated with different ChE inhibitors discussed in Chapter 12.

References

1. Loewi O, Navratyl E. Ueber den Mechanismus der Vaguswirkung von Physostygmine und Ergotamin. *Pflueger's Arch Ges Physiol* 1926; **214:** 689–696.
2. Bignami G, Rosic N, Michalek H, Milosevic M, Gatti GL. Behavioral toxicity of anticholinesterase agents: methodological, neurochemical, and neuropsychological aspects. In: Weiss B, Laties VG, eds. *Behavioral Toxicology*. New York: Plenum, 1975: 155–210.
3. Warburton DM, Wesnes K. Historical overview of research on cholinergic system and behavior. In: Singh MM, Warburton DM, Lal H, eds. *Central Cholinergic System and Adaptive Dysfunctions*. New York: Plenum, 1985: 1–35.
4. Davies P, Maloney AJR. Selective loss of cholinergic neurons in Alzheimer's disease. *Lancet* 1976; **ii:** 1403.
5. Bartus RT, Dean RL, Beer B, Lippa AS. The cholinergic hypothesis of geriatric memory dysfunction. *Science* 1982; **217:** 408–417.
6. Giacobini E, Quadra G. Second and third generation cholinesterase inhibitors: from preclinical studies to clinical efficacy. In: Giacobini E, Becker R, eds. *Alzheimer Disease: Therapeutic Strategies*. Boston: Birkäuser, 1994: 155–171.
7. Giacobini E. From molecular structure to Alzheimer therapy. *Jpn J Pharmacol* 1997; **74:** 225–246.

8. Dubois KP, Doull PR, Salerno PR, Coon JM. Studies on the toxicity and the mechanism of action of *p*-nitrophenyldiethylthionophosphate (parathion). *J Pharmacol Exp Ther* 1949; **95:** 79–91.
9. Giarman NJ, Pepeu G. Drug-induced changes in brain acetylcholine. *Br J Pharmacol* 1962; **19:** 226–234.
10. Pepeu G, Pazzagli A. Amnesic properties of scopolamine and brain acetylcholine in the rat. *Int J Neuropharmacol* 1964; **4:** 291–299.
11. Wecker L, Mobley PL, Dettbarn W-D. Central cholinergic mechanisms underlying adaptation to reduced cholinesterase activity. *Biochem Pharmacol* 1977; **26:** 633–637.
12. Nielsen JA, Mena EE, Williams IH, Nocerini MR, Liston D. Correlation of brain levels of 9-amino-1,2,3,4-tetrahydroacridine (THA) with neurochemical and behavioral changes. *Eur J Pharmacol* 1989; **173:** 53–64.
13. Giacobini E, Zhu X-D, Williams E, Sherman KA. The effect of the selective reversible acetylcholinesterase inhibitor E2020 on extracellular acetylcholine and biogenic amine levels in rat cortex. *Neuropharmacology* 1996; **35:** 205–211.
14. van der Staay FJ, Hinz VCh, Schmidt BH. Effect of metrifonate, its transformation product dichlorvos, and other organophosphorus and reference cholinesterase inhibitors on Morris water escape behavior in young-adult rats. *J Pharmacol Exp Ther* 1996; **278:** 697–708.
15. Scali C, Giovannini MG, Bartolini L *et al.* Effect of metrifonate on extracellular brain acetylcholine and object recognition in aged rats. *Eur J Pharmacol* 1997; **325:** 173–180.
16. Garrone B, Luparini MR, Tolu M, Magnani M, Landolfi C, Milanesi C. Effect of subchronic treatment with the acetylcholinesterase inhibitor heptastigmine on central cholinergic transmission and memory impairment in aged rats. *Neurosci Lett* 1998; **245:** 53–57.
17. Giovannini MG, Scali C, Bartolini L, Schmidt B, Pepeu G. Effect of subchronic treatment with metrifonate and tacrine on brain cholinergic function in aged F344 rats. *Eur J Pharmacol* 1998; **354:** 17–24.
18. Russel RW, Carson VG, Booth RA, Jenden DJ. Mechanism of tolerance to the anticholinesterase DFP: acetylcholine levels and dynamics in the rat brain. *Neuropharmacology* 1981; **20:** 1197–1201.
19. Testylier G, Dykes RW. Acetylcholine release from the frontal cortex in the waking rat measured by microdialysis without acetylcholinesterase inhibitors: effect of diisopropylfluorophosphate. *Brain Res* 1996; **740:** 307–315.
20. Bushnell PJ, Kelly KL, Ward R. Repeated inhibition of cholinesterase by chlorpyrifos in rats: behavioral, neurochemical and pharmacological indices of tolerance. *J Pharmacol Exp Ther* 1994; **270:** 15–25.
21. Raiteri M, Marchi M, Paudice P. Adaptation of presynaptic acetylcholine autoreceptors following long-term drug treatment. *Eur J Pharmacol* 1981; **74:** 109–110.
22. Testylier G, Maalouf M, Butt AE, Miasnikov AA, Dykes RW. Evidence for homeostatic adjustment of rats somatosensory cortical neurons to changes in extracellular acetylcholine concentrations produced by iontophoretic administration of acetylcholine and by systemic diisopropylfluorophosphate treatment. *Neuroscience* 1999; **91:** 843–870.
23. Bernard V, Laribi O, Levey AI, Bloch B. Subcellular redistribution of m2 muscarinic acetylcholine receptors in striatal interneurons in vivo after acute cholinergic stimulation. *J Neurosci* 1998; **18:** 10 207–10 218.
24. Cuadra G, Summers K, Giacobini E. Cholinesterase inhibitor effects on neurotransmitters in rat cortex in vivo. *J Pharmacol Exp Ther* 1994; **270:** 277–284.
25. Mori F, Cuadra G, Giacobini E. Metrifonate effects on acetylcholine and biogenic amines in rat cortex. *Neurochem Res* 1995; **20:** 1081–1088.
26. Beani L, Tanganelli S, Antonelli T, Bianchi C. Noradrenergic modulation of cortical acetylcholine release is both direct and γ-aminobutyric acid-mediated. *J Pharmacol Exp Ther* 1986; **236:** 230–236.
27. Cuadra G, Giacobini E. Coadministration of cholinesterase inhibitors and idazoxan: effects of neurotransmitters in rat cortex in vivo. *J Pharmacol Exp Ther* 1995; **273:** 230–240.
28. Cuadra G, Giacobini E. Effects of cholinesterase inhibitors and clonidine co-

administration on rat cortex neurotransmitters in vivo. *J Pharmacol Exp Ther* 1995; **275:** 228–236.

29. Chu DCM, Penney JB, Young AB. Quantitative autoradiography of hippocampal $GABA_B$ and $GABA_A$ receptor changes in Alzheimer's disease. *Neurosci Lett* 1987; **82:** 246–252.
30. Cowburn R, Hardy J, Roberts P, Briggs R. Presynaptic and postsynaptic glutamatergic function in Alzheimer's disease. *Neurosci Lett* 1988; **86:** 109–113.
31. Pazzagli M, Corsi C, Latini S, Pedata F, Pepeu G. In vivo regulation of extracellular adenosine levels in the cerebral cortex by NMDA and muscarinic receptors. *Eur J Pharmacol* 1994; **254:** 277–282.
32. Deutsch JA. The cholinergic synapse and the site of memory. In: Deutsch JA, ed. *The Physiological Basis of Memory.* New York: Academic Press, 1973: 13–16.
33. McGaugh JL. Drug facilitation of learning and memory. *Ann Rev Pharmacol Toxicol* 1973; **13:** 229–341.
34. Sansone M, Castellano C, Palazzesi S, Battaglia M, Ammassari-Teule M. Effects of oxiracetam, physostigmine, and their combination on active and passive avoidance learning in mice. *Pharmacol Biochem Behav* 1993; **44:** 451–455.
35. Aigner TG, Mishkin M. The effects of physostigmine and scopolamine on recognition memory in monkeys. *Behav Neural Biol* 1986; **45:** 81–87.
36. van der Staay FJ, Hinz VCh, Schmidt BH. Effect of metrifonate on escape and avoidance learning in young and aged rats. *Behav Pharmacol* 1996; **7:** 56–64.
37. Smith RD, Kistler MK, Cohen-Williams M, Coffin VL. Cholinergic improvement of a naturally-occurring memory deficit in the young rat. *Brain Res* 1996; **707:** 13–21.
38. Pedata F, Slavikova J, Kotas A, Pepeu G. Acetylcholine release from rat cortical slices during postnatal development and aging. *Neurobiol Aging* 1983; **4:** 31–35.
39. Davies KL, Mohs RC, Tinklenberg JR, Pfefferbaum A, Hollister LE, Kopell BS. Physostigmine: improvement of long term memory processes in normal human. *Science* 1978; **201:** 272–274.
40. Wetherell A. Effects of physostigmine on stimulus encoding in a memory-scanning task. *Psychopharmacology (Berlin)* 1992; **109:** 198–202.
41. Pepeu G, Marconcini Pepeu I, Casamenti F. The validity of animal models in the search for drugs for the aging brain. *Drug Design Delivery* 1990; **7:** 1–10.
42. Sarter M, Hagan J, Dudchenko P. Behavioral screening for cognition enhancers: from indiscriminate to valid testing: Part I. *Psychopharmacology (Berlin)* 1992; **107:** 144–159.
43. Dawson GR, Bentley G, Draper F, Rycroft W, Iversen SD, Pagella PG. The behavioral effects of heptyl physostigmine, a new cholinesterase inhibitor, in tests of long-term and working memory in rodents. *Pharmacol Biochem Behav* 1991; **39:** 865–871.
44. Riekkinen PJ, Schmidt B, Stefanski R, Kuitunen J, Riekkinen M. Metrifonate improves spatial navigation and avoidance behavior in scopolamine-treated, medial septum-lesioned and aged rats. *Eur J Pharmacol* 1996; **309:** 121–130.
45. Itoh A, Nitta A, Katono Y *et al.* Effects of metrifonate on memory impairment and cholinergic dysfunction in rats. *Eur J Pharmacol* 1997; **322:** 11–19.
46. Wang T, Tang XC. Reversal of scopolamine-induced deficits in radial maze performance by (−)-huperzine A: comparison with E2020 and tacrine. *Eur J Pharmacol* 1998; **349:** 137–142.
47. Riekkinen M, Riekkinen PJ. Effects of THA and physostigmine on spatial navigation and avoidance performance in mecamylamine and PCPA-treated rats. *Exp Neurol* 1994; **125:** 111–118.
48. McGaughy J, Sarter M. Sustained attention performance in rats with intracortical infusions of 192-Ig–saporin-induced cortical cholinergic deafferentation: effects of physostigmine and FG 7142. *Behav Neurosci* 1998; **112:** 1519–1525.
49. Beninger RJ, Wirsching BA, Mallet PE, Jhamandas K, Boegman RJ. Physostigmine, but not 3,4-diaminopyridine, improves radial maze performance in memory-impaired rats. *Pharmacol Biochem Behav* 1995; **51:** 739–746.
50. Kesslak JP, Korotzer A, Song A, Kamali K, Cotman CW. Effects of terhydroaminoacridine

(THA) on functional recovery after seqential lesion of the entorhinal cortex. *Brain Res* 1991; **557:** 57–63.

51. Pepeu G, Giovannelli L. The central cholinergic system during aging. *Prog Brain Res* 1994; **100:** 67–71.
52. Ingram DK. Analysis of age-related impairment in learning and memory in rodent models. *Ann NY Acad Sci* 1985; **444:** 312–331.
53. Bartus RT, Dean RL, Beer B. Memory deficits in aged Cebus monkeys and facilitation with central cholinomimetics. *Neurobiol Aging* 1980; **1:** 145–152.
54. Scali C, Giovannini MG, Prosperi C, Bartolini L, Pepeu G. Tacrine administration enhances extracellular acetylcholine in vivo and restores the cognitive impairment in aged rats. *Pharmacol Res* 1997; **36:** 463–469.
55. Kronforst-Collins MA, Moriearty PL, Schmidt B, Disterhoft JF. Metrifonate improves associative learning and retention in aging rabbits. *Behav Neurosci* 1997; **111:** 1031–1040.
56. Oh MM, Power JM, Thompson LT, Moriearty PL, Disterhof JF. Metrifonate increases neuronal excitability in CA1 pyramidal neurons from both young and aging rabbit hippocampus. *J Neurosci* 1999; **19:** 1814–1823.

11

Synaptic, behavioral, and toxicological effects of cholinesterase inhibitors in animals and humans

Alexander G Karczmar

Introduction

This chapter concerns the sequelae of the 16th century story of the Calabar bean and its active agent, physostigmine, and the studies of these materials in Edinburgh and Germany (see Chapter 1). Subsequent to this story, investigations of anti-cholinesterases (anti-ChEs) followed a number of lines. First, after the demonstration by Dale, Loewi, Feldberg, Eccles, and others of the cholinergic transmission at the periphery and in the central nervous system (CNS), investigations were initiated concerning the actions of anti-ChEs on the central neurons and synapses. Then, following the work of Macht, Gantt, Freile, Funderburk, and Case, cholinergic correlates of overt, conditioned, and cognitive behavior were explored (see also Chapter 10). Parallel studies were addressed at the reflexogenic, respiratory, and other CNS functions. In addition, following the pioneering work of Henderson and Wilson, experiments were begun concerning the effects of anti-ChEs and cholinergic agents on the electroencephalogram (EEG) and sleep.[1–4] The development by the German investigators of the use of the organophosphorus (OP) anti-ChEs as pesticides and potential war gases[1] and the subsequent research with OP agents constitutes yet another line of the anti-ChE research. This chapter describes these various lines of investigation.

Neuronal post- and presynaptic effects of anti-ChEs

Neuronal sites of the action of anti-ChEs

Anti-ChEs exert effects at a number of synaptic, cholinergic sites. First, they act at postsynaptic cholinoceptive receptors innervated by a cholinergic neuron and located at either a cholinergic or a non-cholinergic neuron; this site is generally, but not always, excitatory in nature. Then, they may affect a number of presynaptic cholinoceptive, nerve terminal receptors, whether of cholinergic neurons or non-cholinergic neurons; these receptors are referred to as autoreceptors and heteroreceptors, respectively. The receptors in question are nicotinic or muscarinic, and they differ with regard to the ionic channel and second-messenger phenomena that they engender;[5,6] there are several subtypes of nicotinic and muscarinic receptors.

Released acetylcholine (ACh) acts at all the cholinoceptive sites. If an anti-ChE acted only as an inhibitor of ChEs, its effect would be to augment or potentiate the actions of ACh. Thus, the anti-ChE would combine its presynaptic action, which is concerned with the regulation of ACh release, with its postsynaptic action, which would involve the activation or inhibition of the cholinergic or non-cholinergic neuron. Indeed, early in the research in this area it was envisaged that anti-ChEs, particularly of the OP type, exert this single mechanism of action;[4] more recently, actions not related to the inhibition of ChEs have been described (see Chapter 13).[7] In the present context, the sensitizing and desensitizing effect on the receptors, the direct channel actions, and the neurotoxic effects constitute important aspects of this type of anti-ChE phenomenon (see below).

Presynaptic action of anti-ChEs: muscarinic receptors

As shown in the 1950s and 1960s by Frank MacIntosh, John Szerb, Robert Polak, and Hans Kilbinger,[4,8,9] acute stimulation of peripheral as well as central muscarinic presynaptic autoreceptors by muscarinic agonists and OP or carbamate anti-ChEs generally (for exceptions, see Caulfield[5]) diminishes the release of ACh. This diminution occurs in vitro[10] as well as in vivo; in the latter case, ACh release was monitored at the cerebral surface or in ventricles and deep cerebral structures.[8,11–14] In these experiments anti-ChEs were frequently used to prevent hydrolysis of ACh in the effluent prior to the biochemical or chemical assay, thus obscuring their effect at the nerve terminal; however, measurement of radiolabeled choline originating from the release of ACh partially resolved this problem.[8] In rodents this diminution of release occurred in the cortex, thalamus, hippocampus, corpus striatum and the caudate, brainstem, and superior colliculus; this phenomenon could be established in freely moving animals.[8,15,16] This effect occurs regardless of whether the ACh release is evoked electrically, by K^+, behaviorally, or sensorily, and is subject to tolerance which may be due the desensitization of central presynaptic muscarinic receptors.[17] It is mediated by the m1 and m4 receptor subtypes[5,8] (H Kilbinger, personal communication).

Presynaptic nicotinic autoreceptors, receptors, and muscarinic and nicotinic heteroreceptors

Nicotinic autoreceptors are present at several central sites, including the hippocampus.[18,19] Interestingly, nicotinic autoreceptors seem to facilitate ACh release,[20–22] and the mechanism that underlies the difference between the effect of muscarinic and nicotinic autoreceptor stimulation on ACh release is not clear. A similar facilitatory effect is exercised at muscarinic and nicotinic heteroreceptors on the central release of a number of transmitters, such as norepinephrine, GABA, and glutamate, and bioactive substances, such as hormones.[23–25]

Postsynaptic responses of central nicotinic and muscarinic receptors to anti-ChEs

Postsynaptic excitatory responses of muscarinic and nicotinic receptors are facilitated by carbamate and OP anti-ChEs. As the cholinergic pathways are widespread (see Chapter 8), these central responses are obtained ubiquitously.[2–4] For example, the subthreshold nicotinic excitatory potential of the Renshaw cell, whether evoked by electrophoretic application of ACh or via presynaptic electric stimulation, is converted into a spike by several carbamate and OP anti-ChEs, and the train of spikes that follows such stimulation is prolonged by these agents.[26,27] Similarly, physiological responses — including retinal potentials elicited by light and inhibition of auditory nerves via activation of the cholinergic olivocochlear bundle — are potentiated by anti-ChEs.[20,28,29] Sometimes, the actions of non-hydrolyzable cholinergic agonists were also potentiated. The pertinent information is incomplete with respect to the effect of anti-ChEs at cholinoceptive inhibitory sites, both nicotinic and muscarinic.

Finally, anti-ChEs exhibit direct synaptic actions. Quaternary carbamates do so via the effect of their quaternary head. Then, maintained levels of ACh produce desensitization or receptor inactivation; in particular, OP anti-ChEs contribute to this desensitization via accumulation of ACh as well as other mechanisms.[7] This desensitization is due to structural receptor changes evoked by phosphorylations.[30,31] The opposite process is that of sensitization induced by certain oxamide anti-ChEs and NaF.[7] Also, anti-ChEs exert a direct inhibitory channel effect at the nicotinic receptors.[23,32] These direct actions may contribute to toxic or functional effects of anti-ChEs.

Central functional, endocrine, and behavioral effects of anti-ChEs

The wide distribution of the central cholinergic pathways (see Chapter 8) predicates that every function and behavior of animals and humans exhibits cholinergic correlates,[7,33] and that anti-ChEs exert widespread central behavioral and functional effects. However, it must be stressed that several other ACh transmitters and bioactive, endogenous substances are also involved in the regulation of these phenomena.

Functional effects of anti-ChEs

Reflexogenic effects

The pertinent studies produced equivocal data. Even in the case of monosynaptic extensor reflexes, such as the patellar reflex, both facilitation and attenuation of the reflex was seen in animals with carbamate and OP anti-ChEs.[1,4,34–37] It is of interest that more than half a century ago Kremer found that intrathecal administration to humans of eserine and neostigmine attenuated the patellar reflex.[4] Results were also variable with respect to polysynaptic flexor and extensor reflexes and the polysynaptic spinal reflex that generates the dorsal root potential (DRP) and which can be evoked by antidromic ventral root stimulation or the stimulation of the dorsal root adjacent to the DRP (VR-DRP and DR-DRP reflex, respectively[37,38]). The variability in the results may be due to the involvement of supraspinal anti-ChE actions as well as to auto- and heteroreceptor effects.

Motor effects

Anti-ChEs and muscarinic agonists, whether administered locally to a number of central sites or systemically, induce in unanesthetized animals behavioral arrest as well as some degree of muscle relaxation, while atropinics increase the motor activity.[37,39] However, application of these agents at certain neighboring sites evokes free motor behavior.[40] These anti-ChE effects may be mediated via either muscarinic or nicotinic sites.

Cataleptic immobility, classically induced in rodents by morphine and cannabinoids, is evoked by intraventricular administration, bilateral application to the basal ganglia, or application to the reticular formation of cholinergic agonists and anti-ChE drugs, and these three classes of drug act additively.[37,41,42] Cholinergic catalepsy is accompanied by arousal on the EEG, some rigidity, and increased muscle tonus.

Muscarinic agonists and anti-ChEs induce 'adversive syndrome' or circling behavior[43] following their unilateral injection into the basal ganglia and the nigrostriatal sites. Hintgen and Aprison[43] speculated that the direction of circling is due to the endogenous assymetry of the cholinergic system of rats, as reflected by the levels of acetylcholinesterase (AChE) at the pertinent sites, and that the direction of circling is contralateral with respect to the level of the cholinergic activity. Circling behavior may be considered as a stereotypic or compulsive behavior; other compulsive behaviors evoked by muscarinic agonists and anti-ChEs include gnawing, self-biting, and purposeless, repeated motor activity.[4,44] Anti-ChEs also produce writhing, tremor, palsy-like jerks of the limbs and head, and tongue protrusions.[37] These effects resemble those that occur in Parkinson's disease.

Respiratory effects of anti-ChEs

Although in the 1950s the Nobel Prize winner and respiratory physiologist Corneille Heymans deprecated the action of OP anti-ChEs on respiration,[4] such an action of both OP and carbamate anti-ChEs was clearly demonstrated subsequently. This action is very complex, as both nicotinic and muscarinic receptors are involved and as there is an interaction between the effects of the compounds in question on the bulbar respiratory center and its subcenters, chemoreceptor, baroreceptor and tegmental sites, carotid sinus, and reticular formation; there is also reflex regulation of respiration, and an anti-ChE effect on the motor phrenic and intercostal nerves and the smooth muscle of the respiratory airways.[46–48] The net respiratory effect of muscarinic and nicotinic agonists and anti-ChEs results from the interplay between initial cholinergic stimulation of the medullary centers and activation of the cholinoceptive inhibitory sites abutting on the respiratory center,[49] including sites involved in the generation of rapid eye movement (REM) sleep (see below). Indeed, depression of respiration may occur in humans following anti-ChE induced REM sleep.[50] In addition, acting on the medullary centers, anti-ChEs increase the bronchomotor tone (as well as stimulate the bronchiolar smooth muscle directly).[51]

Used at small, subtoxic doses, anti-ChEs that are capable of central penetration, such as physostigmine and the OP compounds, augment the rate and depth of respiration via their medullary as well as chemoreceptor action; this occurs initially with the toxic doses of these drugs, which ultimately cause respiratory depression (see below). Given systemically, certain cholinergic agonists which cannot penetrate the CNS, such as neostigmine, may indirectly block the respiratory center by activating inhibitory afferents, and depress respiration by constricting bronchiolar musculature. Ultimate respiratory depression induced with high systemically administered doses of tertiary carbamate or OP agents, or inhalation of war gases is due to their central actions rather than to their stimulant action on the bronchi[52] (see below).

Emesis

Emetic function involves reflexogenic pathways originating in the intestine and vestibular complex, chemoreceptor trigger zone located in the area postrema, the vomiting center located in the reticular formation, and the vagal efferents;[53,54] all these sites contain muscarinic receptors, mostly of the m1 subtype, as well as nicotinic receptors[54,55] (see Chapter 8). Central activation of emesis by muscarinic and nicotinic agonists and anti-ChEs has been amply demonstrated, and the OP agents produce emesis upon systemic administration or inhalation[46] (see below). However, anti-ChEs may produce emesis in man and animals, not only centrally but also reflexly; thus, even quaternary agents such as pyridostigmine produce emesis.[56]

Hypothalamically evoked anti-ChE effects: endocrine, including sexual, actions

The hypothalamicopituitary axis includes a system that releases oxytocin and vasopressin from neurohypophysis (posterior pituitary), and a system that liberates regulatory releasing and release-inhibiting hormones. The releasing hormones in turn release male and female gonadotrophins, thyroid stimulating hormone (TSH), adrenocorticotrophic hormone (ACTH), growth hormone (GH), and melanocyte-stimulating hormone from adenohypophysis (anterior pituitary); the latter cause the release of other hormones such as glucocorticoids, thyroid hormones, and sex hormones. Hypothalamic sites of these two systems contain cholinergic pathways (see Chapter 8) and both muscarinic and nicotinic receptors. Indeed, ACh is released from the hypothalamus by stress, a number of physiological parameters, and by K^+, and such release is accompanied by expected hormonal and metabolic changes. Accordingly, nicotinic and muscarinic agonists and anti-ChEs facilitate the hypothalamicohypophyseal function. Thus, anti-ChEs of both carbamate and diisopropylphosphorofluoridate (DFP) type cause the release of oxytocin and vasopressin,[20,57] glucocorticoids, gonadotropins, and other hypothalamic-releasing hormones. On the other hand, cholinergic agonists and anti-ChEs inhibit rather than facilitate the release of prolactin, by releasing either inhibitory regulatory hormones or dopamine.[51]

The release of gonadotropins leads to overt sexual and mating phenomena in both males and females of several species; these effects include, in the rat, lordosis, solicitation and 'attractive' behavior, and ejaculation. These anti-ChE effects[58] involve predominantly m1 or m3 receptors;[59,60] the effect of nicotinic agonists on mating behavior is controversial. Interestingly, anti-ChEs such as tacrine and donepezil augment sexual activity in patients with Alzheimer's disease (AD); however, these effects may be mediated, not via the gonadotropic phenomena, but by the general behavioral improvement of the patients by these agents. Peculiarly, Dorner[61] found that the quaternary anti-ChE, pyridostigmine, administered neonatally permanently augmented rat adult male sexual activity.

The anti-ChE facilitation of sexual activity via peripheral effects on the sex organs may also occur via the action of gonadal hormones on the central, hormonoceptive sites; it also may relate not to the direct anti-ChE actions on the hypothalamus, but to their indirect actions on the hypothalamus via the thalamus, the limbic system or the reticular formation (see Chapter 8), or the spinal cord.[57]

Hypothalamically evoked anti-ChE effects: cardiovascular phenomena

In humans and animals cholinergic agonists and anti-ChEs induce bradycardic and cardiovascular effects via both central and peripheral mechanisms, as these effects are evoked by the intraventricular administration of the agents.[4,62] These actions when induced by the systemic administration of the agents are mediated, at least in part, centrally.[63] The Yugoslavian team of Varagic and Kristic[64] established that the hypothalamus and its m2 receptors[65] are the main site of the effects in question; the hypothalamic pressor actions are mediated via the hypothalamic sympathetic outflow and/or the release of vasopressin, while the bradycardic pressor action is mediated by the vagal nuclei.[66]

Actually, depressor and pressor or biphasic phenomena may be obtained with anti-ChEs and cholinergic agonists, the actual effect depending on the hypothalamic site that is involved; in fact, other central sites such as the medullary, reticular, and spinal sites may mediate anti-ChE cardiovascular action.[67–69] Also, the direction of the cardiovascular effect may be species- or anti-ChE-dependent.[70] Then, the cardiovascular responses to anti-ChEs may depend on the extent of the nicotinic versus the muscarinic action of the accumulated ACh.[65] Finally, anti-ChEs may act directly on the cerebral (as well as peripheral) blood vessels, causing their dilatation and the corresponding changes in vascular tone.

Hypothalamically evoked anti-ChE effects: thermocontrol

Thermocontrol is provided by the hypothalamus and involves cholinergic correlates.[71–73] Anti-ChEs and muscarinic agonists produce sharp hyperthermia when applied to the anterior hypothalamus and preoptic sites, while they produce hypothermia when given systemically or applied to the posterior hypothalamus. Aberrant effects of anti-ChEs may depend on their nicotinic actions, as nicotinic agonists produce hyperthermia when applied to the posterior hypothalamus.[74]

Hypothalamically evoked anti-ChE effects: control of thirst and hunger

In rats, muscarinic and anti-ChE activation of several hypothalamic areas causes polydypsia and an increase in food intake,[71,75] which is consistent with the notion of the role of cholinergic hypothalamic pathways in the control of food and water intake (see Chapter 8). However, different actions may exist in cats and monkeys, and the nicotinic contribution to the effects of anti-ChEs on appetite is not clear.

CNS seizures

In large doses cholinergic agonists as well as carbamate and OP anti-ChEs induce convulsions of the grand-mal type and facilitate strychnine-evoked seizures, whether in animals or humans.[76–78] These effects were obtained upon systemic, intracarotid, and intraventricular administration, or upon topical application of these drugs to the cortex and limbic structures, such as the amygdala and hippocampus;[76] in fact, these agents induce seizure-like activity in isolated preparations.[79] Anti-ChEs induce seizures following nearly complete inhibition of AChE and a several-fold accumulation of brain ACh.[80] Both the limbic muscarinic receptors (m1 subtype) and the cortical nicotinic receptors[77] are involved, the former inducing a spike and wave seizure pattern, while the latter evokes a spike pattern; thus, anti-ChE seizures include both patterns.[77,81,82]

What causes the convulsant action of anti-ChEs? First, in large doses these agents (as well as muscarinic and nicotinic agents) induce respiratory and cardiovascular collapse that may induce or contribute to, seizures; however, this notion is probably untenable.[77,79] Second, while the generation of the kindling and epileptic discharge occur at sites that are rich in cholinergic receptors (see Chapter 8), and while the cholinergic system is mobilized during seizures invoked by convulsants or appropriate electric stimulation,[80] the discharge may be caused via the cholinergic link with amino acid excitatory and inhibitory transmitters.[83,84] Interestingly, in small doses anti-ChEs (and cholinergic agonists) may antagonize convulsants such as picrotoxin or strychnine, since the desynchronization produced by anti-ChEs (see below) may prevent the induction of hypersynchrony, which underlies seizures.[76,80] ACh accumulation caused by large doses of anti-ChEs may desensitize and block the postsynaptic cholinergic sites, and thus prevent the desynchronizing, anticonvulsant role of anti-ChEs.

Behavioral and related actions of anti-ChEs

Effects of anti-ChEs on EEG and brain rhythms

It has been known since the 1930s that administration of muscarinic and nicotinic agonists and anti-ChEs evokes a characteristic shift toward higher EEG frequencies and desynchronization (also referred to as EEG arousal; see below), as well as hippocampal theta rhythms. These phenomena occur in all animal species, including humans; desynchronization was noted in the course of anti-ChE therapy of AD patients.[85,86] Desynchronization is evoked by anti-ChEs whether applied systemically or locally at many sites, extending from the pontine and bulbar reticular formation to the hypothalamus, limbic sites, and the cortex.[7,76] The common denominator in this extensive distribution of cholinoceptive desynchronizing sites is the cholinergic radiations originating in the forebrain, including in particular the nucleus basalis magnocellularis (see Chapter 8), although lesions of midbrain reticular formation and the thalamus do not obliterate, or block only transiently, the cholinergic, anti-ChE, or electrically induced desynchronization.[87]

Muscarinic and anti-ChE induced desynchronization is due to inhibition of rhythmic firing bursts, thalamic spindles and α and δ waves, an increase in the spike frequency, and blockade of such synchronizing EEG activities as recruitment and postreinforcement synchronization; the underlying mechanism may be the interplay between the effects of ACh and GABA.[88,89] Overall, cholinergic desynchronization may be looked upon as antagonistic to the slow EEG rhythms, and as one arm of the dipole between alerting and synchronizing phenomena, the other arm being constituted by the thalamico-cortical axis.[90] The anti-ChE induced antagonism of the slow rhythms is due to the activation of the cholinergic mesopontine nuclei, dorsal geniculate nucleus, and reticular nucleus, and the resultant high-frequency response of the thalamic cells, which spreads to the cortex, and/or to the activation of the gigantocellular tegmental field.[3,90] The anti-ChE induced EEG phenomena involve both muscarinic and nicotinic elements,[3,76,91] and both depolarizing and hyperpolarizing effects appear to be involved.[90] The muscarinic actions are more widely spread than are the nicotinic actions, the latter involving particularly midbrain reticular formation, thalamus, and the habenula and tegmental areas.[91] The action of nicotine may be initiated by an EEG slowing, followed by desynchronization; hence, anti-ChE evoked EEG effects may be biphasic.

A unique slow, synchronized rhythm, the hippocampal theta pattern, is generated or facilitated by anti-ChEs;[92] anti-ChE evocation of the theta rhythm includes nicotinic and muscarinic actions and embraces their different patterns. Like anti-ChEs, learning, attention, and sensory stimulation generate both theta rhythms and EEG desynchronization, and this EEG pattern, however evoked, is frequently referred to as 'EEG arousal'[90,93] (see Chapter 10). Yet, anti-ChE evoked EEG desynchronization and theta pattern differ from that accompanying behavioral arousal, as established by power spectrum analysis;[76,94] the significance of this finding is commented upon below.

Anti-ChEs and phases of sleep

In the 1940s and 1950s, Giorgio Moruzzi, Horatio Magoun, Harold Himwich, and Franco Rinaldi established the notion of the dipole of behavioral arousal or awakening, on the one hand, and of sleep, on the other, and the importance of the cholinergic system for the regulation of these phenomena.[4] Indeed, ACh release from the cortex is increased markedly during REM sleep.[95]

The two main phases of sleep are REM sleep, or deep sleep, and the slow-wave or synchronized (SW) sleep. REM sleep is characterized by the presence of the pontogeniculate occipital (PGO) spikes, eye movement, muscle atonia, EEG desynchronization and theta rhythms, and the presence of dreams. Hence, REM sleep shows paradoxical activities with respect to its deep-sleep nature, and this is also referred to as 'paradoxical sleep' or 'dream sleep'. REM sleep also mediates autonomic phenomena such as phasic changes in blood pressure and cardiac rate, respiratory depression accompanied by augmentation of respiratory volume, and diminution of thermocontrol.[96] The cholinergic sites controlling the two forms of sleep include the brainstem and its tegmental areas, the limbic forebrain and reticular formation with its pontine giant cells, and the mesencephalic neurons[97] (see Chapter 8).

While the control of both forms of sleep is multitransmitter in nature, they both have significant cholinergic correlates and may be evoked by ACh, cholinergic agonists, and anti-ChEs.[37,76,88,98,99] The induction of REM sleep in humans by OP and carbamate anti-ChEs is well documented; anti-ChEs advance the onset and increase the frequency of REM sleep episodes.[90,100,101] In animals, REM sleep is caused by systemic administration or local application of these agents to pertinent tegmental and pontine sites. The nicotinic receptors and various muscarinic receptor subtypes may evoke different patterns of REM sleep;[100,102,103] thus, anti-ChE evocation of REM sleep includes these various elements. It illustrates the importance of the cholinergic system in the generation of REM sleep that anti-ChEs induced REM sleep in animals which were experimentally deprived of norepinephrine and serotonin.[104] Cholinergic facilitation of REM sleep is due to cholinergic excitatory neuronal effects[3,90] and to the action of the second messengers, including nitric oxide.

As REM sleep is a component of the sleep–wakefulness dipole it may relate to the circadian rhythms, which also exhibit cholinergic correlates,[37] and to their regulation by hypothalamic mechanisms. It should be stressed that the cholinergicity and the anti-ChE induction of REM sleep reflect a special cholinergic behavioral syndrome[90] (see below).

Effects of anti-ChEs on aggression

Aggression is an ethological behavior directed at inflicting damage or presenting a threat within a species or across species and genera. Laboratory models of aggression involve social isolation, footshock, limbic lesions, exposure of mice to muricidal rats, observation of small rodents in pseudo-natural habitats, injection of irritants, and administration of appropriate drugs;[45,96,105–107] they represent predatory, defensive (as in the case of territorial or pup defense), or affective (rage) forms of aggression. These various forms of aggression can be induced by systemic administration of anti-ChEs and cholinergic (particularly muscarinic) agonists, their localized application to the hypothalamus, thalamus, and limbic system, to the efferent pathways such as the dorsal periventricular–periaqueductal region and the midbrain central gray,[108] and to the rostral ventrolateral medulla and the nucleus of solitary tract;[96] all these sites are rich in cholinergic pathways[40] (see Chapter 8). Anti-ChE induced aggression exhibits typical motor activities (advance in the case of predatory aggression, and escape in that of defensive aggression), emotions, and autonomic and endocrine effects, as may be expected in view of the central sites that are involved in evoking aggression and in view of their efferent tracts to the spinal cord, preganglionic neurons, and adrenal medulla.[96,109] Endocrine phenomena involve sex hormones, which on their own facilitate male aggression. It is of interest that affective aggression has been induced in humans following accidental or industry-related exposure to anti-ChEs,[57,110] or occasionally following therapeutic treatment of AD patients with anti-ChEs.

Aggression is a multitransmitter phenomenon; yet, well-known pro-aggression paradigms, including administration of dopamine or antiserotonin drugs, facilitate certain but not other types of aggression, while anti-ChEs and muscarinic agents induce all types of aggression. Other illustrations of strong cholinergic correlates of aggression abound: 'killer rats' exhibit high levels of choline acetyltransferase (CAT) and ACh in the amygdala and/or diencephalon as compared to non-killers;[111] mice strains demonstrating high levels of spontaneous, ethological aggression show higher turnover of brain ACh as compared to the non-aggressive strains;[112] and rats bred for cholinergic supersensitivity show markedly increased aggression.[113]

Antinociception

Antinociception or analgesia involves either decreased pain perception or interference with nociceptive mechanisms prior to the arrival of the pain stimuli to the levels of perception. Early, Pellandra[114] and Slaughter and Gross[115] showed that given systemically carbamate anti-ChEs (even the quaternary neostigmine; sic!) are analgesic and/or increase the analgesic effect of morphine and codeine. These findings were amply confirmed in animals and extended to OP and carbamate anti-ChEs as well as nicotinic and muscarinic agonists.[4,37,116] It should be added that, sporadically, some of these drugs are used as analgesic drugs in man.[117]

In animals, analgesic effects were obtained by application of these drugs to many basal forebrain sites, including the thalamic, pontomedullary, and reticular sites, intraventricularly, as well to the spinal cord.[117,118] All these sites exhibit cholinergic pathways[119] (see Chapter 8); cholinergic correlates of pain and analgesia are also evidenced by the findings that the cholinergic system within these sites is activated by pain processes and by acupuncture.[119] Cholinergic analgesia involves postsynaptic m1 and m2 receptor subtypes at the supraspinal levels and nicotinic receptors at both supraspinal and spinal levels.[120,121] Thus, anti-ChEs may block nociception either at the origin of the pain stimuli (at the dorsal horn or in the thalamus) or at the level of pain perception (at the somatosensory cortex).

Several transmitter systems may mediate cholinergic analgesia, including the peptidergic systems, monoamines, and indole amines.[117,122,123] It is not clear whether or not endogenous endorphins (and morphinoids) are involved in cholinergic analgesia. While the morphinoids diminish ACh release from several brain parts and ACh turnover,[7] both morphinoid and cholinergic analgesia are antagonized by naloxone and other opioid antagonists, and enkephalin synthesis may be increased by muscarinic agonists.[116] However, there is evidence that militates against the notion of a close relation between the two mechanisms. Thus, there is controversy about the presence of cross-tolerance between the nociceptive actions of opioids and cholinergics.[116,121] Also, atropine antagonizes cholinergic but not morphinergic antinociception.[116,118] Furthermore, the structure–activity relationship of the opioid anatagonists with respect to their attenuation of opiate versus cholinergic analgesia is widely different.[116,121] Finally, each system may be able to act on its own, as morphinoids but not cholinergic agonists and anti-ChEs are analgesic in primates,[124] while the opposite is true in amphibians.[7,125] Cannabinoid–cholinergic interaction should be mentioned in this context. Cannabinoids are analgesic and exert several effects on the kinetics of CNS ACh.[126] However, cholinergic–cannabinoid interaction which obtains with regard to motor function (see above), has not been investigated with respect to analgesia.

Other sensory phenomena

Anti-ChEs, ACh, and cholinomimetics mimic the inhibition by the olivocochlear bundle of the responses of the auditory nerve terminal.[20,29,37] These effects involve cholinergically mediated release of an inhibitory transmitter[29] and modulate the transfer of auditory energy to inner hair cells.[127] Anti-ChEs and cholinergic agonists also act on the afferent neurons of the semicircular canal and the sacculae[27] and affect the responses of the primary sensory cortex to auditory stimulation.[128,129] These effects of anti-ChEs imply that the cholinergic system plays a role in audition.

Cholinergic synapses between the amacrine and ganglion cells mediate the cholinoceptive response of the retinal ganglion cells, modulate the interaction between ganglionic on- and off-units, and are involved in dark adaptation.[37] These phenomena underlie the facilitation of the dark adaptation in man by anti-ChEs[130] and the action of endogenous photochromic ChE inhibitors.[131] Anti-ChEs may also act on the cholinergic structures of the brain associative and integrative areas.[28,132]

Addiction

Alcoholic, opiate, cocaine, and nicotine addiction that may be demonstrated in animals (e.g. the animals may be induced to self-adminster these agents, a phenomenon which, in the case of nicotine, induces a c-for expression[133]) all exhibit cholinergic correlates.[7,37] That the cholinergic system is involved in the reward–punishment axis is relevant to all four addictions. Furthermore, the cholinergic correlates of all these four conditions are evidenced by the effects of drug-seeking and self-administration behavior, dependence and abstinence on ACh turnover, levels, release, and receptors,[11,134–139] and by the effects of cholinergic agonists and antagonists on dependence and withdrawal.[140] However, the findings are controversial and it is difficult to suggest a coherent picture of the state of the cholinergic system in the course of the development of dependence on these four types of agent and during withdrawal. Furthermore, only a few studies have been reported[141,142] concerning the action of anti-ChEs on these four addictions. In view of the evidence for the mediation by dopamine of nicotine reinforcement it is paradoxical that an anti-ChE, fasciculin-2, blocked the release of dopamine by nicotine in the rat brain.[143]

Other behavioral effects of anti-ChEs

As already stated, every behavior, whether in humans (perhaps with the exception of moral behavior) or in animals exhibits cholinergic correlates and is affected by anti-ChEs. Besides those described, overt fear, imprinting, animal schizoid behavior, and startle phenomena are facilitated by anti-ChEs.[33,37,45]

Anti-ChEs and the consciousness of self

As suggested by Karczmar,[98] in their totality the functional and behavioral correlates of the cholinergic system and the pertinent actions of anti-ChEs present unique features that are significant for the relationship of a vertebrate organism to its environment and survival. In this sense the cholinergic regulation of motor activity, sensory and nociceptive function, reward–punishment phenomena, certain behaviors such as aggression, fear phenomena and cognition (see Chapter 10), and relevant EEG phenomena such as the cholinergic EEG arousal (see above) should be considered. It should be emphasized that EEG arousal is induced by anti-ChEs rather than by either muscarinic or nicotinic agonists, i.e. it depends on the whole range of ACh actions. There are additional illustrations of cholinergic correlates of the organism–environment interaction and its survival significance for the organism: the cholinergic system controls and the anti-ChEs facilitate the habituation to trivial stimuli[140] and the return of the animal to the awareness of environment following the 'blinding' act of reinforcement.[88]

All these phenomena relate to a specifically cholinergic behavioral syndrome that Karczmar named the cholinergic alert non-mobile behavior (CANMB).[33,76,98] This syndrome constitutes an awaken version of REM sleep and it may be considered as representing a cholinergic 'reality check'. It is consistent with this notion that anti-ChEs block schizoid behavior in animal models of schizophrenia;[33] actually, muscarinic agonists and anti-ChEs were used to cause temporary 'awakening' in hebephrenic schizophrenics.[37]

Is it possible to go beyond the definition of the role of the cholinergic system as a 'reality check' and the basis for successful animal–environment interaction? In the course of the recent attempts at resolving the perennial problem of consciousness and the awareness of self (the cartesian problem of mind–body dichotomy), Penrose[145] and others[146] have described self-awareness as the quantal process of 'objective reduction' that may occur at the subcellular levels of the neuronal microtubules, while Woolff[147] has presented evidence for the presence of a cholinergic link at the microtubules of the basal forebrain and speculated about a cholinergic basis of consciousness.

Toxic effects of anti-ChEs

The early, detailed recognition of anti-ChE toxicity was due to the work of German investigators[148] (see Chapter 1), and led ultimately to the development of OP war gases. At the time, the German, as well as the Russian, investigators also recognized the pesticidal potential of both carbamate and OP anti-ChEs, and their use as such became another source of human danger.[33]

Strategic sites of anti-ChE toxicity and lethal actions in man and animals

Anti-ChEs exert toxic and fatal actions due either to their inhibition of ChEs, particularly of AChE and accumulation of ACh, or to other mechanisms. As to the effects ascribable to AChE inhibition, central toxic effects of OP and carbamate anti-ChEs in humans and animals represent an exaggerated form of the actions described above. This statement refers to tertiary anti-ChEs and non-quaternary OP drugs such as DFP, war gases (sarin, soman, and tabun), and insecticides including malathion and parathion that readily penetrate the CNS. Actually, OP drugs may damage the blood–brain barrier, thus facilitating their own CNS entry and that of other compounds;[7] conditions such as stress may additionally facilitate CNS entry (A Friedman, H Soreq, personal communication). Accordingly, in animals toxic doses of anti-ChEs induce central motor effects such as early immobility followed by tremors and jerks, circling motions and ataxia, respiratory stimulation and then respiratory depression, and seizures.[1,4,46]

Anti-ChEs also induce, via AChE inhibition, peripheral effects, hitherto not discussed in this chapter. Toxic peripheral actions, whether in humans or animals, include initial fasciculations followed by ataxia and muscle weakness, and a whole gamut of autonomic effects, such as bradycardia, miosis, rhinorrhea, hypo- or hypertension, the SLUD syndrome (salivation, lachrymation, urination, and defecation), sweating, bronchiolar spasm, and emesis. In humans additional central effects may be distinguished: anxiety, dizziness, tremulousness, confusion, paresthesia, impaired accommodation, anorexia and nausea, headache, sensation of chest constriction, cough, and dyspnea. If the subject survives, with longer acting drugs such as OP agents there are also sleep and mood disturbances, nightmares, hallucinations, and psychotic episodes or personality changes. These latter effects were reported in human volunteers and subjects exposed to OP drugs, whether accidentally or in a war[46,149–151] (see below). The toxic actions of anti-ChEs due to the accumulation of ACh require considerable inhibition of AChE; the relationship between the inhibition that occurs in blood (a measurement frequently carried out in the course of the examination of anti-ChE poisoning) and that occurring in the nervous tissue (particularly in the CNS) depends on the tissue distribution of the agent in question; some 50% inhibition of the blood AChE is pathognomic and it may correspond to either higher or lower levels of AChE inhibition in the CNS.[46,152] It should be stressed that carbamate and OP anti-ChEs differ significantly in their ChE inhibitory potency, the target ChE (i.e. their preference for AChE or butyrylcholinesterase (BuChE)) and their toxicity.[1,4]

Neuropathies may follow initial subtoxic or toxic actions of anti-ChEs. A neuromyal and CNS neuropathy and damage may be related to cholinergically induced hyperactivity, Ca^{2+} mobilization, and formation of free radicals.[153–155] Direct actions of the OP drugs that induce receptor or channel effects independently of the inhibition of AChE may also be involved in these neuropathies[69,152] (see above). Another type of neuropathy induced by some OP drugs is the organophosphorus ester induced delayed neurotoxicity (OPIDN). This neuropathology, which occurs both in humans and animals, involves the brain and spinal ascending and descending sensory and motor tracts, the anterior horns being the particular target. The resultant paralysis is termed 'ginger-Jake' paralysis, as the original human OPIDN, at the end of the 19th century, was due to the consumption of tetra-*o*-cresyl phosphate (TOCP) adulterated extract of ginger.[156] Since the late 19th century, perhaps some 50 000 cases of OPIDN have been reported in humans. While certain OP compounds (classically, TOCP) which cause OPIDN are not potent anti-ChEs, many potent OP anti-ChEs, whether insecticides or potential war gases, may induce OPIDN. OPIDN may be caused either by OP inhibition of a 'neurotoxic esterase',[157] or via interference by the OP compounds with the Ca^{2+} calmodulin kinase II and resulting hyperrelease of Ca^{2+}.[157] Then, there is the delayed cognitive toxicity (DCT). Duffy and Burchfiel[158] reported that industrial workers who were exposed to the OP agents during the course of their manufacture exhibited EEG, sleep, memory, and personality changes sometimes two years after the exposure. DCT, which obviously is not correlatable with inhibition of ChEs, may be duplicated in primates and possibly in humans[152,158,159] (see below). Other unconventional actions of anti-ChEs are described in Chapter 13.

While chronic toxic effects of anti-ChE may be due to OPIDN, DCT, and related effects, acute life-threatening and fatal consequences of exposure to anti-ChE agents are due to AChE inhibition and ACh accumulation. The strategic site of lethality is species dependent. While rodents and dogs die following their exposure to anti-ChEs by 'asthmatic' death due to peripheral effects upon the smooth muscle of the bronchi and the skeletal respiratory musculature, cats and humans die because of the block of the medullary respiratory centers.[46,52,77] This block needs not follow a seizure. It is caused, early, by prolonged ACh-induced depolarization, which is then converted into desensitization or receptor inactivation[7] (see above). A desensitizing receptor or direct blocking channel effect of OP agents may also contribute to these phenomena (see above).

Comparison of toxicities due to carbamates and OP agents

Only tertiary carbamate anti-ChEs and non-quaternary OP agents are capable of the whole range of toxic effects, including those exerted on the CNS; OP drugs, particularly OP war gases are 'highly toxic and extremely dangerous'.[160] These two types of anti-ChE differ in several respects. First, OP war gases such as tabun, sarin, cyclosarin, soman, and VX, are dispensed as gases and become toxic due to absorption via the skin and lung. This is not the case for the carbamates and related agents. Therefore, these two types of agent must differ in their kinetics. OP drugs act rapidly and progressively and, being irreversible inhibitors (see Chapter 7), their duration of action is very long. Carbamate effects are slower to develop than those of the OP drugs, yet they are relatively short-acting drugs; also, they are reversible ChE ligands and their actions wane within a few hours[161] (see Chapters 7 and 12). The return of function after OP intoxication (when the subject does not die) depends partially on 'natural' reactivation and mainly on regeneration of the enzyme, which may take days (unbound OP drugs are rapidly hydrolyzed, thus, while their effect is long lasting, their free presence is limited). Furthermore, the OP-induced phosphorylation of ChEs is subject to 'aging', i.e. to a molecular rearrangement or dealkylation of the inhibitor which renders the phosphorylated enzyme refractory to reactivators[161] (see Chapter 7). Finally, OP drugs rather than carbamates cause OPIND and DCT, which are only indirectly dependent on AChE inhibition.

Treatment and antidoting of anti-ChE toxicity

Since the 1850s atropine has been employed as the symptomatic antagonist of physostigmine toxicity.[1] With the advent of research concerning OP war gases and insecticides, additional means of prevention and treatment of anti-ChE toxicity were explored. Besides newer atropinics, nicotinolytics were employed to antagonize respiratory and other CNS actions of anti-ChEs, curarimimetrics (to restore neuromyal function), benactyzine and other compounds (to reverse spinal and reflexogenic actions), and anticonvulsants (to deal with both seizures and respiratory collapse.[4,46] These agents were not very effective. Atropine can antagonize OP-induced bronchiolar constriction, gastrointestinal symptoms, blood pressure changes, and, in particular, CNS effects, including seizures and respiratory depression. Even so, used alone, atropinics are not capable of increasing the LD_{50} of OP agents by more than two-fold.

The discovery by Hobbiger,[162] Rosenberg, and Wilson[163] (Chapter 7) of the reactivating action on phosphorylated ChEs of hydroxylamines, hydroxamic acid, and oximes, and Koster's[164] demonstration of the protective effect of physostigmine against phosphorylation of ChEs by OP drugs, introduced a new era of antidoting to anti-ChE toxicity. A number of quaternary and tertiary carbamates and of quaternary and bis-quaternary nucleophilic compounds were tested as protectors of AChE and reactivators, respectively. The mono-quaternary oximes, particularly 2-PAM (pralidoxime) are considered on practical grounds as most useful reactivators, while the quaternary carbamate, pyridostigmine, is the mainstay of protective therapy. Of course, their quaternary nature means that they are not effective at the CNS level (although there is some indication that quaternary oximes may cross the blood–brain barrier), yet their combined prophylactic use may protect against ten LD_{50} doses of OP drugs. In the field (e.g. in the case of US troops engaged in 1991 in the Persian Gulf War) pyridostigmine is employed prior to perceived exposure of the personnel, while oximes and atropine are given in the treatment of poisoning after it has occurred. While oximes are relatively devoid of side actions, atropinics and pyridostigmine exhibit, at the doses used, significant side-effects. Several other means of prevention or treatment of anti-ChE toxicity have been recommended, such as the use of 'sponge' compounds to ligate or 'sink' OP drugs, enzymes capable of hydrolyzing the OP agents, and immunizing agents.[33,148,151,165,166]

Toxic actions of insecticides and war gases: Persian Gulf syndrome

Anti-ChE insecticides and pesticides

In the 1950s the development and manufacture of these compounds became a major industry in the USA, Germany, Switzerland, Japan, Australia, Russia, Canada, and elsewhere. Incredibly, as early as the late 1950s over 50 000 such compounds had already been synthesized and tested.[167] In the USA alone, some 100 000 lb of OP insecticides are produced every year. Their use is even more extensive in underdeveloped countries wishing to increase agricultural yields, including Russia.[33] The drugs in question may be used as insecticides, fungicides, and herbicides, as well as pesticides and rodenticides.[152,167] Most of the insecticide carbamate and OP drugs exhibit potent anti-AChE effects; the effectiveness of these compounds is exerted via the inhibition of the insect or pest (e.g. worms) nervous system AChE.[168]

Soon after their introduction into agriculture, reports of their accidental toxicity in animals such as sheep, and occasionally of epidemiological toxicity in man, were forthcoming from all areas of the world. At present some 100 cases of accidental toxicity are caused every year in the USA by the agents in question (K Hamernik, B Chen, personal communication). The toxic effects or death appear to be due to inhibition of AChE, although there may be no strict relationship between the ChE inhibition and the human toxicity of these compounds;[152] tolerance to low levels of AChE may develop, and this phenomenon may obscure the relationship between AChE level and toxicity. OPIDN has occurred in several cases of OP poisoning.[156]

At this time, more than 40 anti-ChE OP insecticides are licensed for agricultural use in the USA.[152,169] To be licensed the compounds must pass a battery of animal tests defined in 1991 by the US Environmental Protection Agency (EPA). Furthermore, the Food Quality Protection Act (FQPA) of 1996 requires the US EPA 'to perform a combined risk assessment for chemicals that produce adverse effects by a common mechanism of toxicity', as such compounds may cause additive or potentiative toxicity.[152] This mandate refers to the use in agriculture of a combination of insecticides or pesticides, sometimes including OP and carbamate agents.[152] The assessment of the potential toxicity of such combinations is not easy.[152] Thus, carbamates and OP drugs may antagonize each other in humans (see Chapter 7); on the other hand, additive effects or even potentiation of toxicity may occur in humans from joint exposure to several carbamates or several OP agents. This arises when one OP drug interferes with the detoxification, via esterases, of another OP drug, as in the case of the potentiation of malathion toxicity by joint application of ethylnitrophenylphenylphosphothionate (EPN).[167] However, this potentiation also arises independently of the detoxification phenomena.[170] This important problem is currently being addressed.[171]

War gases and the Persian Gulf War syndrome

While the USA, Germany, Russia, and the UK were capable in the 1940s of using OP compounds, they were not employed in the Second World War. Yet these four countries continued their development of new OP war gases and their antidotes.[148] There was a volunteer program in the USA concerning field use of carbamate protectors of AChE from phosphorylation, large doses of atropine, and additional agents; some behavioral toxicity (DCT) that persisted for a long time occurred occasionally. Also, manufacturing and pesticidal testing of OP drugs led to accidents involving humans and animals. Then, there is the terrorist use of OP drugs. It is relatively easy to manufacture war gases, and there may have been several instances of their use by terrorists. Best documented are the two incidents of the use of sarin by a religious fanatic group in Japan.[172,173] The more tragic of these two episodes, which occurred in 1995 in a Tokyo subway, involved aproximately 5500 people. There was a number of dead; the 640 people admitted to hospital exhibited low blood ChE (AChE?) values and peripheral symptomatology that was typical of anti-ChE poisoning. Eighteen cases (9 men, 9 women) were examined 6–8 months following the exposure. There were some EEG changes, psychomotor peformance was decreased, and there was evidence of post-traumatic stress disorder (PTSD), fatigue, and general health deterioration.[173]

Finally, there is the use of these agents in combat. There is anecdotal information concerning their use by Iraq against the Kurds and in Iran, and in the course of the recent conflict in Yugoslavia[33,174] (EJ Hogendoorn, personal communication). In addition, there is the Operation Desert Shield, or the Persian Gulf War, that lasted from August 1990 to late February 1991, and the Persian Gulf War syndrome or 'mystery illness'. The total number of US military personnel involved in this operation amounted to almost 700 000.[175,176] Soon thereafter a significant number of returning veterans complained of a variety of symptoms, the most common being joint pain, fatigue, memory loss, and sleep disturbances.[175–177] Accordingly, the US Department of Defense and the Department of Veterans' Affairs established in 1993 a registry program for the Operation Desert Shield veterans in the VA Hospitals of the USA. The program involves screening tests of various depths; some 80 000 Operation Desert Shield veterans were registered. In addition, an intensive research program, both in humans and animals, was initiated. This program includes the study of: the possible role of pyridostigmine, both alone or combined with stress or other agents; neurotoxic phenomena; interaction between OP drugs and non-cholinergic transmitter systems; stress and PTSD; pulmonary abnormalities; viruses, bacteria, and parasites; herbicides and other toxicants; the possible role of fumes resulting from burning oil wells; and immunological and endocrine factors.[33,175,176] The results of these studies and of the analysis of the screening program do not lead to any clear conclusions as to the Persian Gulf War syndrome. A number of clinical studies support the hypothesis that Persian Gulf War syndrome is real and delayed in nature, and that it represents a discrete spectrum of neurological injury involving the central, peripheral, and autonomic nervous systems.[33,178] However, contrary opinions also exist (see, for example, the report of a NIH Technology Assessment Workshop Panel[179]). Also, some doubts were raised as to the reliability and openness of the information provided by the US Department of Defense with respect to certain aspects of the matter.[149,180] Altogether, in the USA the Persian Gulf War syndrome assumes a multifactorial nature: a plethora of committees deal with the syndrome, media coverage is intense, and there are complex political pressures. The net outcome is that, in 1998, a new survey of Persian Gulf veterans and their families was initiated (some 8 years after Operation Desert Shield) in the USA, but it will take some time, if ever, for the problem to be resolved.

Conclusions

The study of anti-ChEs is vital to the understanding of synaptic function, the activities of neuronal systems, behavior, and consciousness. In addition, the applied uses of anti-ChEs, whether in disease or as pesticides, are of great significance. However, to be balanced against these factors are the toxic aspects of their action. Thus anti-ChEs constitute not only a cornucopia of plenty, but also a Pandora's box of calamities.

Acknowledgments

Some of the research from this laboratory referred to in this paper was supported by NIH grants NS6455, NS15858, RR05368, NS16348, and GM77; VA grant 4380; grants from Potts, Fidia, and ME Ballweber Foundations; and Guggenheim (1969) and Senior Fullbright (1985) Fellowships. In addition, help afforded by CARES, Chicago, and AMVETS of Illinois is gratefully acknowledged.

References

1. Holmstedt B. Pharmacology of organophosphorus anticholinesterase agents. *Pharmacol Rev* 1959; **11:** 657–688.
2. Eccles JC. *The Physiology of Synapses.* New York: Springer-Verlag, 1964.
3. Krnjevic K. Central cholinergic mechanisms and function. In: Cuello AC, ed. *Cholinergic Neurotransmission: Function and Dysfunction.* Amsterdam: Elsevier, 1993: 285–293.
4. Karczmar AG. Pharmacologic, toxicologic and therapeutic properties of anticholinesterase agents. *Physiolog Pharmacol* 1967, **3:** 163–322.
5. Caulfield MP. Muscarinic receptors — characterization, coupling and function. *Pharmacol Therapeut* 1993; **59:** 319–379.
6. Changeux JP, Revah F. The acetylcholine receptor: allosteric sites and the ion channel. *TINS* 1987; **10:** 245–250.
7. Karczmar AG. Brief presentation of the story and present status of studies of the vertebrate cholinergic system. *Neuropsychopharmacology* 1993; **9:** 181–199.
8. Starke K, Goethert M, Kilbinger H. Modulation of neurotransmitter release by presynaptic autoreceptors. *Physiol Rev* 1989; **69:** 864–989.
9. Hoss W, Ellis J. Muscarinic receptor subtypes in the central nervous system. *Int Rev Neurobiol* 1985; **26:** 152–199.
10. Szerb JC. Characterization of presynaptic muscarinic receptors in central cholinergic neurons. In: Jenden DJ, ed. *Cholinergic Mechanisms and Psychopharmacology.* New York: Plenum, 1977: 49–60.
11. Pepeu G. The release of acetylcholine from the brain: an approach to the study of the central cholinergic mechanisms. In: Kerkut GA, Phillis JW, eds. *Progress in Neurobiology.* Oxford: Pergamon, 1974: 257–288.

12. Pepeu G. Overview and future of CNS cholinergic mechanisms. In: Cuello AC, ed. *Cholinergic Neurotransmission: Function and Dysfunction.* Amsterdam: Elsevier, 1993: 455–458.
13. Marchi M, Raiteri M. Nicotinic autoreceptors mediating enhancement of acetylcholine release become operative in conditions of 'impaired' cholinergic presynaptic function. *J Neurochem* 1996; **67:** 1974–1981.
14. Karczmar AG. Conference on dynamics of cholinergic function: overview and comments. In: Hanin I, ed. *Dynamics of Cholinergic Function.* New York: Plenum, 1986: 1215–1259.
15. Garrone B, Luparini MR, Tolu L, Magnani M, Landolfi C, Milanese C. Effect of subchronic treatment with the acetylcholinesterase inhibitor heptastigmine on central cholinergic transmission and memory impairment in aged rats. *Neurosci Lett* 1998; **254:** 53–57.
16. Liu J, Pope CN. Comparative presynaptic neurochemical changes in rat striatum following exposure to chlorpyrifos or parathion. *J Toxicol Environ Hlth* 1998; **53:** 531–544.
17. Hamilton SE, Schlador ML, McKinnon LA, Chmelar RS, Hamilton NM. Molecular mechanisms for the regulation of the expression and function of muscarinic acetylcholine receptors. In: Massoulie J, ed. *From Torpedo Electric Organ to Human Brain: Fundamental and Applied Aspects.* Paris: Elsevier, 1998: 275–278.
18. Fuder H, Muscholl E. Heteroreceptor-mediated modulation of noradrenaline and acetylcholine release from peripheral nerves. *Rev Physiol Biochem Pharmacol* 1995; **126:** 265–412.
19. Wonnacott S. Presynaptic nicotinic receptors. *TINS* 1997; **20:** 92–98.
20. Koelle GB. Cytological distributions and physiological functions of chlinesterases. In: Koelle GB, ed. *Handbuch der exper. Pharmakol., Erganzungswk.,* Vol 15, *Cholinesterases and Anticholinesterase Agents.* Berlin: Springer-Verlag, 1963: 189–298.
21. Lapchak PA, Araujo DM, Quirion R, Collier B. Effect of chronic nicotine treatment on nicotinic autoreceptor function and N-[^{3}H]methylcarbamylcholine binding sites in the rat brain. *J Neurochem* 1989; **52:** 483–491.
22. Lukas RJ, Bencherif M. Heterogeneity and regulation of nicotinic acetylcholine receptors. *Int Rev Neurobiol* 1992; **34:** 26–131.
23. Albuquerque EX, Pereira EFR, Braga MFM, Alkondon M. Contribution of nicotinic receptors to the function of synapses in the central nervous system: the action of choline as a selective agonist of α_7 receptors. In: Massoulié J, ed. *From Torpedo Electric Organ to Human Brain, Fundamental and Applied Aspects.* Amsterdam: Elsevier, 1998: 309–316.
24. Gioanni,Y Rougeot C, Clarke PB, Lepouse C, Thierry AM, Vidal C. Nicotinic receptors in the rat prefrontal cortex: increase in glautamate release and facilitation of medio-dorsal thalamo-cortical transmission. *Eur J Neurosci* 1999; **11:** 18–30.
25. Allal C, Lazartigues E, Tran M *et al.* Central cardiovascular effects of tacrine in the conscious dog: a role for catecholamine and vasopressin release. *Eur J Pharmacol* 1998; **348:** 191–198.
26. Eccles JC, Katz B, Koketsu K. Cholinergic and inhibitory synapses in a pathway from motor-axon collaterals to motoneurones. *J Physiol (London)* 1954; **126:** 524–562.
27. Koketsu K. Restorative action of fluoride on synaptic transmission blocked by organophosphorous anticholinesterases. *Int J Neuropharmacol* 1966; **5:** 247–254.
28. Celesia GG. Visual evoked potentials. In: Barber C, Taylor M, eds. *Evoked Potentials Review.* London: IEPS, 1991: 45–46.
29. Guth PS, Perrin P, Norris CH, Valli P. The vestibular hair cells: post-transductional signal processing. *Prog Neurobiol* 1998; **54:** 193–247.
30. Greengard P. Neuronal phosphoproteins. *Neurobiology* 1987; **1:** 81–119.
31. Katz EJ, Cortes VI, Eldefrawi ME, Eldefrawi AT. Chlorpyrifos, parathion and their oxons bind to and desensitize a nicotinic receptor: relevance to their toxicities. *Toxicol Appl Pharmacol* 1997; **146:** 227–236.
32. Eldefrawi AT, Eldefrawi ME. Neurotransmitter receptors. In: Reuhl KR, Lawndes HE, eds. *Comprehensive Toxicology.* Amsterdam: Elsevier, 1997: 139–153.

33. Karczmar AG. Invited review. Anticholinesterases: dramatic aspects of their use and misuse. *Neurochem Intl* 1998; **32:** 401–411.
34. Kissel JW, Domino EF. The effects of some possible neurohumoral agents on spinal cord reflexes. *J Pharmacol Exp Ther* 1959; **125:** 168–177.
35. Yoshioka K, Sakuma M, Otsuka M. Cutaneous nerve-evoked cholinergic inhibition of monosynaptic reflex in the neonatal rat spinal cord: involvement on m2 receptors and tachikininergic primary afferents. *Neuroscience* 1990; **38:** 195–203.
36. Deshpande SB, DasGupta S. Diisopropylfluorophosphate-induced depression of segmental monosynaptic transmission in neonatal spinal cord is also mediated by increased axonal activity. *Toxicol Lett* 1997; **90:** 177–182.
37. Karczmar AG. Central actions of acetylcholine, cholinomimetics, and related compounds. In: Goldberg AM, Hanin I, eds. *Biology of Cholinergic Function.* New York: Raven, 1976: 395–449.
38. Koketsu K, Karczmar AG, Kitamura K. Acetylcholine depolarization of the dorsal root nerve terminals in amphibian spinal cord. *Intl J Neuropharmacol* 1969; **8:** 329–336.
39. Pradhan SN, Dutta SN. Central cholinergic mechanisms and behavior. *Intl Rev Neurobiol* 1971; **14:** 173–231.
40. Brudzynski SM, Kadishevitz L, Fu XW. Mesolimbic component of the ascending cholinergic pathways: electrophysiological–pharmacological study. *J Neurophysiol* 1998; **79:** 1675–1686.
41. Reid MS, Tafti M, Gewary JN *et al.* Cholinergic mechanisms in canine narcolepsy. I. Modulation of catalepsy via local drug administration into the pontine reticular formation. *Neuroscience* 1994; **59:** 511–512.
42. Elazar Z, Peleg N, Paz M, Ring G. The striatal dopaminergic catalepsy mechanism is not necessary for expression of pontine catalepsy produced by carbachol injections into the pontine reticular formation. *Naunyn Schmiedebergs Arch Pharmacol* 1995; **352:** 187–193.
43. Hingtgen JN, Aprison MH. Behavioral and environmental aspects of the cholinergic system. In: Goldberg AM, Hanin I, eds. *Biology of Cholinergic Function.* New York: Raven, 1976: 515–566.
44. Davis M, Nonneman A, Glickman SE. Chemical stimulation of the midbrain reticular system. *Psychol Rep* 1967; **20:** 507–509.
45. Karczmar AG, Scudder CL, Richardson DL. Interdisciplinary approach to the study of behavior in related mice types. *Neurosci Res* 1973; **5:** 159–244.
46. Wills JH. Toxicity of anticholinesterases and treatment of poisoning. In: Karczmar AG, ed. *International Encyclopedia of Pharmacology and Therapy*, Vol 1, *Anticholinesterase Agents.* Oxford: Pergamon, 1970: 356–469.
47. Widdicombe JG. Lung reflexes. *Bull Physiopathol Respir* 1972; **8:** 723–725.
48. Feldman JL. Neurophysiology of breathing in mammals. In: Montcastle VB, Bloom FE, Geiger R, eds. *Handbook of Physiology*, Vol 4, *Intrinsic Regulatory Systems of the Brain.* Bethesda: Physiological Society, 1986: 463–524.
49. Nersesian ON, Baklavadzhian OG. Mikroionoforeticheskoe issledovanie vliianiia kholinerichskikh veshchestv na aktivnost' medullinykh dykhtatel'nykh neironov. *Fiziol Zh SSSR* 1989; **75:** 948–954 [in Russian].
50. Kok A. REM sleep pathways and anticholinesterase intoxication: a mechanism for nerve-agent-induced respiratory failure. *Med Hypotheses* 1993; **41:** 141–149.
51. Haselton JR, Padrid PA, Kaufman MP. Activation of neurons in the rostral ventrolateral medulla increases bronchomotor tone in dogs. *J Appl Physiol* 1991; **71:** 210–216.
52. Rickett DL, Beers E. Differentiation of medullary and neuromuscular effects of nerve agents. In: Dun NJ, Perlman RL, eds. *Neurobiology of Acetylcholine. A Symposium in Honor of AG Karczmar.* New York: Plenum, 1986: 535–550.
53. Borison HL, Wang SC. Physiology and pharmacology of vomiting. *Pharmacol Rev* 1953; **5:** 193–215.
54. Pedigo NW, Brizzee KR. Muscarinic cholinergic receptors in area postrema and brainstem regulating emesis. *Brain Res Bull* 1985; **14:** 169–177.

55. Beleslin DB, Krstic SK. Further studies on nicotine-induced emesis: nicotinic mediation in area postrema. *Physiol Behav* 1987; **39:** 681–686.
56. Almog S, Winkler E, Amitai Y, Dani S, Shefi M, Tirosh M, Shemer J. Acute pyridostigmine overdose: a report of nine cases. *Isr J Med Sci* 1991; **27:** 259–263.
57. Karczmar AG. Drugs, transmitters and hormones, and mating behavior. In: Ban TH, Freyhan, eds. *Drug Treatment of Sexual Dysfunction.* Basel: Karger, 1980: 1–76.
58. Soulairac ML, Soulairac A. Monoaminergic and cholinergic control of sexual behavior in male rat. In: Sandler M, Gessa GL, eds. *Sexual Behaviour.* New York: Raven, 1975: 99–116.
59. Kow LM, Tsai YF, Weiland NG, McEwen BS, Pfaff DW. In vitro electropharmacological and autoradiographic analyses of muscarinic receptor subtypes in rat hypothalamic ventromedial nucleus: implications for cholinergic regulation of lordosis. *Brain Res* 1995; **694:** 29–39.
60. Retana-Marquez S, Salazar ED, Velazquez-Moctezuma J. Muscarinic and nicotinic influences on masculine sexual behavior in rats: effects of oxotremorine, scopolamine, and nicotine. *Pharmacol Biochem Behav* 1993; **44:** 913–917.
61. Dorner G. Sexual differentiation and behavior. *CIBA Foundation Symposium on Sexual Behavior.* Amsterdam: Excerpta Medica, 1970: 81–101.
62. Hornykiewicz O, Kobinger W. Ueber den Einfluss von Eserin, Tetraethylpyrophosphat (TEPP), und Neostigmin auf den Blutdruck und die pressorischen Carotissinusreflexe der Ratte. *Arch Exp Pathol Pharmakol* 1956; **228:** 493–503.
63. Lazartigues E, Freslon JL, Tellioglu T *et al.* Pressor and bradycardic effects of tacrine and other acetylcholinesterase inhibitors in the rat. *Eur J Pharmacol* 1998; **361:** 61–71.
64. Varagic V, Krstic M. Adrenergic activation by anticholinesterases. *Pharmacol Rev* 1966; **18:** 796–800.
65. Murugaian J, Sundaram K, Sapru H. Cholinergic mechanisms in the ventrolateral medullary depressor area. *Brain Res* 1989; **502:** 287–295.
66. Brezenoff HE, Xiao YF. Acetylcholine in the posterior hypothalamus is involved in the elevated blood pressure in the spontaneously hypertensive rat. *Life Sci* 1989; **45:** 1163–1170.
67. Sundaram K, Murugaian J, Krieger A, Sapru H. Microinjection of cholinergic agonists into the intermediolateral cell column of the spinal cord at T1–T3 increase heart rate and contractility. *Brain Res* 1989; **503:** 22–31.
68. Lacombe P, Sercombe R, Verrechia C, Philipson V, MacKenzie ET, Seylaz J. Cortical blood flow increases, induced by stimulation of the substantia innominata in the unanesthetized rat. *Brain Res* 1989; **491:** 1–14.
69. Kubo T, Ishizuka T, Fukumori R, Asari T, Hagiwara V. Enhanced release of acetylcholine in the rostral ventrolateral medulla of spontaneously hypertensive rats. *Brain Res* 1995; **686:** 1–9.
70. Van Meter WG, Karczmar AG, Fiscus RF. CNS effects of anticholinesterases in the presence of inhibited cholinesterases. *Arch Int Pharmacodynam* 1978; **231:** 249–360.
71. Myers RD. *Drug and Chemical Stimulation of the Brain.* New York: Van Nostrand Reinhold, 1974.
72. Lomax P, Jenden DJ. Hypothermia following systemic and intracerebral injection of oxotremorine in the rat. *Intl J Neuropharmacol* 1966; **5:** 353–359.
73. Yasumatsu M, Yazawa T, Otokawa M, Kuwasawa K, Hasegawa H, Aihara Y. Monoamines, aminoacids and acetylcholine in the preoptic area and posterior hypothalamus of rats: measurements of tissue extracts and in vivo dialysates. *Comp Biochem Physiol* 1998; **121:** 13–23.
74. Hall GH, Myers RD. Temperature changes produced by nicotine injected into the hypothalamus of conscious monkey. *Brain Res* 1977; **37:** 241–251.
75. Brito NA, Brito MN, Kettelhut IC, Migliorini RN. Intra-medial hypothalamic injection of cholinergic agents induces rapid hyperglycemia, hyperlactatemia and gluconeogenesis activation in fed, conscious rats. *Brain Res* 1993; **626:** 339–342.
76. Karczmar AG. Brain acetylcholine and animal electrophysiology. In: Davis KL, Berger

PA, eds. *Brain Acetylcholine and Neuropsychiatric Disease.* New York: Plenum, 1979: 265–310.
77. Glenn JF, Hinman DJ, McMaster SB. Electroencephalographic correlates of nerve agent poisoning. In: Dun NJ, Perlman RL, eds. *Neurobiology of Acetylcholine. A Symposium in Honor of AG Karczmar.* New York: Plenum, 1987: 503–534.
78. Cavalheiro EA, Fernandez MJ, Turski L, Naffah-Mazzacoratti MG. Spontaneous recurrent seizures in rats: amino acid and monoamine determination in the hippocampus. *Epilepsia* 1994; **35:** 1–11.
79. Lebeda FJ, Rutecki PA. Organophosphorus anticholinesterase-induced epileptiform activity in the hippocampus. In: Dun NJ, Perlman RL, eds. *Neurobiology of Acetylcholine. A Symposium in Honor of AG Karczmar.* New York: Plenum, 1987: 437–450.
80. Karczmar AG. Brain acetylcholine and seizures. In: Fink M, Kety S, McGaugh J, Williams TA, eds. *Psychobiology of Convulsive Therapy.* New York: Wiley, 1974: 251–270.
81. Cruikshank JW, Brudzynski SM, McLachlan RS. Involvementof m1 muscarinic receptors in the initiation of cholinergically induced epileptic seizures in the rat brain. *Brain Res* 1994; **643:** 125–129.
82. Maslanski JA, Powelt R, Deirmengiant C, Patelt J. Assessment of the muscarinic receptor subtypes involved in pilocarpine-induced seizures in mice. *Neurosci Lett* 1994; **168:** 225–228.
83. Ryan GP, Hackman JC, Davidoff RA. Spinal seizures and excitatory acid-mediated synaptic transmisison. *Neurosci Lett* 1984; **44:** 161–168.
84. Starr MS, Starr BS. The new competitive NMDA receptor antagonist CGP 40116 inhibits pilocarpine-induced limbic motor seizures and unconditioned motor behavior in the mouse. *Pharmacol Biochem Behav* 1994; **47:** 137–131.
85. Alhainen KJ, Riekkinen PJ Jr, Soininnen HS, Reinikainen KJ, Riekkinen PJ Sr. Responders and nonresponders in experimental therapy of Alzheimer's disease. In: Becker R, Giacobini E, eds. *Cholinergic Basis for Alzheimer Therapy.* Boston: Birkhauser, 1991: 231–237.
86. Nunez PL. Mind, brain and encephalography. In: Nunez PL, ed. *Neocortical Dynamics and Human EEG Rhythms.* Oxford: Oxford University Press, 1995: 133–194.
87. Semba K. The cholinergic basal forebrain: a critical role in cortical arousal. In: Napier TC, Kalivas PW, Hanin I, eds. *The Basal Forebrain.* New York: Plenum, 1991: 197–218.
88. Marczynski TJ. Neurochemical mechanisms in the genesis of slow potentials: a review and some clinical implications. In: Otto D, ed. *Multidisciplinary Perspectives in Events Related to Brain Potential Research.* Washington, DC: US Environmental Protection Agency, 1978: 23–25.
89. Radek RJ. Effects of nicotine on cortical high voltage spindles in rats. *Brain Res* 1993; **625:** 23–28.
90. Steriade M. Modulation of information processing in thalamicocortical systems: Chairman's introductory remarks. In: Cuello AC, ed. *Cholinergic Function and Dysfunction.* Amsterdam: Elsevier, 1993: 341–342.
91. Kawamura H, Domino EF. Differential actions of m and n cholinergic agonists on the brain stem activating system. *Intl J Neuropharmacol* 1969; **8:** 105–115.
92. Longo VG, Loizzo A. Effects of drugs on hippocampal theta-rhythm. Possible relationship to learning and memory processes. *Proc Intl Congr Pharmacol* 1973; **4:** 46–54.
93. Longo VG. Mechanism of the behavioral and electroencephalographic effects of atropine and related compounds. *Pharmacol Rev* 1966; **18:** 965–996.
94. Fairchild MD, Jenden DJ, Mickey MR. An application of long-term frequency analysis in measuring drug-specific alterations in the EEG of the cat. *Electroencephalogr Clin Neurophysiol* 1975; **38:** 337–348.
95. Jasper HH, Tessier J. Acetylcholine liberation from the cerebral cortex during paradoxical (REM) sleep. *Science* 1971; **172:** 601–603.
96. Benarroch EE. *Central Autonomic Network: Functional Organization and Clinical Correlations.* Armonk, NY: Futura, 1997.
97. Leonard CS, Llinas R. Serotonergic and cholinergic inhibition of mesopontine cholin-

ergic neurons controlling REM sleep: an in vitro electrophysiological study. *Neuroscience* 1994; **59:** 309–330.

98. Karczmar AG. Cholinergic substrates of cognition and organism–environment interaction. *Prog Neuro-Psychopharmacol Biol Psychol* 1995; **19:** 187–211.
99. Muhlethaler M, Serafin M. Thalamic spindles in an isolated and perfused preparation in vivo. *Brain Res* 1990; **524:** 17–21.
100. Gillin JC, Salin-Pascual R, Velazquez-Moctezuma J, Shiromani P, Zoltoski R. Cholinergic receptor subtypes and REM sleep in animals and normal controls. In: Cuello AC, ed. *Cholinergic Function and Dysfunction.* Amsterdam: Elsevier, 1993: 379–387.
101. Hobson JA, Datta S, Calvo JM, Quatrocchi J. Acetylcholine as a brain state modulator: triggering and long-term regulation of REM sleep. In: Cuello AC, ed. *Cholinergic Function and Dysfunction.* Amsterdam: Elsevier, 1993: 389–404.
102. Angeli P, Bianchi S, Imeri L, Mancia M. The influence of the muscarinic receptor subtypes on the sleep–wake cycle. *Farmaco* 1993; **48:** 1197–1206.
103. Datta S, Calvo JM, Quatrocchi JJ, Hobson JA. Long-term enhancement of REM sleep following cholinergic stimulation. *Neuroreport* 1991; **2:** 619–622.
104. Karczmar AG, Longo VG, Scotti de Carolis A. Pharmacological model of paradoxical sleep: the role of cholinergic and monoamine system. *Physiol Behav* 1970; **5:** 175–182.
105. Moyer KE, ed. *Physiology of Aggression.* New York: Raven, 1976.
106. Valzelli L. Human and animal studies on the neurophysiology of aggression. *Prog Neuro-Psychopharmacol* 1978; **2:** 591–610.
107. Allikmets LH. Cholinergic mechanisms in aggressive behavior. *Med Biol* 1974; **52:** 19–30.
108. Bandler RJ, Flynn JP. Neural pathways form thalamus associated with regulation of aggressive behavior. *Science* 1974; **183:** 96–98.
109. Loewy AD, Spyer KM, eds. *Central Regulation of Autonomic Function.* New York: Oxford University Press, 1990.
110. Devinsky O, Kernan J, Bear DM. Aggressive behavior following exposure to cholinesterase inhibitors. *J Neuropsychiatry Clin Neurosci* 1992; **4:** 189–194.
111. Ebel A, Mack G, Stefanovic V, Mandel P. Activity of choline acetyltransferase and acetylcholinesterase in the amygdala of spontaneous mouse-killer rats and in rats after olfactory lobe removal. *Brain Res* 1973; **57:** 248–251.
112. Karczmar AG, Kindel GH. Acetylcholine turnover and aggression in related three strains of mice. *Prog Neuro-Psychopharmacol* 1981; **5:** 35–48.
113. Pucilowski V, Eichelman TS, Overstreet DH, Rezwani AH, Janowsky DS. Enhanced affective aggression in genetically bred hypercholinergic rats. *Neuropsychobiology* 1990; **24:** 34–41.
114. Pellandra CL. La geneserine-morphine adjuvant de l'anesthesie generale. *Lyon Med* 1933; **151:** 653.
115. Slaughter D, Gross EG. Some new aspects of morphine action. Effect on intestine and blood pressure; toxicity studies. *J Pharmacol Exp Ther* 1940; **68:** 104–112.
116. Koehn GL, Henderson G, Karczmar AG. Diisopropyl phosphofluoridate-induced antinociception: possible role of endogenous opiates. *Eur J Pharmacol* 1980; **61:** 167–173.
117. Wiklund L, Hartvig P. Cholinergic agents in clinical anesthesiology. In: Aquilonius SM, Gillberg PG, eds. *Cholinergic Neurotransmission: Functional and Clinical Aspects.* Amsterdam: Elsevier, 1990: 399–405.
118. Naguib M, Yaksh T. Characterization of muscarinic receptor subtypes that mediate antinociception in the rat spinal cord. *Anesth Analg* 1997; **85:** 847–853.
119. Zemlan FP, Behbehan MM. Nucleus cuneiformis and pain modulation: anatomy and behavioral pharmacology. *Brain Res* 1988; **453:** 389–392.
120. Gillberg PG, Askmark H, Aquilonius SM. Spinal cholinergic mechanisms. In: Aquilonius SM, Gillberg PG, eds. *Cholinergic Neurotransmission: Functional and Clinical Aspects.* Amsterdam: Elsevier, 1990: 361–370.
121. Pedigo NW, Dewey WL. Comparison of the antinociceptive activity of intravenously administered acetylcholine to narcotic

antinociception. *Neurosci Lett* 1981; **26:** 85–90.
122. Rogers DT, Iwamoto ET. Multiple spinal mediators in parenteral nicotine-induced antinociception. *J Pharmacol Exp Ther* 1993; **267:** 341–349.
123. Smith MD, Yang X, Nha JY, Buccafusco JJ. Antinociceptive effect of spinal cholinergic stimulation: interaction with substance P. *Life Sci* 1989; **45:** 1255–1261.
124. Pert A. The cholinergic system and nociception in the primate: interactions with morphine. *Psychopharmacology* 1975; **44:** 131–137.
125. Nistri A, Pepeu GC, Cammelli E, Spina L, De Bellis AM. Effects of morphine on brain and spinal cord acetylcholine levels and nociceptive threshold in the frog. *Brain Res* 1974; **80:** 199–209.
126. Dewey WL. Cannabinoid pharmacology. *Pharmacol Rev* 1986; **38:** 151–178.
127. Wikstrom MA, Lawoko G, Heilbronn E. Cholinergic modulation of extracellular ATP-induced cytoplasmic calcium concentrations in cochlear outer hair cells. In: Massoulie J, ed. *From Torpedo Electric Organ to Human Brain: Fundamental and Applied Aspects.* Amsterdam: Elsevier, 1998: 345–349.
128. Metherate R, Weinberger NM. Cholinergic modulation of responses to single tonus produces tone-specific receptive field alterations in the cat auditory cortex. *Synapse* 1990; **6:** 133–145.
129. Shapovalova KB, Pominova EV, Diubkacheva TA. The role of activation of the intralaminar thalamic nuclei in regulating the participation of the neostriatal cholinergic system in the differentiation of sound signals by dogs. *Zh Vyssh Nerv Deiat Im IP Pavlova* 1994; **44:** 849–852.
130. Rubin LS, Goldberg MN. Effect of sarin on dark adaptation in man: threshold changes. *J Appl Physiol* 1957; **11:** 439–444.
131. Friedman AH, Marchese AL. Reversal by bicuculline of the simulated state of light adaptation produced by GABA in the dark-adapted isolated eye perfused through its ophthalmic artery. *Pharmacologist* 1973; **15:** 162.
132. Rasmusson DD. Cholinergic modulation of sensory information. In: Cuello AC, ed. *Cholinergic Function and Dysfunction.* Amsterdam: Elsevier, 1993: 357–364.
133. Merlo-Pich E, Chiamulera C, Tessari M. Neural substrate of nicotine addiction as defined by functional brain maps of gene expression. In: Massoulie J, ed. *From Torpedo Electric Organ to Human Brain: Fundamental and Applied Aspects.* Paris: Elsevier, 1998: 225–228.
134. Casamenti F, Pedata F, Corradetti R, Pepeu GC. Acetylcholine output from the cerebral cortex, choline uptake and muscarinic receptors in morphine-dependent, freely-moving rats. *Neuropharmacology* 1980; **19:** 597–605.
135. Antkiewicz-Michaluk L, Michaluk J, Romanska I, Vetulani J. Cortical dihydropyridine binding sites and behavioral syndrome in morphine-abstinent rats. *Eur J Pharmacol* 1990; **180:** 129–135.
136. Collins AC. Interaction of ethanol and nicotine at the receptor level. *Recent Dev Alcoholism* 1990; **8:** 221–231.
137. Hodges H, Allen Y, Sinden J *et al.* The effects of cholinergic drugs and cholinergic-rich foetal neural transplants on alcohol-induced deficits in radial maze performance in rats. *Behav Brain Res* 1991; **43:** 7–28.
138. Wilson JM, Carroll ME, Lac ST, DiStefano LM, Kish SJ. Choline acetyltransferase activity is reduced in rat nucleus accumbens after unlimited access to self-administration of cocaine. *Neurosci Lett* 1994; **180:** 29–32.
139. Pietila K, Lahde T, Attila M, Ahtee L, Nordberg A. Regulation of nicotinic receptors in the brain of mice withdrawn from chronic oral nicotine treatment. *Naunyn-Schmiedeberg's Arch Pharmacol* 1998; **357:** 176–182.
140. Holland LN, Shuster LC, Buccafusco JJ. Role of spinal and supraspinal muscarinic receptors in the expression of morphine withdrawal symptoms in the rat. *Neuropharmacology* 1993; **32:** 1387–1395.
141. Way EL, Iwamoto ET, Bhargava HN, Loh HH. Adaptive cholinergic–dopaminergic responses in morphine dependence. In: Mandfell AJ, ed. *Neurobiological Mechanisms of Adaptation and Behavior.* New York: Raven, 1975: 169–187.

142. Stone WS, Rudd RJ, Gold PE. Glucose and physostigmine effects on morphine- and amphetamine-induced increases in locomotor activity in mice. *Behav Neurol Biol* 1990; **54:** 146–155.
143. Dajas-Bailador F, Costa G, Dajas F, Emmett S. Effects of α-erabutoxin, α-bungarotoxin, α-cobratoxin and fasciculin on the nicotine release of dopamine in the rat striatum in vivo. *Neurochem Int* 1998; **33:** 307–312.
144. Carlton PL. Cholinergic mechanisms in the control of behavior. In: Efron DH, ed. *Psychopharmacology — A Review of Progress. PHS Publication No. 836.* Washington, DC: US Government Printing Office, 1968: 125–135.
145. Penrose R. *Shadows of the Mind.* Oxford: Oxford University Press, 1994.
146. Karczmar AG. Sir John Eccles, a testimonial. *Perspectives Biol Med* 1999; in press.
147. Woolff NJ. A possible role for cholinergic neurons of the basal forebrain and pontomesencephalon in consciousness. *Consciousness Cognition* 1997; **6:** 574–596.
148. Karczmar AG. History of the research with anticholinesterase agents. In: Karczmar AG, ed. *International Encyclopedia of Pharmacology and Therapy*, Vol 1, *Anticholinesterase Agents.* Oxford: Pergamon, 1970: 1–44.
149. Ward JR, Gosselin R, Comstock J, Stagg J, Blanton BR. Case report of a severe human poisoning by GB. *MLRR 151.* Edgewood, MD, 1952.
150. Gershon S, Shaw F. Psychiatric sequelae of chronic exposure to organophosphorus insecticides. *Lancet* 1961; **i:** 1371–1374.
151. Karczmar AG. Acute and long lasting central actions of organophosphorus agents. *Fundament Appl Toxicol* 1984; **4:** S1–S17.
152. Mileson BE, Chambers JE, Chen WL *et al.* Common mechanism of toxicity: a case study of organophosphorus pesticides. *Toxicol Sci* 1998; **41:** 8–20.
153. Petras JM. Soman neurotoxicity. *Fundament Appl Toxicol* 1981; **1:** 242.
154. Dettbarn WD. Pesticide-induced muscle necrosis: mechanisms and prevention. *Fundament Appl Toxicol* 1984; **4:** 18–26.
155. Yang ZP, Dettbarn WD. Lipid peroxidation and changes in cytochrome C oxidase and xanthine oxidase activity in organophosphorus anticholinesterase induced myopathy. In: Massoulié J, ed. *From* Torpedo *Electric Organ to Human Brain: Fundamental and Applied Aspects. J Physiol Paris,* Elsevier Paris, 1998; **92:** 157–161.
156. Abou-Donia MB, Lapadula DM. Mechanisms of organophosphorus ester-induced delayed neurotoxicity: type I and type II. *Annu Rev Pharmacol Toxicol* 1990; **30:** 405–440.
157. Johnson MK. The target for initiation of delayed neurotoxicity by organophosphorus ester: biochemical studies and toxicological applications. *Rev Biochem Toxicol* 1982; **4:** 141–202.
158. Duffy FH, Burchfiel JL. Long term effects of the organophosphate Sarin on EEG in monkeys and humans. *Neurotoxicology* 1980; **1:** 667–689.
159. Steenland K, Jenkins B, Ames RG, O'Malley M, Chrislip D, Russo J. Chronic neurological sequelae to organophosphate pesticide poisoning. *Am J Publ Hlth* 1994; **84:** 731–736.
160. Moore DH. Long term health effects of low exposures to nerve agent. In: Massoulie J, ed. *From Torpedo Electric Organ to Human Brain: Fundamental and Applied Aspects.* Paris: Elsevier, 1998: 325–328.
161. Usdin E. Reactions of cholinesterases with substrates, inhibitors and reactivators. In: Karczmar AG, ed. *International Encyclopedia of Pharmacology and Therapy*, Vol 1, *Anticholinesterase Agents.* Oxford: Pergamon, 1970: 47–354.
162. Hobbiger F. Effect of nicotinhydroxamic acid methiodide on human plasma cholinesterase inhibited by organophosphate containing a dialkylphosphate group. *Br J Pharmacol* 1955; **10:** 356–362.
163. Wilson IB. Acetylcholinesterase. XI. Reversibility of tetraethyl pyrophosphate inhibition. *J Biol Chem* 1951; **190:** 111–117.
164. Koster R. Synergisms and antagonisms between physostigmine and diisopropylfluorophosphate in cats. *J Pharmacol Exp Ther* 1946; **88:** 39–46.
165. Karczmar AG. *Workshop on New Conceptual Approaches to Prophylaxis in Therapy of Organophosphorus Poisoning, Port St Lucie,*

FL, 1984. Washington, DC: USA Department of Defense.
166. Gunderson CH, Lehmann CR, Sidell FR. Nerve agents: a review. *Neurology* 1992; **42:** 940–950.
167. Hayes WJ. *Pesticides Studied in Man*. Baltimore: Williams & Wilkins, 1982.
168. Chadwick LE. Actions on insects and other invertebrates. In: Koelle GB, ed. *Cholinesterases and Anticholinesterase Agents*. Berlin: Springer-Verlag, 1963: 741–798.
169. Carlock L, Chen WL, Gordon E *et al. Risk Assessment for Cholinesterase Inhibiting Pesticides: Less than Life Exposures*. Washington, DC: Acute Cholinesterase Risk Assessment Work Group, 1997: 1–126.
170. Karczmar AG, Awad O, Blachut K. Toxicity arising from joint intravenous administration of EPN and malathion to dogs. *Toxicol Appl Pharmacol* 1962; **4:** 133–137.
171. Mileson BE, Chambers JE, Chen WL *et al.* Common mechanism of toxicity: a case study of organophosphorus pesticides. *Toxicol Sci* 1998; **41:** 8–20.
172. Morita H, Yanagisawa N, Nakajima T *et al.* Sarin poisoning in Matsumoto. *Lancet* 1995; **346:** 290–293.
173. Yokoyama K, Araki S, Murata K *et al.* Chronic neurobehavioral and central and autonomic nervous system effects of Tokyo subway sarin poisoning. In: Massoulie J, ed. *From Torpedo Electric Organ to Human Brain: Fundamental and Applied Aspects*. Paris: Elsevier, 1998: 317–323.
174. Anon. *Int Jane's Defence Weekly* 21 August 1993; 5.
175. Department of Veterans' Affairs. A working plan for research on Persian Gulf veterans' illnesses. Persian Gulf Veterans' Coordinating Board. Project no. VA46. Washington, DC: US Government Printing Office, 1996.
176. Department of Veterans' Affairs. Annual report to Congress. Federally sponsored research on Persian Gulf veterans' illnesses, I13 10–40. Washington, DC: US Government Printing Office, 1997.
177. *VA Cooperative Study*. [Confidential.]
178. Haley RW, Kurt TL, Hom J. Is there a Gulf War syndrome? Searching for syndromes by factor analysis of symptoms. *JAMA* 1997; **277:** 215–222.
179. National Institutes of Health Technology Assessment Workshop on Persian Gulf Experience and Health, April 1994, Bethesda, MD, USA.
180. Gerrity T. *Health Effects of Low-Level Chemical Warfare Nerve Agent Exposure*. Cincinnati, OH: Department of Veterans' Affairs, 1997.

12

Cholinesterase inhibitors: from the Calabar bean to Alzheimer therapy

Ezio Giacobini

Development of cholinesterase inhibitors in Alzheimer's therapy: a hundred-year history

Cholinesterase (ChE) inhibitors represent the drugs of choice in the treatment of Alzheimer's disease (AD) (Fig. 12.1). Following the introduction in the 1980s of a first generation of non-specific drugs such as physostigmine and tacrine, a second generation of more suitable compounds was developed in the 1990s. The latter drugs are clinically more efficacious and produce less severe side-effects at effective doses. Data originating from numerous clinical trials on large patient populations demonstrate that the maintenance of clinical gains for one year or more is a feasible target in one-third of patients.

A considerable amount of basic knowledge related to the cholinergic system anatomy, biochemistry, physiology, pharmacology and molecular biology has been accumulated during the last 50 years of intensive fundamental research in the field of acetylcholine (ACh), and has led to the award of four Nobel Prizes (H Dale, O Loewi, J Eccles and B Katz) and greatly contributed to the therapeutic success of ChE inhibitors. Physostigmine was the first active compound to be isolated as a natural product and used in therapy. Its clinical use in 1877

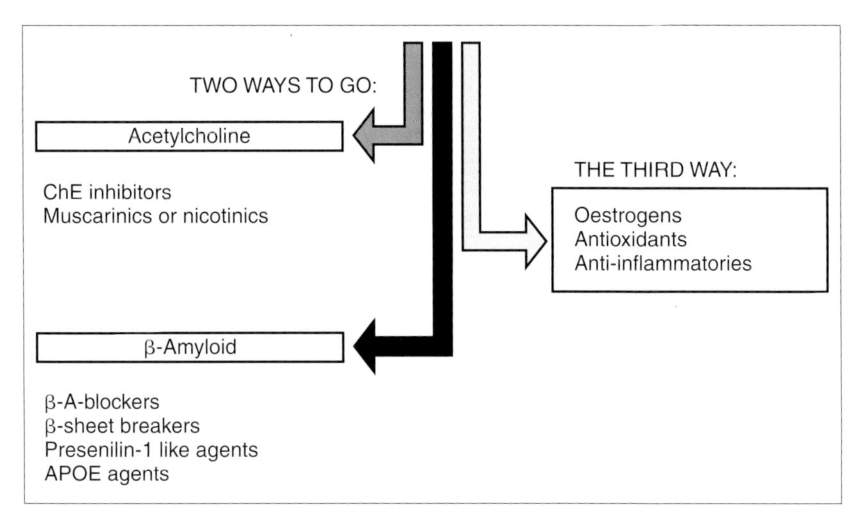

Figure 12.1 *Pharmacological treatment of AD: three possible routes.*

	Year	Discovery	Ref.
Physostigmine	1864	Isolation *(Jobst and Hesse)*	1
	1877	Therapeutic use *(Laquer)*	1
Acetylcholine	1867	Synthesis *(Baeyer)*	1
	1914	A neurotransmitter *(Dale)*	2
Cholinesterases	1932	Isolation and name *(Stedtman)*	3
	1935	Found in the brain *(Stedtman and Stedtman)*	3
Cholinesterase	1979	Physostigmine intravenous *(Davis and Moss)*	4
	1983	Physostigmine oral *(Thal and Fuld)*	5
Inhibitors as anti-Alzheimer drugs	1988	Physostigmine i.c.v. *(Giacobinl* et al*)*	6
	1983	Tacrine oral *(Summer* et al*)*	7
	1988	Metrifonate oral *(Becker and Giacobini)*	8
	1989	Galantamine oral (Rainer et al)	9

i.c.v., intracerebral ventricular.

Table 12.1
ChE inhibitors: a long history.

anteceded by almost half a century the discovery of ACh as a brain neurotransmitter in 1914[1–3] (Table 12.1) (see Chapter 1).

The use of physostigmine provided crucial experimental support to the demonstration of ACh as a an active ester and to the postulation of an enzyme contributing its inactivation. ChEs were isolated from the mammalian brain and purified in 1932, i.e. 20 years after the discovery of ACh as a neurotransmitter (see Table 12.1). The introduction of ChE inhibitors as antidementia (specifically as anti-Alzheimer) drugs represents the most recent neuropharmacological application of these agents in neurology and psychiatry. This trend started in the 1950s with the first neuroleptic (chlorpromazine), continued into the 1960s and 1970s with antidepressants and was followed by the development of anti-Parkinsonian agents. Unlike the discovery of other neurotransmitter-based central nervous system (CNS) drugs such as neuroleptics, tricyclic antidepressants and anxiolytics, the clinical application of ChE inhibitors in the treatment of cognitive deficits in AD was neither accidental nor serendipitous. Its rationale was founded on data derived from experimental physiology and behavioural pharmacology of the cholinergic system in animals and humans. Clinical results on the effect of these drugs on cognition (memory, attention and concentration) and, more recently, on behavioural symptoms in AD (apathy, hallucinations and motor agitation) confirmed predictions of their potential clinical efficacy based on laboratory data.

The first ChE inhibitor to be used against AD was physostigmine administered under various modes;[4–6] this was followed by oral tacrine[7] (see Table 12.1). Subsequently, metrifonate[8] and galantamine[9] were tested orally (see Table 12.1). ChE inhibitors, particularly second generation ones (i.e. those discovered after physostigmine and tacrine), affect corti-

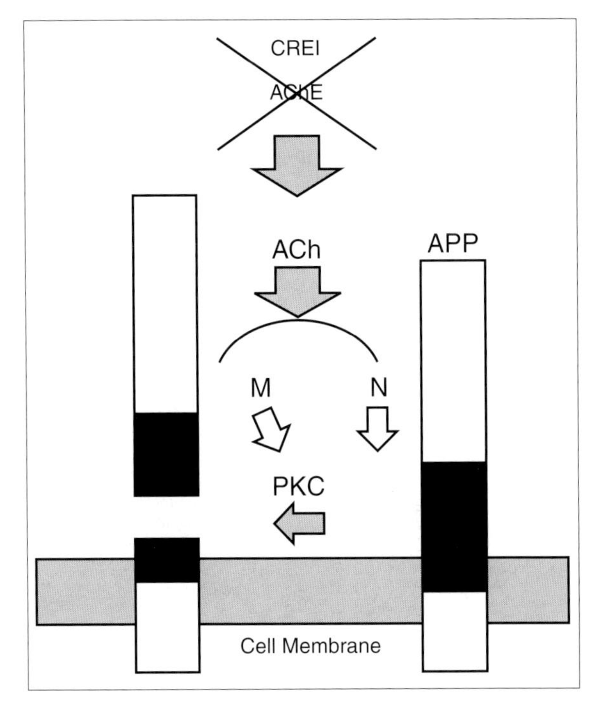

Figure 12.2
Secretion of the APP can be increased in brain by activation of m1 muscarinic receptors (M) and subsequent protein phosphorylation (protein kinase C, PKC) or through direct activation of nicotinic receptors (N). Increased basal release of non-amyloidogenic amino-terminal soluble derivatives has been demonstrated by increasing ACh levels in brain cortex with AChE inhibitors.[11] This effect might concomitantly decrease secretion of potentially amyloidogenic βA peptides (solid intramembrane segments) and exert neuroprotection. CREI, cell membrane.

cal as well as subcortical neurotransmitters other than ACh.[10] Their selective effects on norepinephrine (NE) and dopamine (DA) cortical release are of particular clinical interest with regard to behavioural symptoms related to biogenic amines.[10] Another feature of ChE inhibitors is their ability to enhance the release of non-amyloidogenic-soluble derivatives of β-amyloid precursor protein (APP) from cortex, both in vitro and in vivo.[11] This specific effect could slow down the formation of amyloidogenic compounds in AD brain[11] (Fig. 12.2). The same mechanism may also explain the stabilizing effect on cognitive deterioration seen during the first 6–12 months of ChE inhibitor therapy.[12] Cholinergic alternatives other than the use of ChE inhibitors are being actively investigated, e.g. use of nicotinic and muscarinic agonists. However, receptor non-selectivity and severe side-effects (mainly cardiovascular and gastrointestinal) have prevented therapeutic application of these compounds as complementary cholinergic approaches to ChE inhibitors in AD therapy. As a next development, the combination of cholinergic therapies with oestrogens or anti-inflammatories (or antioxidants) is envisioned (see Fig. 12.1). This application depends on the demonstration of clinical effects in ongoing prospective trials. The search for compounds believed to influence more directly the development of the disease by influencing the processing of the β-amyloid (β-sheet breakers and compounds acting on presenilins, secretases, APOE-4, etc.) has not yet been translated into therapies. The result of such a 'causal' therapy could be classified as structural or 'disease modifying'. In this chapter an attempt is made to demonstrate that, based on clinical and experimental results, we may define ChE-inhibitor-based treatment as a functional and 'disease-stabilizing' therapy.

Changes in ChE activity in AD brain and CSF

Enzymatic variations occurring during disease progression

The brain of mammals contains two major forms of ChE: acetylcholinesterase (AChE)

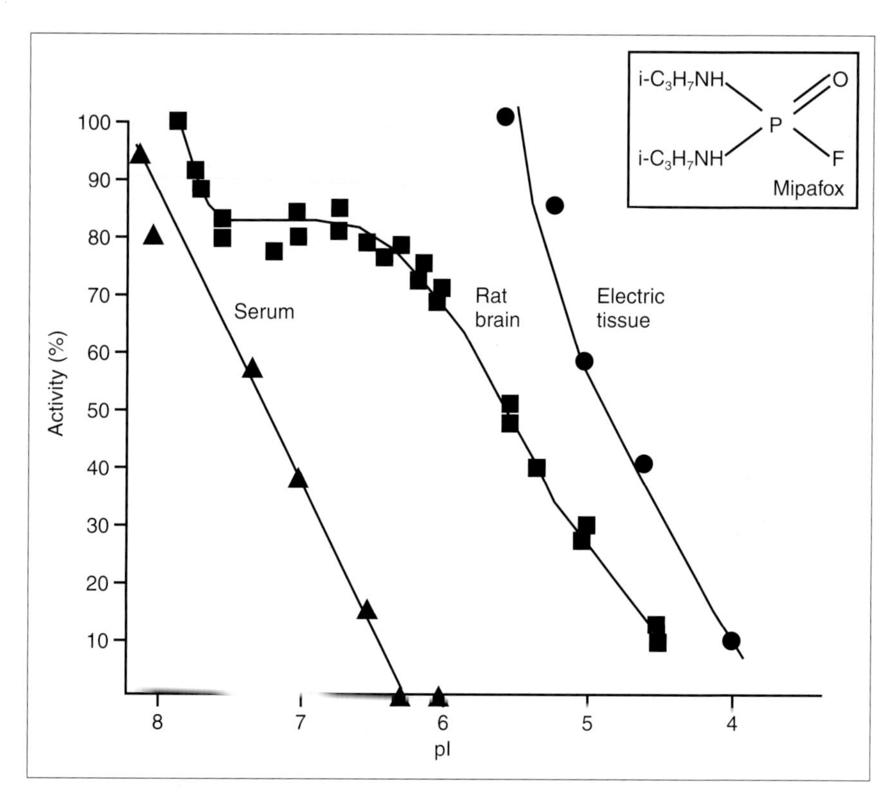

Figure 12.3
Inhibition curve for rat brain homogenate obtained using a selective AChE inhibitor (Mipafox), two purified enzymes and a selective substrate. The activity values are percentages of the ChE activity of the control. pI, −log (molar concentration of inhibitor). Approximately 80% is AChE activity and approximately 20% is BuChE activity.[14]

and butyrylcholinesterase (BuChE).[13] Using specific substrates and selective inhibitors, it has been demonstrated that, unlike AChE, brain BuChE preferentially hydrolyses butyrylcholine.[14] Butyrylcholine (BuCh) is not a physiological substrate in human brain which makes it difficult to interpret the function of BuChE. It should be kept in mind that AChE is present in excess concentrations in the mammalian brain. Therefore, it is not a limiting factor in ACh metabolism unless it reaches very low concentrations, such as at very advanced stages of the disease or at high levels of inhibition (<90%).

By combining specific substrates with selective inhibitors it has been demonstrated that in rat brain approximately 80% of ChE activity is due to AChE and 20% to BuChE[14] (Fig. 12.3). In rat brain, BuChE activity is mainly localized in the glial cells, while AChE activity is mainly concentrated in nerve cells[15,16] (Table 12.2). In the human brain, both AChE and BuChE are found in neurons and glia and, in AD patients, in neuritic plaques and tangles as well.[17] Due to its prevalently neuronal distribution, AChE activity is higher in human brain than is BuChE activity. Depending on the region, human-brain AChE activity is 1.5-fold (temporal and parietal cortex) to 60-fold (caudate nucleus) higher than BuChE activity.[18–24]

An important feature distinguishing BuChE present in serum from AChE present in neurons and erythrocytes is the kinetics of the

Enzyme	Location	Activity (% of normal)	Molecular form
AChE	Neuronal	10–15	50–70% decrease, mainly G4
BuChE	Glia-plaques	120	20% decrease, mainly G1
ChAT	Neuronal	10–15	

Table 12.2
The variation in cholinergic enzyme activity in the cortex of AD patients (late stages) and normal controls.

former toward concentrations of ACh (Fig. 12.4). An excess of the substrate of the order of micromoles will inhibit only AChE but not BuChE.[13,14] Due to the difference in K_m of the two enzymes, glial BuChE is less efficient in hydrolysing ACh at low substrate concentrations (below micromolar) than is neuronal AChE. The concentration of ACh in the cholinoceptive synapse (determined at the neuromuscular junction) is believed to reach low micromolar levels and to be increased two- to eight-fold in the CNS following topical application of a ChE inhibitor (physostigmine). The mechanism of inhibition caused by

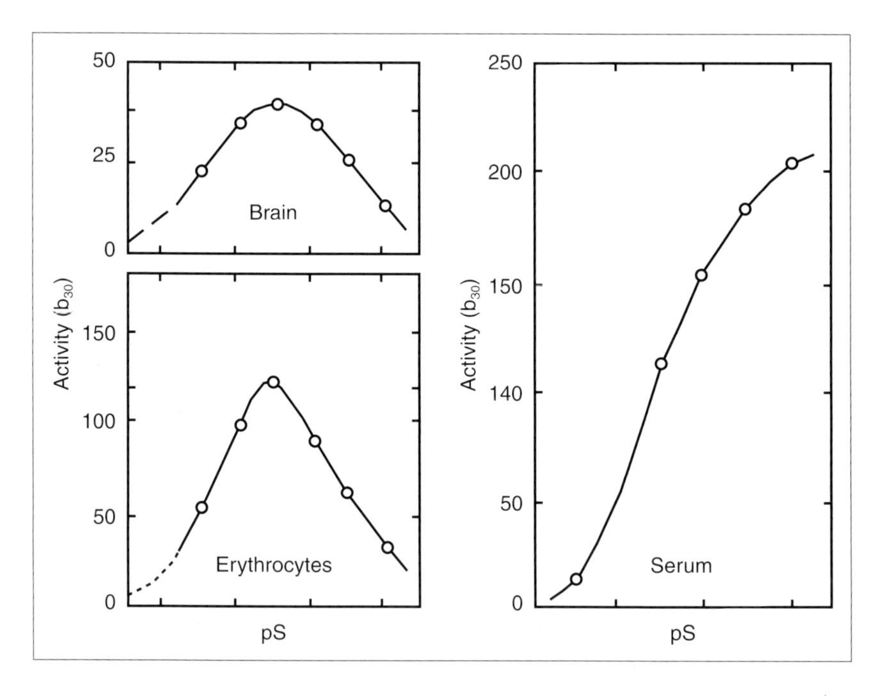

Figure 12.4
Effect of ACh concentration (pS) on the activity of AChE (brain and erythrocytes) and BuChE (serum).

the excess of substrate (ACh) has been clarified by Shafferman *et al.*[25] It is related to a change in conformation of the Tyr337 amino acid which lines the upper part of the catalytic gorge of the enzyme, representing in part the peripheral site overlapping the substrate inhibitory site. A change in conformation of Tyr337 or a mutation in a single amino acid will induce allosteric changes in the peripheral site which are responsible for this inhibition.[25] In the case of BuChE, in which alanine replaces the tyrosine of AChE, no substrate inhibition is observed.

No significant decrease in AChE activity has been seen in cerebral cortex with normal ageing. Conversely, a decrease in cortical AChE activity has been seen both biochemically and by positron emission tomography (PET) in AD patients (see Chapter 14). The proportion of the two ChEs present in human brain is also strongly altered over the course of AD. There are correlations between decreases in AChE activity and dementia severity as well pathology related to substantial losses of cholinergic innervation of the cortex at advanced stages of the disease.[18–21] In the cortex of patients affected by AD, AChE activity decreases progressively in certain brain regions from mild to severe stages of the disease, finally reaching only 10–15% of normal values, while BuChE activity is unchanged or even increased by a maximum of 20% (see Table 12.2).[18–24] In spite of the general reduction in brain AChE activity, the enzyme appears to be increased within and around neuritic plaques. In the plaques, AChE is closely associated with β-amyloid (βA). As examples of regional difference in changes, the BuChE/AChE ratio increases from 0.6 to 0.9 in the frontal cortex and from 0.6 to 11 in the entorhinal cortex. This change may reflect a combination of reactive gliosis following severe neuronal damage (glial cells containing mainly BuChE) and of an accumulation of BuChE in neuritic plaques (which contain both AChE and BuChE).[17] Due to disease progression, in the presence of a strongly decreased concentration of synaptic ACh and AChE (particularly the membrane-anchored G4 form),[26–28] a ChE inhibitor may increase ACh concentrations to micromolar levels, which may inhibit AChE activity. This increase in substrate concentration may trigger glial BuChE to hydrolyse ACh. This could represent a compensatory mechanism for the loss of neuronal AChE activity. Given the close spatial relationship between glial cell protoplasm and the synaptic gap, it is likely that extracellularly diffusing ACh may come into contact with glial BuChE and be effectively hydrolysed, as demonstrated in experiments using intracerebral microdialysis in the rat.[29,30]

ChE molecular forms in human brain: are there selective inhibitors?

Human brain AChE exists in multiple molecular forms defined by their different sedimentation coefficients. Due to their different shapes, collagen-tailed asymmetric forms (An) and globular forms (Gn) can be separated.[27,28] Studies on whole brain fractions suggest that 60–90% of the tetrameric (G4) form is extracellular and membrane located while 90% of the monomeric (G1) form is intracellular and cytoplasmic.[27,28,31] In the rat skeletal muscles four molecular forms are found, with both G1 and G4 globular forms being represented. Ideally, the most effective inhibitor would be one that selectively inhibits brain AChE forms and has no effect on peripheral tissues such as skeletal or cardiac muscle. The differential effect of certain inhibitors may be related primarily to the location of the enzyme and the penetration of the inhibitor rather than to pharmacological or tissue selectivity.

Selective loss of the membrane-bound G4 form has been reported in AD, suggesting a presynaptic location of this form.[28] In severe AD patients, the membrane-bound G4 form is decreased in frontal (by 71%) and parietal (by 45%) cortex and in the caudate putamen (by 47%) as compared with control levels (see Table 12.2).[28] Heptylphysostigmine (eptastigmine), a carbamate physostigmine analogue, shows preferential inhibition for the G1 form of rat and human brain. Similarly, rivastigmine, also a carbamate compound, preferentially inhibits the G1 form.[31] Conversely, edrophonium inhibits the G4 form more potently than the G1 form, while physostigmine, tacrine and metrifonate inhibit both forms with similar potency.[28] Characteristic changes in the density and distribution of binding with a G4 selective inhibitor could be useful diagnostically. Accumulation of BuChE, and possibly G1 AChE, in neuritic plaques has been described.[32]

Differences in allosteric effects or differential affinities for G4 and G1 could also explain the inhibitor selectivity. Since the active sites of all four molecular forms is similar, it is difficult to explain a differential inhibition based on isoform selectivity. Functional and location differences (synaptic vs intracellular) of the two molecular forms (G4 and G1) make the therapeutic application of isoform-selective AChE inhibitors difficult to interpret. Since the AChE monomeric form which is preserved in AD is primarily cytoplasmic, it is unlikely that a selective inhibition of this molecular form would have a significant effect on synaptic ACh. On the other hand, inhibition of the tetrameric ACh-metabolizing form would prolong the synaptic action of ACh and be more beneficial to function.

Is there a function for brain BuChE?

In order to determine a possible function of glial BuChE, and in particular its role in the regulation of ACh extracellular concentrations, the highly BuChE-selective carbamate inhibitor MF-8622 was perfused intracortically in the rat (Table 12.3) at concentrations varying between 15 and 170 μM.[33] Simultaneous with the perfusion the extracellular ACh levels were measured,[33] without using a second ChE inhibitor, by means of a sensitive microdialysis method.[29,30] At the highest MF 8622 concentration (170 μM) the ACh level increased 15-fold (from 5 nM at baseline to 75 nM). This would raise the ACh concentration in and around cortical cholinergic synapses from nanomolar to low micromolar levels.[29] Such values approach the inhibitory concentrations for AChE[34] and may affect glial BuChE activity. For comparison the carbamate eptastigmine, a non-selective ChE inhibitor, was tested under similar conditions, and it was found that at 15 μM eptastigmine the ACh level was elevated 20-fold (from 5 nM at baseline to 100 nM). Thus, the dose-dependent steady-state elevation of cortical ACh seen with the BuChE-selective inhibitor MF 8622 is comparable in magnitude to the effect of a mixed AChE–BuChE inhibitor such as eptastigmine. However, to obtain the same elevation in ACh a 10-fold higher concentration of MF 8622 is necessary (170 μM).[33] This suggests that, as expected, under normal conditions, AChE inhibition is much more efficient than BuChE inhibition in elevating cortical ACh. It is interesting to note that in spite of the high ACh elevation in brain seen after the BuChE inhibitor, the animals perfused with MF 8622 did not show typical cholinergic side-effects.

These results represent the first study in

Compound	IC_{50} (nM) AChE*	BuChE†	BuChE/AChE
BW 284 C51	18.8	48.000	2.553
Donepezil	5.7	7139	1.252
Phenserine	22.2	1552	70
DDVP (Metrifonate)	800	18.000	22.5
Physostigmine	5.4	35	6.5
Galantamine	0.35	18.6	5.3
Rivastigmine	48.000	54.000	1.1
Tacrine	190	47	0.25
Eptastigmine	20	5	0.025
Hetopropazine	260.000	300	0.001
Bambuterol	30.000	3	0.001
MF 8622	100.000	9	0.00009

*Human erythrocytes.
†Human plasma.

Table 12.3
In vitro activity of ChE.

which a selective BuChE inhibitor has been shown to elevate significantly the extracellular levels of cortical ACh without producing side-effects in the animal. The results of this experiment support the concept of two pools of functional ChE in rat brain: one neuronal (AChE), acting mainly under physiological conditions; and one glial (BuChE), acting under conditions of decreased AChE activity such as in AD brain. The two pools show different kinetic properties with regard to regulation of brain ACh and can be separated by selective inhibitors. It is possible that selective BuChE inhibitors could be used to treat more severe cases of AD, which show higher BuChE levels. Since BuChE is present in the neuritic plaques together with AChE, a drug inhibiting BuChE could, in theory, reduce the formation of βA.

From a clinical point of view, this implies that a potent selective BuChE inhibitor may produce significant increases in brain ACh without triggering peripheral or central cholinergic side-effects. From the experimental point of view, a wide range of inhibition (measured in rat brain and plasma) of the two types of ChE inhibitor is seen for various ChE inhibitors (Table 12.4). As an example, the BuChE/AChE ratio of inhibition in rat brain is around 1000 for donepezil and huperzine A and around 1 for tacrine and metrifonate. Several other inhibitors (e.g. physostigmine, galantamine and rivastigmine) show intermediate values.

Particularly interesting from the clinical point of view is the relative rate of inhibition shown by several ChE inhibitors toward human erythrocytes (AChE) and plasma

Compound	IC_{50} (nM) AChE*	BuChE†	BuChE/AChE
Donepezil	5.7	7138	1252
Huperzine A	82	74.400	908
Physostigmine	0.68	8.1	11.9
Galantamine	2000	12.600	6.3
Rivastigmine	57.000	16.000‡	3.6
DDVP (Metrifonate)	1600	1500	0.9
Tacrine	80.6	73.0	0.9

*Striatum.
†Plasma.
‡Heart.

Table 12.4
In vitro activity of ChE inhibitors in the rat.

(BuChE) (see Table 12.3). It can be seen that most inhibitors presently used in AD therapy are not selective for AChE; however, they all show significant clinical efficacy. Considering the drastic decrease in AChE activity that occurs in certain brain regions of patients with advanced AD (at autopsy of severe cases, the AChE levels are 5% in some brain areas) and the large pool of BuChE available in glial neurons and neuritic plaques, it may not be an advantage for a ChE inhibitor to be selective for AChE. On the contrary, a good balance between AChE and BuChE inhibition might result in optimal therapeutic efficacy of a non-selective ChE inhibitor to be used in moderate to severe cases of AD.

The present knowledge of the molecular configuration of the two enzymes makes it possible to design compounds possessing either AChE or BuChE selectivity or well-balanced AChE–BuChE specificity, high CNS penetration and low peripheral and central cholinergic toxicity. Some of the second-generation ChE inhibitors have already demonstrated these characteristics and display particularly low peripheral toxicity. An additional feature is that, unlike AChE, binding of ACh to the site outside the catalytic site in BuChE can cause an increase in activity called 'substrate activation' (see Chapter 7). Activation of BuChE activity in serum, liver or brain could theoretically increase its efficacy.

ChEs in CSF: origin and characteristics

Two hypotheses can be suggested with regard to the origin of AChE in cerebrospinal fluid (CSF): it may originate either from the spinal cord and brain as a result of a secretory process, or from blood plasma by a process by formation in the choroid process (Fig. 12.5).

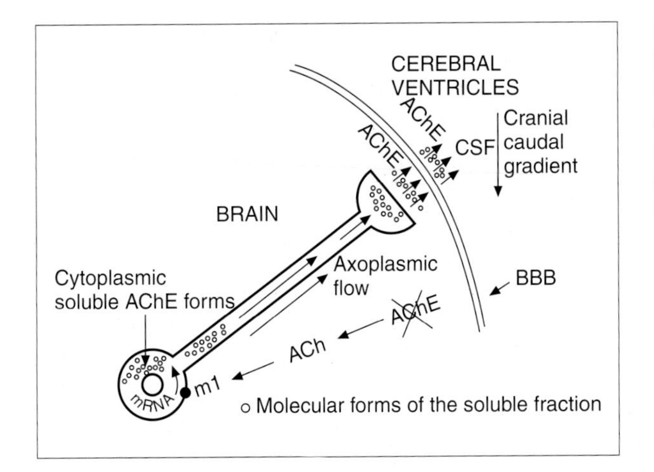

Figure 12.5
CNS synthesis and secretion of AChE in the CSF. Muscarinic ACh (m1) receptors activate AChE synthesis in central cholinergic neurons.[43] Cytoplasmic-soluble AChE forms are transported to the synapse and secreted in the CSF.[35,36] BBB, blood–brain barrier.

Observations based on the distribution of selective molecular forms in dog CSF ruled out the latter hypothesis, demonstrating that AChE originates primarily from the spinal cord and brain as a result of a secretory process.[35] A combined origin from both blood plasma with synthesis in the liver and brain (astrocytes?) is probable for BuChE.

The activity of AChE and BuChE and the molecular forms G1 and G4 were measured in the lumbar CSF of AD patients, with severity of disease varying from Clinical Dementia Rating (CDR) I (mild) to II and III (moderate and severe), and age-matched controls.[36] BuChE activity was found to represent approximately 5% of the total CSF ChE activity. The mean lumbar AChE activity in AD patients did not differ significantly from that of controls, and there was no significant decline when CSF was analysed at 6-month intervals over 12 months. Neither molecular form (G1 and G4) showed a selective decline. However, in another study,[37] the G4 form was found to be significantly decreased. BuChE activity was not significantly changed in the case of the AD patients, and the AChE/BuChE ratio was not found to be significantly different in AD patients compared with controls.[36]

It has also been observed that CSF AChE activity increases linearly as a function of age and more than doubles between 2 and 80 years of age.[36] Some workers have found a reduction of up to 60% in CSF AChE activity in histologically confirmed AD patients,[36] but the bulk of the literature suggests that if such declines do exist they are too small and variable to be of diagnostic use for AD.

With regard to the relationship between ChE activity and time of onset of AD, Kumar *et al.*[38] found a significantly lower CSF AChE activity in early-onset patients but not in late-onset patients compared with age-matched controls. This may indicate severe cholinergic damage in early-onset cases. On the contrary, choline levels were found to be more elevated in the CSF of AD patients than in controls.[39] This finding could relate to neuronal membrane breakdown and reduced choline uptake by cholinergic neurons.

With regard to the ACh content of the CSF,

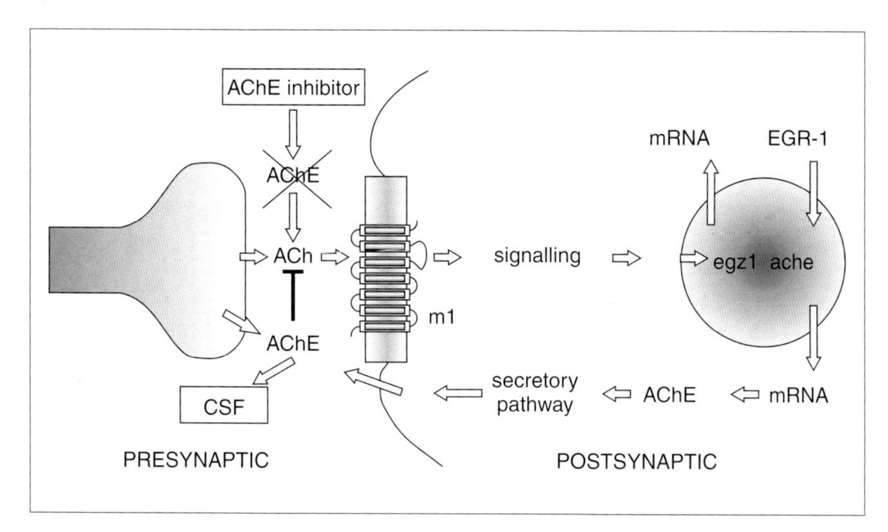

Figure 12.6
Mechanism of cholinergic control of AChE expression.[43] Increased levels of ACh resulting from AChE inhibition stimulate m1 ACh receptors, increasing Egr-1 mRNA and Egr-1 protein binding to and activation of the AChE gene promoter. AChE is then secreted into the synapse and CSF.[35,36] (Adapted from Nitsch et al.[43])

the literature is highly equivocal, but favours the idea that no sizeable CSF measurements distinguish AD patients from normal subjects.[26,40,41] The observation[42] that CSF AChE activity was significantly increased (36–80%) in dogs administered physostigmine i.v. may have some clinical implications with regard to the use of ChE inhibitors. The increase in activity was dose dependent and could be blocked with pretreatment with atropine, suggesting a muscarinic-modulated mechanism. This observation could not be explained until the recent demonstration of a gene-activated, muscarinic-regulated increase in AChE biosynthesis following inhibition (Fig. 12.6).[43]

The variations in CSF AChE activity in various neurological disorders and in age-matched controls are shown in Table 12.5. It

Disorder	*Decrease vs controls*	*Ref.*
AD	Not significant	36
Familial AD vs non-familial AD	No difference	44
Cerebellar ataxia	Not significant	45
Huntington's chorea	Not significant	46
Early-onset AD vs late-onset AD	Decrease in early-onset AD	38
Agraphia in AD	No difference	47
Parkinson's disease	Not significant	48

Table 12.5
Variations in CSF AChE activity in neurological disorders and controls.

	ChAT (μmol/h per g protein)	AChE (μmol/h per g protein)	Ratio	ACh synthesis (dpm/min per mg protein)	ACh levels in CSF, extracellular (nM)
Controls	6	0.25	24	6.2	33–300
AD patients	2.3	0.18	13	2.9	15–152
Decrease (%)	62	28	46	53	48

Table 12.6
Cholinergic parameters in the frontal cortex of AD patients and controls.

can be seen that only patients with early-onset disease show a significant decrease in CSF AChE activity; this could be related to a more rapid loss of cholinergic synapses in early-onset AD.[44–48]

Decline in cholinergic function with progression of AD: premises for an effective cholinergic therapy

During a 15- to 20-year period of progression of AD, a continuous loss of cholinergic neurons (50–87%) in the nucleus basalis Meynert (NBM) and of cortical cholinergic synapses is observed.[49] From a total average number of 350 000 neurons in young-adult controls, as few as 72 000 neurons are found in the NBM of AD patients. This profound loss in subcortical nuclei results in a progressive cortical cholinergic denervation.[50] It is still controversial whether or not early decline in cognition is associated with a decrease in cortical choline-acetyltransferase (ChAT) activity or with other changes in cholinergic function such as selective choline uptake, ACh vesicular storage and release, or ACh synthesis (Table 12.6). Due to the vast excess of both AChE and ChAT, enzyme activities measured in brain do not represent, particularly at early stages of the disease, a sensitive marker of cholinergic function such as ACh levels, ACh synthesis or release (which can be determined either in bioptic material or, in particular cases, in vivo). In addition, it is probable that the well-documented loss in basal forebrain cholinergic neurons may be accompanied by enzymatic overexpression of AChE or ChAT in surviving neurons, which would mask cortical neuritic losses.[49,50] The fact that brain AChE is localized not only to cholinergic axons, but also to cholinoceptive perikaria, and in a smaller amount to neuroglia and neuritic plaques (in AD), makes it difficult to interpret the biochemical data. More accurate evaluation of cortical cholinergic denervation is a combination of biochemical and microscopic observations of cortical axons visualized with monoclonal antibodies to AChE and ChAt. Such an approach demonstrates signs of early cholinergic denervation in AD.[50]

The formulation of a cholinomimetic strategy calls for an improvement in ACh effect resulting in enhanced cognitive capacities of the patient. The most efficacious intervention so far has been the use of drugs such as ChE inhibitors, which enhance synaptic concentrations of ACh. One typical example of this effect at the synaptic level is the action of prostigmine at the neuromuscular junction which produces an improvement of neuromuscular transmission. This drug has been used for many decades to treat the defect in cholinergic nicotinic receptors of myasthenic patients (see Chapter 15). At the clinical level, the potential of ChE inhibitors to improve cognition and function is exemplified by changes induced in cerebral glucose metabolism, as measured using PET before and after administration of a single dose of physostigmine. Simultaneously with an increase in cortical glucose metabolism in several brain regions there is a rapid (minutes) improvement in the outcome of a cognitive test (Necker cube task, design a cube).[51] In bioptic brain tissue of age-matched neurological controls (non-AD patients) the synthesis/hydrolysis ratio of ACh was 24[52] (see Table 12.6). The same ratio found in cerebral tissue biopsied from AD patients was 13 (a reduction of 43%). ACh synthesis measured radiometrically in human brain tissue obtained from biopsy was decreased by 53% in early AD patients,[53] and the CSF level of ACh was decreased by 62% compared with controls[41] (see Table 12.6). Levels of extracellular ACh measured directly in human brain by microdialysis and HPLC-ED[54] (see Table 12.6), and indirectly in ventricular CSF samples from awake AD patients or in the lumbar CSF,[6,26] indicate a difference from normal controls of the order of 50%, which is in agreement with the CSF values. ACh levels measured in brain or CSF would represent a more functional marker than enzymatic activity. This approach is limited in humans.

Based on available data, one can hypothesize that doses of ChE inhibitors capable of doubling ACh levels in the cortex of patients with mild to moderately severe AD could re-establish normal levels of the neurotransmitter. Preclinical experimental results in animals and clinical data in humans demonstrate that such an effect is feasible with most second-generation ChE inhibitors without causing severe or irreversible side-effects.[55]

Data relevant to cholinergic function in the human frontal cortex of AD patients and controls are summarized in Table 12.6. These data underline an indicative 46% decrease in the synthesis/hydrolysis ratio of ACh in the cortex of AD patients.

Effect of ChE inhibitors in non-human primates

The search for an effective treatment for memory loss began almost 20 years ago with the development of animal models of age-related memory impairment. However, by the mid-1980s only incomplete information was available. As summarized in the paper by Bartus and Dean,[56] cognitive models in aged rodents and primates were developed almost simultaneously. In the mid-1970s it was demonstrated that aged monkeys suffer significant behavioural and cognitive deficits and most impressively, memory impairment is present.[56] These memory changes resemble those observed in elderly humans. The possibility of a relationship between animal models and a potential treatment for age-related memory impairment has received strong support from the formulation of a cholinergic hypothesis of memory loss.[57] Starting from this observation it was a only a short step to test whether or not the performance of aged monkeys on memory tasks could be improved

Species, age	Test	Drug	Effect on memory	Ref.
Macaca, young adult	Visual recognition	Physostigmine, i.m., 0.06 mg/kg	Increase	59
Cebus, aged	Delayed visual recall	Physostigmine, Tacrine, p.o. several doses	Increase	56
Macaca, adult	Delayed spatial recall Visual recognition	Physostigmine, i.m., 0.03–0.08 mg/kg Tacrine, i.m., 0.8 mg/kg	Increase	58
Macaca, young adult, aged	Delayed match sample	Tacrine p.o., 0.5–2 mg/kg	Increase	60

Table 12.7
The effect of ChE inhibitors on the memory of non-human primates.

by a ChE inhibitor such as physostigmine (Table 12.7). In the 1990s, physostigmine was replaced by tacrine, which had already been demonstrated to have an effect in humans (see Table 12.7). Tests such as visual recognition, delayed visual recall and delayed match sample were used in young-adult, middle-age adult and aged monkeys. Table 12.7 summarizes the results of some representative studies in non-human primates with both drugs. The effect was an enhancement of memory function. In some cases, disruption of performance with a muscarinic antagonist (scopolamine) could be fully reversed by either physostigmine or tacrine.[58–60] These systematic changes in performance support the view that cholinergic mechanisms contribute to the memory process and suggest that cholinomimetic drugs could be of help in treating memory disorders, including AD.

Effect of ChE inhibitors on non-demented subjects: effect in the normal young and adults

Despite the early positive results in animals suggesting that the cholinergic system plays a key role in memory, the earliest clinical experimentation with ChE inhibitors in neuropsychiatry was not on memory dysfunction and dementia but on affective disorders such as mania and depression.[61] The first group of normal young subjects treated with a high dose of physostigmine i.v. showed an impairment of short-term memory and 'became tearful and depressed' (Table 12.8).[61]

In a second trial using a lower dose, storage and retrieval of information from long-term memory improved, while short-term memory was not altered.[62] Other trials involving the same low dose did not produce any clear change in short-term memory (see Table 12.8). These results confirmed the strong dose dependency of the effect (inverted-U response)

Memory	Dose (mg)	Age (years)	No. of subjects	Effect on memory	Ref.
Short term, long term	2–3 i.v.	21–55	13	Impairment	61
Long term	0.5 i.v.	18–35	10	Slight enhancement	61
Short term, long term	1 i.v.	18–35	19	Improvement in storage and retrieval	62
Short term	0.5 s.c.	20–35	10	Marginal	63
Short term	0.5 i.v.	20–35	32	None	64

Table 12.8
The effect of physostigmine on memory in normal young adults.

Memory	Dose (mg)	Age (years)	No. of subjects	Effect on memory	Ref.
Short term	2–3 i.v.	<65	6	Impairment of storage and retrieval	61
Short term	0.8 s.c.	64–82	13	Improvement in storage	63
Short term	0.5 i.v.	<65	3	Improvement	65
Short term	0.5 i.v.	64–77	16	No effect	64

Table 12.9
The effect of physostigmine on memory in aged normal subjects.

seen in animals, and made evident the difficulty of demonstrating a significant improvement in memory function in normal young subjects. The effects of physostigmine on cognition in young humans are subtle and trials require carefully controlled test conditions and dose ranges, together with highly sensitive neuropsychological tests. No long-term studies of the use of physostigmine in normal adults have been reported. Data related to the acute effects of physostigmine on memory are summarized in Table 12.8.[61–65]

Effects of physostigmine on memory in aged normal subjects

The first encouraging data showing positive effects of physostigmine on the memory of non-demented elderly subjects came from a study of three individuals with an average age of 64 years (Table 12.9).[65] The results of a previous clinical study were negative (see Table 12.9).[61] According to the authors 'the drug improves memory only in a narrow, low dose range that may vary across subjects'. These results were confirmed by two sub-

Diagnosis	*Dose (mg)*	*Age (years)*	*No. of subjects*	*Ref.*
Herpes simplex, encephalitis	0.8 s.c.	1	1	67
Huntington's chorea	0.5 i.v.	63	1	65
Post-trauma amnesia	1.5 p.o.	36	1	68
Post-encephalitic amnesia	2.5/day p.o.	20	1	69
Delirium	0.5 i.v.	79	1	70
Post-ECT amnesia	0.5 i.v.	55	17	71
Progressive supranuclear palsy	0.5–2 p.o.	76	7	72

Table 12.10
The use of physostigmine to improve memory in non-demented neurological patients.

sequent studies (see Table 12.9). The study by Drachman and Sahakian[63] showed a 'consistent trend toward improvement' and 'a considerable increase in memory storage'. Despite the positive and encouraging results, the authors concluded with a pessimistic comment: 'So far, therapeutic trials of choline or physostigmine in patients with obvious dementia have been generally disappointing'. Based on these considerations they recommend the use of cholinergic stimulation in normal ageing memory-impaired subjects rather than in demented patients. From these limited studies one could formulate the hypothesis that minimally cognitively impaired elderly subjects may be better responders to ChE inhibitors than normal aged subjects. The clinical verification of this hypothesis is now being pursued thanks to the availability of ChE inhibitors more efficacious than physostigmine.

ChE inhibitors improve memory of non-demented neurological patients

Central cholinergic agents with muscarinic antagonistic properties (e.g. scopolamine) are known to produce transient memory deficits in humans and animals. Brain damage resulting from herpes simplex or from other encephalitic infections also produces memory impairment. Other types of amnesia occur after trauma, after electroconvulsive therapy (ECT), in cases of progressive supranuclear palsy and in delirium (Table 12.10). It is difficult to recognize a common factor in these disorders, such as that cholinergic function could be altered in all states. However, as reported in Table 12.10, there is a limited number of observations of a positive effect of physostigmine on amnesia, which tends to support such an explanation. Particularly interesting are cases of severe acute amnesic syndromes following drug intoxication or overdosage (thioridazine, amitryptilline and ketamine) in which a reversal effect of tacrine was

Drug	Recent memory	Dose (mg/kg)	No. of patients	Effect of drug	Ref.
Thioridazine	Absent	0.75	1	Reversal	66
Amitryptiline	Decreased	0.75	1	Reversal	66
Ketamine	Delirium	1	49	Reversal	73

Table 12.11
The effect of tacrine on acute amnesic syndromes and delirium following drug intoxication or overdosage.

observed. The use of tacrine in cases of delirium is not accidental as the compound itself was originally developed to treat cases of post-anaesthesia confusion (see Table 12.10).

Treatment of tricyclic antidepressant overdosage with physostigmine has been reported in several studies since the 1970s.[66] It is of interest to note that it was Summers and Kaufman[66] who a few years later pioneered the use of tacrine in AD. The effect of tacrine in early studies on acute amnesic syndromes and delirium following drug intoxication or overdosage is reported in Table 12.11.[66–73]

ChE inhibitors in AD therapy

ChE inhibitors improve behaviour in AD patients

It is known from classical cholinergic toxicology (see Chapters 1 and 11) that accidental intoxication with ChE inhibitors, gene overdose of AChE in transgenic animals (see Chapter 4) or administration of cholinergic antagonists (particularly in elderly subjects) may produce toxicity symptoms characterized not only by cognitive deficits but also by hallucinations, delusions, a high stress level and motor agitation. There is evidence that cholinergic deficits in the limbic or paralimbic structures in AD patients may contribute to the development of behavioural abnormalities in these patients. Both clinical and experimental evidence indicates interactions between the cholinergic system and the biogenic amine system in the cognitive impairment observed in AD.[12] Additional information has been derived from microdialysis experiments in rodents, with simultaneous measurement of ACh and biogenic amine levels in the CNS.[10,12] Changes in extracellular cortical levels of ACh following AChE inhibition are related to significant increases in norepinephrine as well as dopamine.[10,12] These results may help in understanding the behavioural and cognitive effects of ChE inhibitors[12] (see Chapter 13). Human pharmacology data support the concept that stabilization of cholinergic function could improve both behavioural symptoms and cognitive deficits. Early studies with physostigmine and tacrine showed a reduction in psychotic symptoms and improvements in apathy in AD patients and in psychotic symptoms in delirium[66] (see Tables 12.10, 12.11 and 12.12).

Non-cognitive symptoms are common features of AD, occurring in almost 90% of patients. Multidimensional assessments suggest that apathy or indifference is the most common behavioural disturbance (70%) in AD patients. Major depression occurs in 10–17% and mild depressive symptoms in almost 40% of all AD patients. Psychotic symptoms, including delusions, hallucinations and misindentification syndromes, appear in approximately 25–50% of patients with mild or severe AD.[74] The neuropsychiatric symptoms of AD are often responsible for patient distress, motor agitation and increased disability, as well as for an intolerable burden and stress for care-givers. Care-givers may ultimately themselves suffer anxiety disorders, depression and stress-related illness. Biochemical and pharmacological considerations suggest the involvement of a cholinergic deficiency in the mechanism of psychotic symptoms in AD. Furthermore, anticholinergic agents (including anti-parkinsonian and antidepressant drugs) can induce psychosis, while cholinergic agents (ChE inhibitors and muscarinic agonists such as xanomeline) have been shown to ameliorate psychosis and reduce agitation.[75] It is interesting to note that the first clinical study of ChE inhibitors involved affective disorders such as mania and depresssion.[61]

Severe pathology (increased regional plaque and tangle density) is seen in the temporal and frontal lobes, together with decreased cerebral blood perfusion (CBF) and glucose hypometabolism in agitated AD patients. These findings may link memory impairment, agitation and psychotic behaviour with severe cholinergic denervation in AD. As a consequence, ChE inhibitors that ameliorate cholinergic function should also decrease psychotic symptoms and agitation. There have been only a few controlled studies of behavioural treatment in AD, mainly with non-specifically acting neuroleptics. The first ChE inhibitor reported to have beneficial effects on the behaviour of AD patients was tacrine[74] (see Table 12.12). Its effects on behaviour may partially account for the reported delay in nursing-home placement of these patients. Preliminary evidence suggested that donepezil may also improve behavioural symptoms[74] (see Table 12.12).

Two larger studies specifically linking cholinergic effects to behavioural symptoms involved metrifonate[76,77] (see Table 12.12). Both studies show significant positive effects of the drug on behaviour. One of these studies, a 36-week multicentre study of over 1500 patients tested using the Neuropsychiatric Inventory (NPI), showed an effect of metrifonate not only on cognitive and global function but also on behavioural disturbances.[76] The most reduced symptom was hallucination. This is the first clear demonstration that a ChE inhibitor can favourably affect the behavioural function of AD patients. It is also the first study of any drug therapy for AD to show concurrent cognitive, global and behavioural benefit, and thus suggesting a cholinergic link between cognitive and behavioural deficits in AD. The second of these studies, a 26-week study of 450 patients, showed that 60% of the metrifonate-treated patients experienced no deterioration in their psychiatric and behavioural symptoms (NPI score) and 40% significantly improved in their symptoms.[77] Metrifonate treatment was particularly effective in decreasing agitation, hallucinations, depression, apathy and aberrant motor behaviour.[77] The attenuation of such symptoms may reduce care-giver burden and significantly decrease the costs of patient care.

This antipsychotic effect of the ChE inhibitor is unique from the psychopharmaco-

Drug	Delusions, hallucinations	Apathy	Motor agitation	Depression, anxiety	Ref.
Metrifonate	++	++	++	++	76, 77, 149
Physostigmine	+	–	+	–	147
Donepezil	+	+	+	+	74
Tacrine	+	+	+	+	74, 75

++, 0.05 vs placebo.

Table 12.12
Effects of ChE inhibitors on behavioural symptoms in AD.

logical point of view. No neuroleptic drug effect can be linked to one single neurotransmitter in the way that the effect of a ChE inhibitor is related to ACh. One practical advantage of using ChE inhibitors to treat behavioural symptoms in AD patients is that the concomitant use of neuroleptics, and thus their potential extrapyramidal side-effects, can be avoided. More work needs to be done in order to understand the complex effect of ChE inhibitors on behaviour.

ChE inhibitors: how do they work and are there differences between them?

ChE inhibitors that have been tested in clinical trials or are in current (1999) use in Japan, the USA and Europe number less than ten drugs (Table 12.13). Most of these compounds have advanced to clinical Phase III and IV trials, and three (tacrine, rivastigmine and donepezil) are registered in the USA and in Europe.[12] Two other compounds (galantamine and metrifonate) are awaiting registration in both the USA and Europe (see Table 12.13). The second-generation ChE inhibitors (donepezil, rivastigmine and metrifonate), in order to replace tacrine, had to fulfil specific requirements, such as a lower toxicity (hepatic) and easier administration, besides being demonstrably clinically effective.[12] Based on current selection criteria and toxicity, a number of ChE inhibitors, particularly certain carbamates, which were in advanced clinical phases have been discontinued (see Table 12.13). Eptastigmine, a carbamate analogue of physostigmine, was subjected to clinical trials in the USA and Europe, and showed excellent efficacy and low side-effects. However, it was withdrawn from clinical studies because two patients developed aplastic anaemia. The cause of eptastigmine bone marrow toxicity is unknown, but could be related either to the presence of the heptyl chain (physostigmine does not produce this effect) or to haematotoxic effects of the eseroline metabolite. A second carbamate (NX-066) was withdrawn because of similar haematological complications.

There are differences among the presently used compounds with regard to their efficacy, percentage of treatable patients and responders, dropouts, and severity and incidence of side-effects. The effects of eight ChE inhibitors on the AD Assessment Scale–Cognitive sub-

Compound	Country	Company	Clinical phase§	Side-effects comments
Physostigmine, slow release	USA	Forest	III	Moderate
ENA713, Rivastigmine	USA, Europe	Novartis	IV R	Low
Eptastigmine*	USA, Italy	Mediolanum	III	Haematology‡
E2020, Donepezil	USA, Japan	Pfizer, Eisai	IV R	Low
MDL 73,745*	USA, Europe	Marion Merrell Dow	–*	Low
Metrifonate	USA, Germany	Bayer	IV	Low
Tacrine (THA)†	USA, Europe	Warner-Lambert	IV R	Hepatotoxicity
Velnacrine (HP029)*	USA, Europe	Hoechst-Roussel	–*	Haematology‡
Suronacrine (HP128)*			–*	Hepatotoxicity
Galantamine	Germany	Shire-Pharm	III	Low
	USA	Janssen	III	
Methanesulphonyl fluoride	USA	U Texas	II	Low
Huperzine A	China	Chinese Academy of Science	III	NA
NX-066*	England, USA	Astra Arcus	–*	Haematology‡
CP-118,954*	USA	Pfizer	–*	NA
KA-672	Germany	Schwabe	I	NA
NIK 247	Japan	Nikken	III	Low
TAK 147	Japan	Takeda	III	Low
GEN 2819	Italy	Chiesi	II	Low

*Withdrawn.
†Other indications: HIV, tardive dyskinesia.
‡Bone marrow toxicity.
§Clinical Phase in the USA/Europe. R, registered in the USA and/or in some European countries and extra-European countries.
NA, not available.

Table 12.13
Clinical trials of the use of ChE inhibitors in AD.

scale (ADAS-Cog) test, using intention-to-treat (ITT) criteria are compared in Table 12.14.[78–92] Pharmacologically, these drugs represent either reversible (tacrine, donepezil, eptastigmine and galantamine) or pseudo-irreversible or irreversible (rivastigmine, metrifonate, methanesulphonyl fluoride (MSF) and physostigmine) ChE inhibitors. The duration of these Phase III clinical trials varied from 24 to 30 weeks (with the exception of that of MSF) and over 10 000 patients from 26 different countries were tested. The six most extensively clinically tested ChE inhibitors (tacrine, eptastigmine, donepezil, rivastigmine, metrifonate and galantamine) all produced statistically significant improvements in multiple clinical trials using similar standardized and internationally validated measures of both cognitive and non-cognitive function. The structure of ADAS-Cog is such that it measures four cognitive areas (memory, orientation, language and praxis) as well as non-cognitive symptoms. The total number of points assigned to the cognitive area is 70. The mean annual change in ADAS-Cog scores in untreated AD patients has been evaluated in longitudinal studies to be approximately 8–9 points/year. Obviously, there are large individual variations, as the level of change seems to be dependent on the degree of illness. The magnitude of cognitive effects measured with the ADAS-Cog scale for all six drugs either expressed as a difference between drug- and placebo-treated patients or as the difference between drug-treated patients and baseline is similar under present treatment conditions (see Table 12.14). This similarity in size of cognitive improvement at 26–30 weeks of treatment suggests a 'ceiling effect' of approximately 5 points (ADAS-Cog) average for approximately one-third of patients in mild to moderate (CDR 1–1.5) stages of the disease at presently used doses. It should be pointed out that both the clinical and economic significance of an improved ADAS-Cog score becomes more substantial when evaluated over one year (8–9 points or more). The differences in the effect of the various drugs may be related in part to the rate of deterioration of the placebo group, which can vary from trial to trial. These results suggest that the maximal clinical effect has not yet been reached with all the drugs tested at the present dosages and the present degree of severity of the disease (mild to moderate). Analysis of the results suggests that both very mild and more severe cases need to be studied.

The data show wide variations in the magnitude of effect among patients treated with same drug. In some patients the effect can be twice as large as the average effect. Cholinergic side-effects are transient, reversible and very similar for all drugs. They are predictable on the basis of our pharmacological knowledge. A small percentage of the more severe non-cholinergic side-effects seen with some particular drugs (such as hepatotoxicity with tacrine) or the bone-marrow toxicity seen with certain carbamates (velnacrine, eptastigmine, etc.) are less predictable and are difficult to explain. To our knowledge, no deaths directly related to the use of ChE inhibitors in the treatment of AD have been reported. However, the cholinergic toxicity related to the maximum tolerated dose (MTD) suggests a limit in safe achievable levels of ChE inhibition.

Similarity in clinical efficacy is also underlined by similar but not identical effects on global scales such as the Clinician's Interview Based Impression of Change–Plus (CIBIC-plus) of 0.3–0.5 points difference in favour of treatment with donepezil, rivastigmine and metrifonate, drugs with very different chemical structures and pharmacological profiles. The percentage of improved patients varies from 25% (rivastigmine low dose) to 50% (tacrine and donepezil high dose) with an average of 34%. This indicates that more than one-third of treated patients showed a significant clinical response to the ChE inhibitor. This effect can be maintained for four drugs (tacrine, donepezil, rivastigmine and metrifonate) for at least one year, representing a high impact value for patients and caregivers.[98–100] A smaller percentage (about 10–15%) of patients did not improve on the ADAS-Cog score with any of the drugs used, while a second group of patients (5% or more) demonstrated an improvement significantly higher than 5 points. This difference between responders and non-responders may reflect the level of cholinergic damage present in the brain, genetic factors or gender (e.g. APOE-4 alleles) (Table 12.15) and also too low levels of ChE inhibition in brain.[93–97] From the 6-month data presented in Table 12.14 it can

Drug	Dose (mg/day)	Duration of study (weeks)	Treatment difference From placebo*	From baseline†	Improved (%)	Drop-outs (%)	Side-effects (%)
Tacrine[78,79]	120–160	30	4.0–5.3	0.8–2.8	30–50	55–73	40–58
Eptastigmine[80,81]	45	25	4.7	1.8	30	12	35
Donepezil[82–84,147]	5–10	24	2.8–4.6	0.7–1	40–58	5–13	6–13
Rivastigmine[85,86,151]	6–12	24	1.9–4.9	0.7–1.2	25–37	15–36	15–28
Metrifonate[87–90]	25, 75, 80	12–26	2.8, 3.1, 3.2	0.75–0.5	35–40	2, 21, 8	2–12
Metrifonate[91]	60–80	26	3.9	2.2	40	15	7
Galantamine[92,152]	30	12	3.3	1.8	–	33	–
MSF[144]	10–13‡	16	6.8	3.4	80	0	–
Physostigmine[154] (controlled release)	30–36§	24	1.6–2.3	1.8	–	–	–

*Study end-point vs placebo.
†Study end-point vs baseline.
‡3 times weekly, methanesulphoryl-fluoride (MSF).
§ $n = 15$.

Table 12.14
The effects of eight ChE inhibitors as assessed on the ADAS-Cog test (ITT).

be seen that patients treated with the active compound changed little cognitively or behaviourally over the course of the study (Fig. 12.7). This trend suggests a stabilization effect of disease-related deterioration which is clinically more significant than the expected symptomatic improvement. This stabilizing effect is particularly evident for certain ChE inhibitors such as metrifonate, which significantly improved cognitive performance above pretreatment baseline levels at 26 weeks (see Table 12.14 and Fig. 12.7).[87,88]

Recent data suggest the presence of long-term efficacy of ChE inhibitors and indicate that the treatment could be prolonged for up to 2 years during which period the drug is well tolerated and still efficacious (Table 12.16). The long-term effect of ChE inhibitors has been suggested in seven different studies using five different drugs (donepezil, tacrine, metrifonate, galantamine and rivastigmine) extending from 1 to 4.5 years and using various tests. Some of these data can be criticized for not being placebo controlled, as deterioration (expressed in ADAS-Cog, MMSE or CDR scores) was calculated on the basis of an anticipated annualized rate of deterioration. Due to large variations among patients it is difficult to accurately assess long-term effect differences between drugs. As a consequence, studies to evaluate specifically long-term efficacy need to be redesigned, particularly to determine if a protective effect for subjects with minimal cognitive impairment but at risk for conversion to

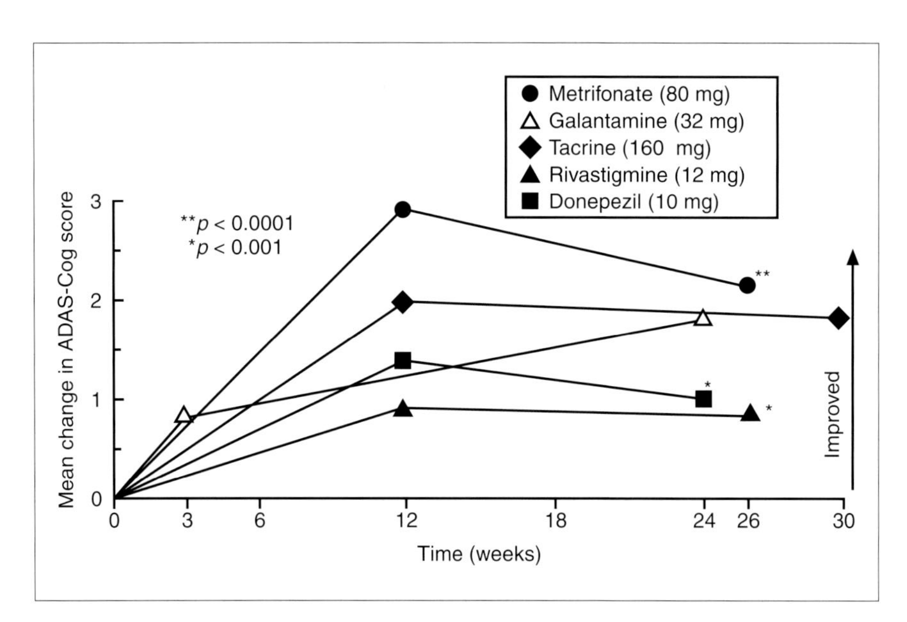

Figure 12.7
Stabilization effect of 6-month treatment with five ChE inhibitors. The patients treated with the active compound changed little cognitively from baseline. This stabilizing effect is particularly evident with metrifonate treatment, which significantly stabilizes cognitive performance above pretreatment levels at 26 weeks.[87,88]

Drug	*Measure*	*Response difference (APOE-4 +ve vs APOE-4 −ve)*
Tacrine[93]	ADAS-Cog	No difference
Metrifonate[94]	ADAS-Cog	Decrease
Tacrine[95]	ADAS-Cog	No difference
Tacrine[96]	ADAS-Cog	Decrease in women
Tacrine[97]	ADAS-Cog	Decrease
Xanomeline94	ADAS-Cog	Decrease

APOE-4, apolipoprotein.

Table 12.15
Effect of APOE-4 on the efficacy of cholinomimetic drugs in AD.

AD can be demonstrated. Evidence of such an effect would modify our present definition of ChE inhibitors from drugs with a symptomatic stabilizing effect to drugs with a neuroprotective action, and extend treatment indication to individuals identified as subjects at risk (familiar cases or carriers of high-risk genes).

Differences in cholinergic effects among ChE inhibitors

The relationship between peripheral ChE inhibition (plasma or red blood cells (RBC)) and cognitive (ADAS-Cog) or global impression of change (CGIC) effect is reported in Table 12.17.[4,5,78–82,92,101,103,104] These data support

Drug	No. of subjects	Max. treatment duration (years)	Test	Benefit difference	Ref.
Donepezil	1600	4, 5	ADAS-Cog	Positive	98
Donepezil*	431	1	ADFACS, CDR	Positive	155
Donepezil*	286	1	GBS, MMSE	Positive	156
Tacrine	25	1	MMSE-EEG	Positive	99
Metrifonate	605	3	ADAS-Cog, MMSE	Positive	90
Rivastigmine	2149	2	ADAS-Cog, MMSE, CIBIC, GDS	Positive	100
Galantamine	44	3	ADAS-Cog	Positive	153

*Prospective, placebo-controlled.

Table 12.16
Long-term efficacy of ChE inhibitors in AD.

basic principles of cholinergic pharmacology, e.g. that brain ChE inhibition relates directly to an increase in synaptic ACh levels which may lead to cognitive and memory improvement. This relationship might vary for each drug, as each compound may produce significant cognitive improvement and therapeutic effect at different levels of ChE inhibition.[5,12,104] There is a correlation between cognitive effects and level of AChE inhibition. This correlation would be best observed in brain or in CSF, in agreement with pharmacological data in animals[12,101] and humans.[12,103,104] The level of peripheral enzyme inhibition seen in the patient as AChE activity in RBC or plasma BuChE activity represents an indirect measure of the drug effect (such as drug concentration). It varies between 30% and 80%, depending on dose and pharmacokinetic characteristics of the compound (see Table 12.17). The ChE inhibition measured in the periphery (plasma or RBC) does not accurately reflect the level of CNS inhibition and relates more to peripheral side-effects than to central cognitive effects. For some drugs (donepezil and metrifonate) the mean level of peripheral ChE inhibition is 65–70%, and could be safely brought to 90%. For other drugs, the practical limit of inhibition can be as low as 30% (physostigmine, tacrine and eptastigmine) and may be increased only at the expense of severe side-effects. As predicted by pharmacological and behavioural data, there is a direct correlation between CSF or brain AChE inhibition and cognitive effect.[10,101–106] Also, as expected from results obtained in animal models, there is no direct correlation between CNS AChE inhibition and the severity of side-effects[6] as many of these effects relate mainly to peripheral inhibition.[8,81,104] As CSF AChE inhibition cannot be readily monitored, it may be practical to correlate cognitive or other effects with peripheral (RBC) AChE inhibition. As an example, following metrifonate treatment, the degree of improvement in ADAS-Cog and CIBIC-plus is directly proportional to AChE inhibition, while side-effects are very mild even at a high level (70–80%) of inhibition.[107,108] A direct clinical implication of this relationship is that drugs producing high levels (70–80%) of central AChE inhibition at low dosage and having short (1–2 h) half-lives will produce only mild peripheral cholinergic side-effects. A large increase in brain ACh may be produced in the patient within a full range of therapeutic potency. For example, rivastigmine at doses of 6–12 mg/day (corresponding to 62% inhibition in CSF at 6 mg) will produce a significantly greater improvement in cognitive function than at a dose of 1–4 mg.[85,86] A further increase in dose may be more effective, but probably would increase side-effects significantly.

What makes ChE inhibitors more effective and less toxic?

The pharmacologically predicted U-shaped curve of the relationship between ChE inhibition and cognitive effect can be observed for some ChE inhibitors within a certain dose range in humans (see Table 12.17). This relationship is explained by the fact that, by increasing the dose of the inhibitor, efficacy is progressively increased until adverse effects become a limiting factor. A second explanation is the specific inhibition kinetics of ChE inhibitors and the substrate-induced saturation effect on the enzyme (substrate inhibition effect of ChEs) (see Fig. 12.4). Within normal conditions, the level of brain ACh is feedback regulated (through its synthesis) in order to maintain steady-state levels, and varies according to AChE inhibition.[12,101,108] With increased brain concentrations of ACh, the substrate inhibition kinetics of enzyme activity become a limiting factor (see Fig. 12.4). This relationship is observed in brain tissue in vitro and is present also in vivo.[101,108]

A plot of the velocity of enzymatic reaction against substrate (ACh) concentration is a bell-shaped curve with a defined peak in the case of AChE activity in brain or RBC, and a sigmoid curve in the case of BuChE activity in plasma (see Fig. 12.4). Thus, a large excess of ACh, caused by a high level of ChE inhibition, can inhibit ChE activity. In other words, an increase in substrate decreases the catalytic potency of the enzyme and subsequently its pharmacological effect. From this relationship it can be predicted that high ChE inhibition reached too rapidly during treatment because of rapid passage of a high dose of drug into the brain and its accumulation in CNS will not further increase efficacy but only augment CNS-dependent side-effects (drowsiness, vomiting, etc.). In addition, ChE inhibition in peripheral organs (gastrointestinal tract, bronchi, heart and muscles) may produce a number of commonly observed adverse effects.

For most ChE inhibitors this effect can be reduced by progressive and slow dose titration. It is advantageous to use a slow-release type of ChE inhibitor such as metrifonate. For practical reasons of administration, in order to reach steady-state levels of brain ACh gradually, an escalating dosage should not be necessary. Slow-release-type administration may also lower the risk of cholinergic receptor down-regulation and tolerance-producing enzyme induction. There is an agreement between clinical and animal data for both physostigmine and tacrine with regard to dose–behaviour relationships. Rupniak *et al.*,[109] using two primate models in rhesus monkeys, found that both tacrine and physostigmine improved visual recognition memory significantly. Both drugs showed a clear inverse U-shaped relationship, with a maximum effect at around 0.0010–0.02 mg/kg i.m. physostigmine and 0.8–1 mg/kg tacrine. Lower or higher doses did not improve drug performance, only side-effects. Central cholinergic side-effects, which may develop early in treatment, are not related directly to brain AChE inhibition but mainly to elevation of CNS ACh levels.[101,108] In the rat, this increase in ACh is correlated not only to AChE inhibition, but also to other synaptic mechanisms.[10] Peripheral side-effects may occur depending on a rapid redistribution of the drug (or of its metabolites) between non-CNS (peripheral organs such as bronchi, gastrointestinal tract, heart and muscles) and CNS compartments and peripheral (BuChE) inhibition.

The occurrence of both peripheral and central side-effects seen with all the compounds tested so far demonstrates that none of the presently available inhibitors is truly 'brain specific'. With very long-acting ChE inhibitors (such as methanesulphonyls and organophosphates) the rate of enzyme activity recovery is limited by the rate of synthesis of new enzyme. This rate of synthesis is different for each tissue. In rat brain, the synthesis of new enzyme occurs with a mean half-life of approximately 11 days, compared to 1, 3 and 6 days for ileum, heart and muscles, respectively.[110] This ten-fold difference between slow regeneration of brain AChE and rapid re-synthesis of peripheral enzyme produces good brain selectivity in these long-acting compounds, thus favouring therapeutic effect over side-effects. This mechanism explains the low toxicity of metrifonate and methanesulphonyl fluoride even at high (80–90%) levels of AChE inhibition as compared to other ChE inhibitors producing a lower level of inhibition (see Table 12.14).[111]

Tolerance of the effects of ChE inhibitors: is it a problem?

The cognitive effect (ADAS-Cog) seen with most ChE inhibitors becomes statistically significant after 2–3 weeks of treatment. This delay may be partly due to the attenuation of the placebo effect. In most patients, following a period of 30–36 weeks of treatment a decrease in clinical effect is observed. Does this progressive decrease depend exclusively on patient deterioration, or do other factors contribute to lower clinical efficacy?

Tolerance to repeated doses of ChE inhibitors may contribute, combined with patient deterioration, to the attenuation of the clinical effect of the drug. There is a vast literature addressing the phenomenon of tolerance to both single and repeated doses of ChE inhibitors.[8,12,112] Adaptation due to a decreased effect of the ChE inhibitor is supported by behavioural as well as toxicological studies.[112] Tolerance to repeated doses of a ChE inhibitor might be explained by two mechanisms. First, a reduction in sensitivity to ACh depending on a decreased number (down-regulation) of muscarinic and/or nicotinic receptors. This explanation may be valid within experimental conditions, but is more difficult to apply under actual treatment conditions as AD patients show a deficit in both the synthesis and levels of ACh and a decrease in the number of nicotinic (but not muscarinic) receptors.[101,106]

A second, more likely, explanation is induction of enzyme synthesis by increased AChE gene expression at nerve terminals (see Fig. 12.6). The signal for increased gene expression may be through direct stimulation of m1 muscarinic receptors by the ChE inhibitor or by ACh itself (see Fig. 12.6).[43] Von der Kammer *et al.*[113] demonstrated that activation of m1 increased transcription from Egr-dependent promoters, including the AChE promoter. Therefore, the AChE gene can be under the control of extracellular signals coupled to muscarinic receptor activation, such as the direct or indirect effect of ACh, muscarinic agonists or ChE inhibitors. This effect is reflected by an increased level of AChE activity (not an inhibition) in the CSF of patients treated long-term. The AChE activity in CSF originates from spinal cord and brain neurons

Drug	Dose (mg/day)	Steady state (% inhibition)	Optimal ChE† (% inhibition)	ChE inhibition (ADAS-Cog or CGIC scores)	Ref.
Physostigmine	3–16	40–60	BuChE 30–40	U-shaped	4, 5, 103
Rivastigmine	6–12	37	BuChE 30–40	Linear	151
Eptastigmine	30–60	13–54	AChE 30–35	U-shaped	80, 81, 104
Metrifonate	30–70	35–75	AChE 65–80	U-shaped	105, 145, 146
Donepezil	5	64	AChE 60	Linear	82
Tacrine	160	40	BuChE 30	Linear	78, 79
		60	AChE –	–	
Galantamine	20–50	50–60	AChE 50	U-shaped	92
MSF	10–13*	85–90	AChE 85	–	144

*Three times weekly.
†AChE, in RBC; BuChE, in plasma.

Table 12.17
Relationship between percentage ChE inhibition and ADAS-Cog and CGIC scores.

as a result of a secretory process (see Fig. 12.5).[114,115] This secretion shows a tendency to decrease with increasing severity of dementia and disease progression (see Table 12.5).[116]

Stimulation of AChE release from neurons into the CSF by ChE inhibitors has been postulated by Bareggi and Giacobini[114] (see Fig. 12.5). Increased AChE activity in the CSF was demonstrated by Mattio *et al.*[102] in dogs chronically administered high doses of physostigmine intraventricularly. Muscarinic antagonists block the increase in ChE activity. An example of up-regulation of AChE activity in the CSF of tacrine-treated patients (80–160 mg for 12 months) has been reported by Nordberg *et al.*[117] (Table 12.18). In that study no change was observed in AChE inhibition in RBC while a 50% increase (sic) in AChE (but not BuChE) activity was seen in CSF. Several other observations indicate that increased production of both AChE and BuChE may occur in dogs and rats treated with various types of ChE inhibitor (see Fig. 12.5).[110,118] One should note that resaturation of enzyme activity after inhibition follows different mechanisms for reversible (e.g. carbamates and acridines) and irreversible (e.g. organophosphates) inhibitors. Metrifonate, an organophosphate, does not produce AChE up-regulation in the brain or RBC of rats treated for 12 weeks,[118] while physostigmine and analogues such as heptastigmine and MF268 show decreased inhibition after 2 weeks' administration to rats.

Plasma BuChE activity, in particular, might overshoot the basal level for some time after dosing is stopped. Particularly relevant to clinical conditions is the increase in plasma BuChE activity seen in humans receiving long-term doses of organophosphates such as parathion, either orally or in spray form.[119]

Drug	Administration (dose)	AChE activity (%) CSF	AChE activity (%) RBC	Time	Ref.
Physostigmine	1 μg i.c.v.	30	100	5 min	6
Physostigmine	8 μg i.c.v.	15	100	5 min	6
Rivastigmine	3 mg p.o.	60	90	2 h	85
Tacrine	80–160 mg p.o.	150	75	12 months	117

i.c.v., intracerebral ventricular.

Table 12.18
The effect of ChE inhibitors on CSF and RBC AChE activity.

Plasma BuChE activity increases after 2–5 days of treatment and stabilizes between day 7 and 10 of treatment. After exposure is stopped there is a rebound of ChE activity, sometimes reaching twice the normal level.[120] In AD patients, in contrast with those treated with tacrine, RBC AChE inhibition was the same at 2, 8 and 12 weeks following administration of metrifonate (0.65–2.0 mg/day).[121–123] These results in humans confirm the observation of Hinz *et al.*[118] of no apparent enzyme up-regulation in rat brain following administration of this compound; this represents an advantage in terms of clinical application of the drug.

In conclusion, there seems to be a difference in the production of tolerance and enzymatic induction between the different compounds. Is there also a behavioural tolerance to ChE inhibitors? Behavioural adaptation to a number of ChE inhibitors, from organophosphates to carbamates, has been described in numerous publications.[11,112]

Reduction in the sensitivity or the number (down-regulation) of both muscarinic and nicotinic receptors after repeated doses of ChE inhibitors has been demonstrated to be associated with tolerance in several studies.[8,111,112] Can tolerance be avoided or reduced in a clinical setting in order to prolong the effect?

Two strategies to limit tolerance to ChE inhibitors could be used during the course of treatment: progressive dose readjustment with slow titration to higher levels of inhibition; or temporary discontinuation (washout periods) until AChE activity returns to baseline levels. With both approaches control of RBC/CSF AChE activity ratio would be of value. The washout period should be followed by a restart of treatment at the same or at a higher dose. Long-term trials have shown that such a strategy is feasible at the expense of a small reduction in clinical effect.

Pharmacokinetic differences

A summary of the pharmacokinetic properties of six ChE inhibitors currently being tested or used in AD therapy is presented in Table 12.19.[84,85,93,123,124] Several important differences are apparent with regard to metabolism, as well as other characteristics. While tacrine, galantamine and donepezil are metabolized through the hepatic route (P450), rivastigmine and metrifonate are not. Metrifonate, in particular, is transformed non-enzymatically into an active compound (2,2-dichlorovinyl-dimethyl phosphate, DDVP) which is a potent inhibitor of ChE. This difference is clinically important since elderly patients show decreased hepatic metabolism, and therefore drugs that are not hepatically metabolized are preferred.

Another important characteristic is the difference in drug elimination. The half-life is long (73 h) for donepezil, intermediate (5 h) for rivastigmine and short (2 h) for metrifonate. Such a difference is important as a short half-life reduces the time of exposure of the peripheral pool of ChE to the inhibitor, thus decreasing side-effects. Bioavailability is maximal for galantamine and metrifonate (100% and 90%, respectively). Plasma protein binding is lowest (10% and 20%, respectively) for galantamine and metrifonate and highest (96%) for donezepil. Since elderly patients are generally treated with several drugs simultaneously, this factor is of particular interest in relation to drug–drug interactions.

The pharmacokinetic properties of the ChE inhibitors may constitute important differences with regard to the efficacy and the severity of side-effects. In order to maximize therapeutic brain effects and minimize peripheral (bronchial, muscular, gastrointestinal and cardiac) side-effects, the elimination half-life should be short (around 1–2 h). The effective dose should be low but able to produce a substantial CNS enzyme inhibition (60–80%), as well as a steady (small diurnal–nocturnal variations) and long-lasting (several days) level of inhibition. Irreversible ChE inhibitors (such as metrifonate and methanesulphonyl fluoride) satisfy such criteria more closely than do the reversible ones.[110]

Long-term effect of ChE inhibitors

The long-term effects and prolonged clinical efficacy of ChE inhibitors are suggested by two kinds of observation. First, if drug treatment is interrupted, cognitive effect may continue for several weeks even in the absence of ChE inhibition. This suggests that the effect of ChE inhibitors may not be only symptomatic and solely related to an elevation in brain ACh levels. Other mechanisms such as the modification of APP metabolism could be present (see Figs 12.2 and 12.8). β-Amyloid deposition in AD brain has been linked to AChE expression through an AChE–β-amyloid cycle process (see Fig. 12.8), as demonstrated by both in vitro[125–127] and in vivo studies.[11] Long-term studies for periods longer than one year (up to 4.5 years) (see Table 12.16) suggest that in the presence of the drug, clinical efficacy may continue for one year or more during which time the drug is well tolerated. The studies listed in Table 12.16 indicate that benefit differences can be maintained in a number of patients for 1–2 years for five different drugs (donepezil, tacrine, eptastigmine, metrifonate, rivastigmine and galantamine). In terms of improvement in the ADAS-Cog score this may sum up to a total of 15–20 points (2 years difference). Studies to measure specifically the long-term efficacy are being designed. It is particularly important to evaluate a potential protective effect of therapy on minimally cognitively impaired subjects at risk of conversion to AD. As an example, the results of a study demonstrating long-term (60–104 weeks) cognitive benefits with rivastigmine are shown in Fig. 12.9.[100] After 39 weeks of treatment, the drug-treated group of patients was still above baseline. At 2 years (104 weeks) the small decline in ADAS-Cog from baseline seen in patients treated with the

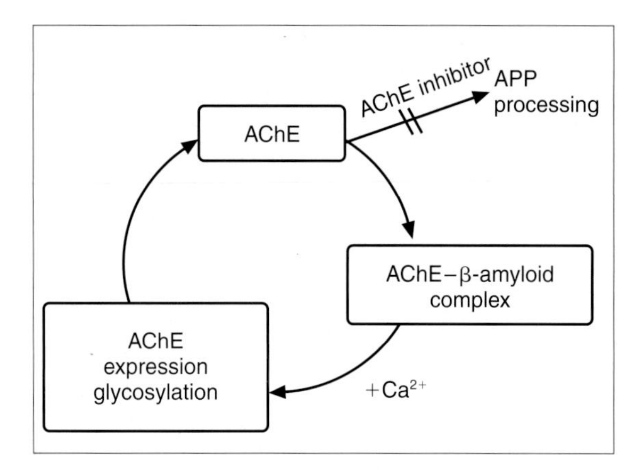

Figure 12.8
Proposed AChE–β-amyloid cycle. AChE co-localizes with β–A and accelerates β-amyloid formation and deposition in AD brain. Reciprocally, β-amyloid protein regulates AChE expression, assembly and glycosylation. As a result, AChE which is expressed around neuritic plaques influences β-amyloid formation which then stimulates AChE formation.[125–127] Inhibition of AChE may influence APP processing and β-amyloid deposition.[11]

drug from day 1 was still above the decline in placebo-treated patients in the first 6 months.[100]

Potential interactions with other drugs and side-effects

Non-selective or selective AChE inhibitors all produce similar adverse effects which are related to cholinergic central and peripheral hyperactivity. Peripheral side-effects are not the result of BuChE inhibition since their incidence is not decreased in drugs that selectively inhibit AChE, such as donepezil.

ChE inhibitors tested for the treatment of AD are generally well tolerated despite the fact that patients are elderly individuals (often over 80 years old) displaying multiple morbidity and being treated with several drugs simultaneously. Side-effects are generally mild to moderate, of short duration and resolve spontaneously or after reducing the dose. Severe side-effects are rare and are generally reversible events. Neutropenia has been seen in a very few patients treated with eptastigmine, and leukopenia or agranulocytosis has occurred following treatment with velnacrine but so far has not been reported with any other ChE inhibitor. The side-effects of ChE inhibitors are mainly cholinergic in nature and predictable. Side-effects occur most often during the initial 1–2 week initial (or titration) phase and tend to decrease in intensity or disappear with time. During the maintenance phase, adverse effects occur at a rate close to that of placebo. Long-term treatment of one year or longer does not produce a higher frequency of side-effects than shorter (6 months) treatment periods. Typical side-effects are nausea, vomiting, dizziness and diarrhoea, which relate mainly to peripheral ChE inhibition. They can be attenuated by muscarinic antagonists that do not penetrate the CNS, such as glycopyrrolate. In animals, complete BuChE inhibition with highly selective inhibitors does not result in an increase in side-effects.

As shown in Table 12.14, the occurrence (as a percentage of treated patients) of side-effects varies from drug to drug, being the highest for tacrine due to its hepatic toxicity. For other drugs occurrence varies at the higher doses from 12% to 35%. As adverse events leading to discontinuation of treatment occur more frequently during dose titration, some drugs (such as rivastigmine) require a slow titration period. Other drugs (donepezil and metrifonate) can be administered at a fixed dose without a phase of slow titration, and thus a safe maintenance dose can be reached

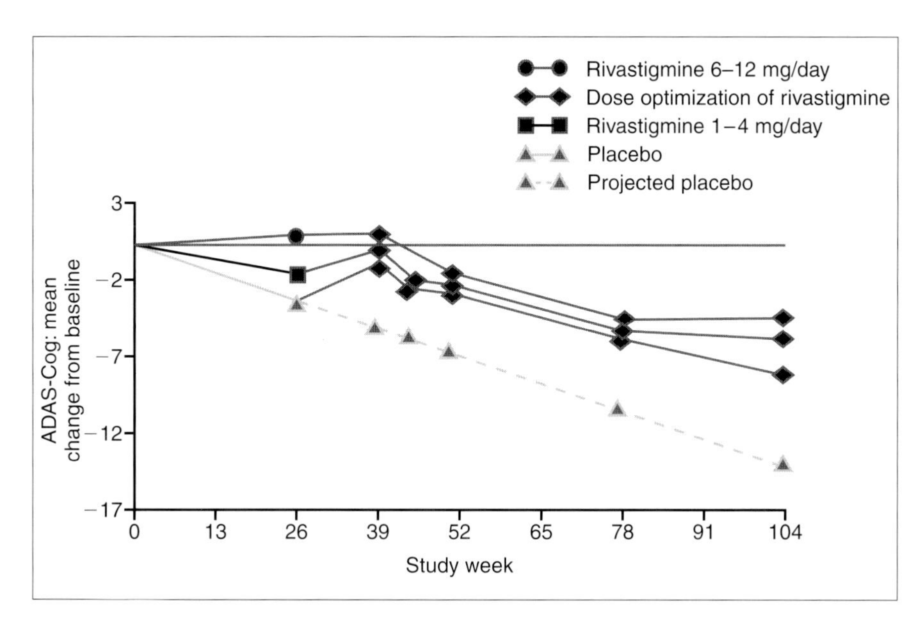

Figure 12.9
Long-term effects of rivastigmine on ADAS-Cog score. All patients were dose-optimized for rivastigmine.[100]

more quickly. Side-effects depend on the rate of inhibition of both central and peripheral ChE. A gradual and smooth increase in brain ChE inhibition and a short-lasting peripheral inhibition represent two key factors in decreasing the intensity of initial side-effects. There are differences in the central/peripheral ratio of inhibition for different ChE inhibitors. Some drugs induce both central (tremor, dizziness and nausea) and peripheral (tracheobronchial constriction, lacrimation, salivation, diarrhoea, muscle weakness, etc.) effects. A better separation between central and peripheral effects is achieved by donepezil, rivastigmine and metrifonate than by tacrine and physostigmine.[12] For irreversible inhibitors such as metrifonate, the difference in the rate of re-synthesis between peripheral and central pools of the enzyme could explain their low peripheral toxicity.[128] With metrifonate it is possible to take advantage of the more rapid synthesis of ChE in peripheral tissues as compared to brain, the long-lasting inhibition and the rapid rate of elimination of the parent drug. Low doses of the active metabolite of metrifonate (DDVP) are sufficient to maintain a high level of central inhibition. Accordingly, a low level of peripheral side-effects is seen despite a high AChE brain inhibition. In rare cases, at high doses an increase in ACh levels at the neuromuscular junction may produce a transient and reversible muscular weakness.

Depending on the pharmacological selectivity of ChE inhibitors there is a potential risk for interaction with other drugs, particularly with those acting indirectly via ChE inhibition or directly (via muscarinic or nicotinic receptors) on the cholinergic system. These interactions are common to all ChE inhibitors and include potentiation of cholinomimetic action, and prolongation or reversal of such an effect (Table 12.20). A typical example is the poten-

Drug	Plasma concentration (μg/l)	Time to peak (R)	Elimination half-life (h)	Metabolism	Ref.
Tacrine	–	1–2	2–4	Hepatic (P450)	93
Donepezil	30–60	3–4	73	Hepatic (P450)	84
Rivastigmine	114	1.7–1	5	Non-hepatic	85–151
Metrifonate	500	0.5	2	Non-hepatic	123
Galantamine	543	0.5	4.4–5.7	Hepatic (P450)	124
Huperzine A	8.4	1.1	4.8	Hepatic (P450)	136

Table 12.19
Comparison of the pharmacokinetic properties of six ChE inhibitors after oral dosing in humans.

tiation of the paralysing effect of succinylcholine through inhibition of plasmatic BuChE. Although rare and reversible there is a possibility for an interaction with histaminic antagonists such as ranitidine, nizatidine and cimetidine which have shown low ChE inhibitory properties in vitro (but not in vivo in patients). Other specific interactions have been observed either in experimental animals or in patients. All of them, such as hypertension, increased gastrointestinal motility, bronchoconstriction and hypomania, can be explained by a cholinergic mechanism (Table 12.21). It should stressed that the side-effects seen during treatment with AChE inhibitors are transient and reversible. They are neither more severe in intensity nor more frequent in number than those produced by antidepressants (particularly of the tricyclic type), anxiolytics of the benzodiazepine type and neuroleptics (particularly haloperidol) in elderly patient populations.

In summary, ChE inhibitors are well tolerated drugs. Most side-effects are mild in severity, short lived (lasting only a few days) and reversible. As expected from experimental pharmacology data in animals, patients tend to become rapidly tolerant to side-effects. However, ChE inhibitors should be used with caution in patients with severe asthma, significant chronic obstructive pulmonary disease, cardiac conduction defects, severe hypertension or clinically significant bradycardia and should not be used in association with drugs that have ChE inhibiting properties (see Tables 12.20 and 12.21).

New indications for ChE inhibitor therapy: subjects at risk

An immediate challenge to the findings of a prolonged effect of ChE inhibitors is to investigate whether or not early treatment may alter the course of the disease and thus delay clinical onset of AD. The possibility of genotyping several members of the same family, which allows diagnosis of the disease presymptomatically, together with the possi-

Interacting drug	*Side-effect*	*Mechanism*
Cholinomimetics	Potentiation	Choline potentiation
Muscarinic agonists	Bronchospasm	
Phenotiazines Tricyclic antidepressants Antihistaminics	Reversal of action	Depressed receptor selectivity
Ranitidine	Potentiation	Rapid reversal, competition
Nizatidine Cimetidine	Prolongation of neuromuscular block during anaesthesia, increased gastrointestinal motility, muscle weakness	AChE inhibition, tubocurarine reversal, block, enhanced ACh release
Metoclopramide	Increased gastric motility	ChE inhibition, enhanced ACh release
Succinylcholine	Potentiation	BuChE inhibition
Beta-blockers Anti-asthmatics Opiates[139]	Bradycardia Bronchospasm Analgesia	Potentiation Potentiation Potentiation

Table 12.20
Potential interactions with ChE inhibitor treatment.

bility of a preventive treatment has increased the interest in identifying individuals at risk as suitable candidates for early treatment.

Asymptomatic at-risk members of familial AD pedigrees present a challenging possibility of early detection and treatment. Familial AD, however, represents a relatively small group (5–10%) of all AD patients. A longitudinal study of presymptomatic individuals at risk belonging to autosomal-dominant early-onset familial AD by Fox *et al.*[129] suggested that measurable cognitive decline is present 2–3 years before the manifestation of symptoms and 4–5 years before individuals fulfil the criteria for probable AD. The most sensitive measure in this study was verbal recall. The detection of individuals with early memory impairment suggests a therapeutic intervention with a ChE inhibitor to evaluate the hypothesis that the drug may slow down disease progression.

A second suitable group of patients is constituted by Down's syndrome patients. Early treatment would allow one to examine the effect of drugs on the development of the characteristic plaque pathology.

Preclinical stages in individuals with minimal cognitive impairment (MCI) who are at risk of AD constitute a suitable target for preventive therapy (see Fig. 12.10).[130] These indi-

ChE inhibitor	*Interacting drug*	*Side-effect*	*Mechanism/site*
Tacrine	Haloperidol	Parkinsonism	Dopamine antagonist, cholinergic hyperactivity/ basal ganglia
Tacrine	Neuromuscular blockers	Prolonged duration of action	Neuromuscular transmission/ muscle contraction
Other ChEIs			
Tacrine, Physostigmine	Nicotinics	Arrhythmia, chronotropic inotropic action	Nicotinic block/ guinea-pig atria
Tacrine, other ChEIs	Hypertensive drugs	Acute hypertension, hypertensive crisis	Cholinergic/adrenergic
Tacrine	Pentamidine	Bronchoconstriction	Muscarinc activity/isolated human bronchi
Tacrine, other ChEIs	Epileptogenic drugs	Generalized seizures	Cholinergic potentiation
Donepezil, other ChEIs	Nortryptiline, fluoxetine, verapamil, beta-blockers	Hypomania–mania	Cholinergic–noradrenergic imbalance drug metabolism
Eptastigmine, other ChEIs	Food intake	Lower ChE inhibition	Reduced availability

Table 12.21
Specific side-effects of ChE inhibitor combination.

viduals demonstrate some cognitive decline with increasing age but are still functioning well. It is implicit in this concept that patients who develop AD start with normal cognitive function and slowly progress to a stage of MCI. Ferris and de Leon[131] found that a test of delayed paragraph recall produced an overall prediction accuracy greater than 90% for conversion from MCI to AD. This suggests the possibility of identifying an 'enriched population' of subjects at risk of converting to probable AD over a 3-year period. A parallel study with volumetric magnetic resonance imaging (MRI) of hippocampal structures correlating imaging with cognitive data could provide evidence of structural changes. Patients with MCI at high risk for conversion represent an ideal population for therapeutic interventions. The ability of a ChE inhibitor to prolong the time to clinical diagnosis is being evaluated in follow-up trials of 3 years. At the same time the effect of a ChE inhibitor in reducing the rate of brain (hippocampal) atrophy is being studied using quantitative MRI.

Dementia with Lewy bodies (DLB) has been found by certain studies to be the second most common cause of dementia after AD, accounting for more than 25% of cases in prospective

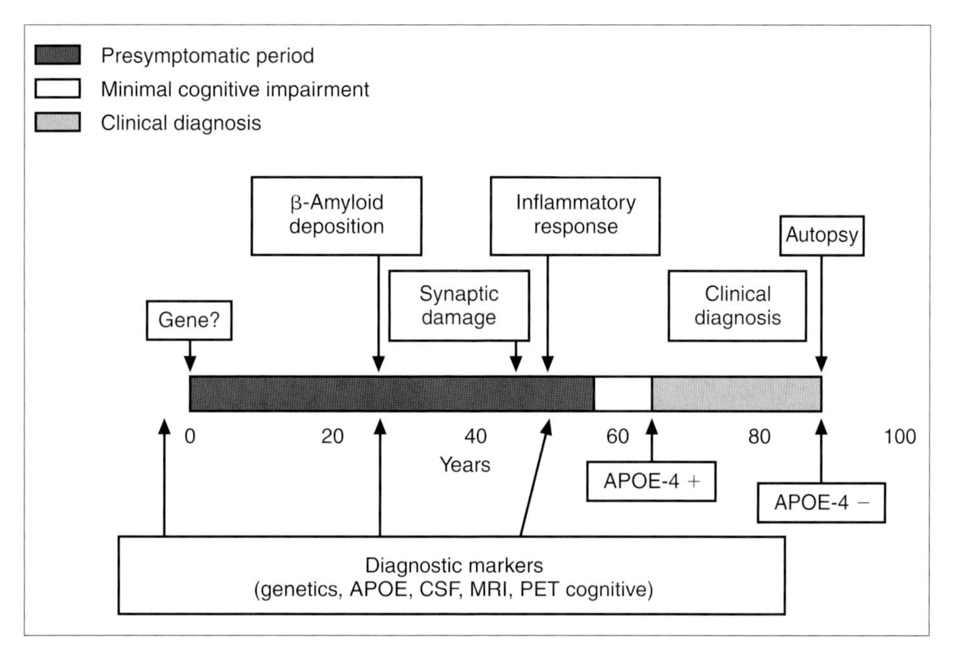

Figure 12.10
Potential periods of AD treatment: the presymptomatic period, the period of minimal cognitive impairment and the period after clinical diagnosis. Pathology, potential diagnostic markers and risk factors (APOE-4) are included. (After Sunderland.[130])

studies.[132] One neurotransmitter system repeatedly implicated in DLB is the cholinergic system.[132] Neocortical cholinergic activity is more severely depleted in DLB than in AD. This transmitter deficit has been related to the incidence of hallucinations that are characteristic of the disease. In DLB patients, not only is cortical cholinergic function affected, but also the caudate nucleus, thalamus and brainstem.[132] Since typical neuroleptics are contraindicated, cholinergic therapy could be attempted. A multicentre trial is in progress of the ChE inhibitor rivastigmine in DLB. The results of a nine case, 12-week study with donepezil showed that by both cognitive testing (BCRS and MMSE) and family reports, cognition and function (IADL) improved in seven DLB patients.[133] Hallucinations decreased in frequency, duration and intensity in eight patients; however, parkinsonism worsened in three.[133] Positive results were also described with tacrine in an earlier report on three cases.[134] A new indication for ChE inhibitors could be the treatment of dementia associated with Parkinson's disease. Hutchinson and Fazzini[135] reported elimination of hallucinations in five of seven cases and a reduction in two. With this indication a possible increase in parkinsonian symptoms should be considered. An attempt to treat vascular dementia with a ChE inhibitor (rivastigmine) showed encouraging results.[136] No other studies have been reported in the literature on this subject. The difficult diagnosis of vascular dementia represents, in many cases, an obstacle for evaluation. The combination of AD and mild ischaemic damage, which is present in a substantial percentage of dementia patients, may be a target for ChE inhibitor therapy, as demonstrated in recent studies with metrifonate.

The benefits of cholinergic therapy with a new ChE inhibitor have been unambiguously demonstrated for AD and myasthenia gravis (see Chapter 15). Huperzine A, a reversible ChE inhibitor and a natural product from China, which is at present being tested in AD patients, has also been tested in the treatment of myasthenia gravis (128 patients) producing a symptomatic benefit with less side-effects than neostigmine.[137] Future studies will show if treatment with ChE inhibitors can be extended to other disorders presenting cholinergic deficits.

Future applications of ChE inhibitors: combination treatment

ChE inhibitors are the first drugs to demonstrate efficacy in the treatment of cognitive symptoms in AD patients within a certain period of the natural history of the disease. The introduction of ChE inhibitors represents a milestone in AD therapy, being a transition from drugs with no demonstrable effect to drugs tested for the first time in randomized, placebo-control studies in vast populations of patients. The improvement is reflected in activities of daily living and global functioning. While cognitive impairment is an early and consistent feature of AD, behavioural disturbances are also typical of the disease. The symptoms have an impact on both the individual and the care-giver, and result in increased costs of care and earlier institutionalization.

Recent clinical trials with ChE inhibitors have shown some of these drugs to produce demonstrable reduction in psychiatric and behavioural symptoms. Forty per cent of metrifonate-treated patients experienced a decrease in certain behavioural symptoms (depression, anxiety, apathy and aberrant motor behaviour) over a treatment period of 26 weeks.[75–77] Future studies will show whether or not this effect is produced by all ChE inhibitors or only some. It is expected that muscarinic and nicotinic agonists with lower side-effects and higher selectivity than those presently available will be developed. Human cholinergic pharmacology suggests that the combination of ChE inhibitors with either muscarinic or nicotinic agonists may potentiate specific effects of the ChE inhibitors, such as orientation, attention and alertness.[139]

An immediate challenge is to investigate whether or not ChE inhibitors may alter the course of the disease, delaying conversion of MCI to clinical AD (see Fig. 12.10). These studies need to be prospective, multicentre, long term (at least 3 years) and involve relatively large numbers of subjects (at least 1000). In addition, selective markers (imaging, such as volumetric MRI, and CSF) of structural changes need to be included in the study.

The combination of ChE inhibitors with non-steroidal anti-inflammatory drugs (NSAIDs), antioxidants (vitamin E or other) and oestrogens represent an interesting new strategy (see Fig. 12.1). With regard to oestrogens, it will be necessary to develop selective agonists for brain receptors that are devoid of carcinogenic activity. Preliminary results of short-term studies support the hypothesis that concomitant use of oestrogens (or oestrogen/progesterone) and a ChE inhibitor may yield greater cognitive benefits than the use of a ChE inhibitor alone. The safety and efficacy of combining oestrogen and a ChE inhibitor need to be examined further; a prospective, randomized, double-blind placebo-controlled trial is currently underway. The lack of non-toxic and effective antioxidants is still preventing large-scale combination trials with ChE inhibitors.

Within the field of anti-inflammatory drugs, the combination of cyclooxygenase inhibitors (COX-2 I), which are presently under clinical evaluation, with ChE inhibitors is an attractive alternative to ChE inhibitor monotherapy. Cyclooxygenase is the enzyme that catalyses the first step in the prostanoid synthesis pathway and is the target of NSAIDs.[138]

Indications of ChE inhibitor treatment of dementia other than in AD, such as in LBD, Down's syndrome[140] and vascular dementia, represent three unchallenged territories of dementia therapy. Other indications might be found outside the field of dementia. For example, cognitive dysfunction is one of the leading causes of disability for individuals with multiple sclerosis. Memory loss is the most common cognitive impairment. Pilot studies have demonstrated memory improvement in multiple sclerosis patients treated with donepezil relative to placebo-treated controls.[141]

The cholinergic hypothesis 12 years later: how correct were we in predicting the effects of ChE inhibitors?

Cholinesterase inhibitors therapy has been developed under the following assumptions:

- the cholinergic system is early and selectively damaged in AD patients;
- a steady-state elevation in synaptic ACh levels produces a symptomatic short-term cognitive benefit;
- this benefit is symptomatic in nature, and therefore the clinical effect of ChE inhibitors cannot be expected to persist beyond cessation of treatment;
- because of the first point above, only early, mildly affected patients will respond to ChE inhibitor therapy; and
- clinical benefits will be limited to cognitive (not to behavioural) improvement.

Clinical data still support the hypothesis that pharmacological enhancement of central cholinergic synaptic transmission improves cognition in AD patients. Experiments with steady-state intravenous administration of either the muscarinic agonist arecoline or the ChE inhibitor physostigmine to AD patients confirm the dose-dependent effect on cognition seen in normal young and aged subjects (see Tables 12.8 and 12.9) and support the idea that the memory enhancement caused by ChE inhibitor is a direct one and is not related to secondary activation of the hypothalamic–pituitary–adrenal axis.[142]

Recent data suggest that some of our early hypotheses on the effects of ChE inhibitors may have been inaccurate. New results show that, although neocortical cholinergic deficits (such as a decrease in ChAT and AChE activity) are characteristic of severely demented patients (see Table 12.6), they are not clearly apparent in individuals with early mild AD.[143] Significant cholinergic enzymatic deficits are not demonstrable until relatively late in the course of the disease.[143] Two conclusions can be drawn from these findings. First, the clinical effect of ChE inhibitors seen at mild stages of the disease suggests that, in spite of the fact that there are no apparent changes in cholinergic enzyme activities, the cholinergic system may be subfunctional with respect to ACh synthesis, storage, release or receptor (nicotinic) mechanisms. As pointed out previously, cholinergic enzymes are present in excess concentrations in brain (see Table 12.6). These findings suggest that patients with more severe disease should be a target for cholinergic treatment. So far such treatment has been reserved for mild and moderately severe cases (see Table 12.14). However, in the future pharmacological intervention could be extended from patients with mild (MCI) disease to those with severe disease. Such an approach could prolong the therapeutic period to span 2–3 years of the approximately 8–9 year natural history of the disease. The development of more effective ChE inhibitors that produce less cholinergic side-effects is a future target.[144] The requirement of high and consistent levels of brain ChE inhibition and ACh increase with once-daily dosage and a high degree of tolerability and efficacy is still valid. The ideal ChE inhibitor would be an irreversible inhibitor that selectively inhibits brain AChE and has little or no effect on peripheral ChEs.[145,146]

Is there a prolonged effect of ChE inhibitors? Recent observations[100,147] have shown that the effect of ChE inhibitors can be maintained for 3–4 weeks following the interruption of the treatment (washout period). In addition, the effect can be seen as an improvement in ADAS-Cog scores when treatment is restarted after being suspended for 3–6 weeks. It is difficult to explain such effects by means of a symptomatic mechanism of action. One explanation could be that ChE inhibitors interfere with disease progression by promoting soluble APP release, as demonstrated by Mori *et al.*[11] in rat brain and recently confirmed by Racchi *et al.*[148] in neuroblastoma cells.

The benefits of ChE inhibitor treatment have been considered to be mainly, if not exclusively, cognitive in nature. However, it has now been demonstrated that improvement involves behavioural as well as cognitive symptoms[149,150] (see Table 12.2). This translates into a stabilizing effect of the general condition of the patient[151] (see Fig. 12.7), an effect which can last for 6–12 months or more (see Fig. 12.9).

ChE inhibitor therapy has generally been considered a short-term intervention (3–6 months). However, recent studies have demonstrated that the drug effect can be seen in some patients for as long as 2 years (see Fig. 12.9), producing a strong clinical and economical impact.[155,156]

Conclusions: ChE inhibitor treatment is not cost neutral

The results of the treatment of over 8000 patients (as summarized in Table 12.14) with six drugs allow us to draw some conclusions with regard to the therapeutic value, toxicity, advantages and limitations and future indications of the use of ChE inhibitors in the treatment of AD. Based on these data and on large clinical studies we can draw some conclusions regarding the pharmaco-economic impact and value of these drugs.

An increased utilization of ChE inhibitors, which improve cognition and influence mood, will not only be a relevant part of future care strategies of AD but will also play a significant economic and social role.

The introduction of ChE inhibitors in AD therapy has triggered a new interest which is likely to have far-reaching effects. It has stimulated improved diagnostic procedures and stronger attention from physicians, which might in itself produce a positive effect on outcomes. Pharmacological treatments that improve quality of life and influence the ability of AD patients to perform activities of daily living reduce the emotional and economic burden of the care-giver and the costs of disease management.[152] Recent European studies have shown a cost saving of 2000 Euro/patient (approximately US$2100) for each MMSE point saved (B. Winblad, unpublished results, 1999). A cost–benefit analysis of 6 months of tacrine treatment in a Scandinavian patient population demonstrated a cost saving of 1–17% (average of 5–6%).[152] The saving is related mainly to the delay to institutionalization and the reduction in care-giver burden (providing the drug does not influence mortality rate). In calculating cost savings one should keep in mind that approximately 75% of total-care costs is for severely demented patients.

Acknowledgements

Mr D Bhantooa, Department of Geriatrics, University of Geneva, is acknowledged for designing the figures.

References

1. Goodman & Gilman. *The Pharmacological Basis of Therapeutics*, 9th edn. New York: McGraw-Hill, 1996: 137, 141, 161, 175.
2. Dale HH. The action of certain esters and ethers of choline, and their relation to muscarine. *J Pharmacol Exp Ther* 1914; **6:** 147–190.
3. Giacobini E. Cholinergic foundations of Alzheimer's disease therapy. *J Physiol (Paris)* 1998; **92:** 283–287.
4. Davis KL, Mohs RC. Enhancement of memory by physostigmine. *N Engl J Med* 1979; **301:** 946–956.
5. Thal J, Fuld PA. Memory enhancement with oral physostigmine in Alzheimer's disease. *N Engl J Med* 1983; **308:** 708–718.
6. Giacobini E, Becker R, Mcilhany M, Kumar V. Intracerebroventricular administration of cholinergic drugs: preclinical trials and clinical experience in Alzheimer patients. In: Giacobini E, Becker R, eds. *Current Research in Alzheimer Therapy.* New York: Taylor & Francis, 1988: 113–122.
7. Summer WK, Viesselman JO, Marsh GM, Candelora K. Use of THA in treatment of Alzheimer-like dementia: pilot study in twelve patients. *Bio Psychiatry* 1981; **16:** 145–153.
8. Becker E, Giacobini E. Mechanisms of cholinesterase inhibition in senile dementia of the Alzheimer type. *Drug Dev Res* 1988; **12:** 163–195.
9. Rainer M, Mark TH, Haushofer A. Galanthamine hydrobromide in the treatment of senile dementia of Alzheimer's type. In: Kewitz T, Thomsen L, Bickel A, eds. *Pharmacological Interventions on Central Mechanisms in Senile Dementia.* Munich: Zuckschwerdt, 1989.
10. Giacobini E, Cuadra G. Second and third generation cholinesterase inhibitors: from preclinical studies to clinical efficacy. In: Giacobini E, Becker R, eds. *Alzheimer Disease: Therapeutic Strategies.* Boston: Birkhäuser, 1994: 155–171.
11. Mori F, Lai CC, Fusi F, Giacobini E. Cholinesterase inhibitors increase secretion of APPs in rat brain cortex. *Neurol Rep* 1995; **6:** 633–636.
12. Giacobini E. Cholinesterase inhibitors do more than inhibit cholinesterase. In: Becker R, Giacobini E, eds. *Alzheimer Disease: From Molecular Biology to Therapy.* Boston: Birkhäuser, 1996: 187–204.
13. Silver A. *The Biology of Cholinesterases.* New York: Elsevier/Agricultural Research Council Institute, 1974: 426–447.
14. Giacobini E, Holmstedt B. Cholinesterase content of certain regions of the spinal cord as judged by histochemical and cartesian diver technique. *Acta Physiol Scand* 1958; **42:** 12–27.
15. Giacobini E. Distribution and localization of cholinesterases in nerve cells. *Acta Physiol Scand* 1959; **45(Suppl 156):** 1–45.
16. Giacobini E. Metabolic relations between glia and neurons studied in single cells. In: Maynard M, Cohen MD, Snider R, eds. *Morpho-*

logical and Biochemical Correlates of Neural Activity. New York: Harper & Row, 1964: 15–38.

17. Wright CI, Geula C, Mesulam MM. Neuroglial cholinesterases in the normal brain and in Alzheimer's disease: relationship to plaques, tangles and patterns of selective vulnerability. *Ann Neurol* 1993; **34:** 373–384.
18. Perry EK, Perry RH, Blessed G, Tomlinson BE. Changes in brain cholinesterases in senile dementia of Alzheimer type. *Neuropathol Appl Neurobiol* 1978; **4:** 273–277.
19. Arendt T, Brückner MK, Lange M, Bigl V. Changes in acetylcholinesterase and butyrylcholinesterase in Alzheimer's disease resemble embryonic development — a study of molecular forms. *J Neurochem Intl* 1992; **21(3):** 381–396.
20. Davies P. Neurotransmitter-related enzymes in senile dementia of the Alzheimer type. *Brain Res* 1979; **171:** 319–327.
21. Atack JR, Perry EK, Bonham JR, Candy JM, Perry RH. Molecular forms of butyrylcholinesterase in the human neocortex during development and degeneration of the cortical cholinergic system. *J Neurochem* 1987; **48(6):** 1687–1692.
22. Atack JR, Perry EK, Bonham JR, Candy JM, Perry RH. Molecular forms of acetylcholinesterase and butyrylcholinesterase in the aged human central nervous system. *J Neurochem* 1986; **47(1):** 263–277.
23. Arendt T, Brückner M, Lange M, Bigl V. Changes in acetylcholinesterase and butyrylcholinesterase in Alzheimer's disease resemble embryonic development — a study of molecular forms. *J Neurochem* 1992; **21(3):** 231–244.
24. Giacobini E, DeSarno P, Clark B, McIlhany M. The cholinergic receptor system of the human brain — neurochemical and pharmacological aspects in aging and Alzheimer. In: Nordberg A, Fuxe K, Holmstedt B, eds. *Progress in Brain Research*. Amsterdam: Elsevier, 1989: 335–343.
25. Shafferman A, Velan B, Ordentlich A *et al.* Substrate inhibition of acetylcholinesterase: residues affecting signal transduction from the surface to the catalytic center. *EMBO J* 1992; **11(10):** 3561–3568.
26. Tohgi H, Abe T, Hashiguchi K *et al.* Remarkable reduction in acetylcholine concentration in the cerebrospinal fluid from patients with Alzheimer type dementia. *Neurosci Lett* 1994; **177:** 139–142.
27. Ogane N, Giacobini E, Messamore E. Preferential inhibition of acetylcholinesterase molecular forms in rat brain. *Neurochem Res* 1992; **17(5):** 489–495.
28. Ogane N, Giacobini E, Struble R. Differential inhibition of acetylcholinesterase molecular forms in normal and Alzheimer disease brain. *Brain Res* 1992; **17:** 307–312.
29. Cuadra G, Summers K, Giacobini E. Cholinesterase inhibitor effects on neurotransmitters in rat cortex *in vivo*. *J Pharmacol Exp Ther* 1994; **270(1):** 277–284.
30. Cuadra G, Giacobini E. Coadministration of cholinesterase inhibitors and idazoxan: effects of neurotransmitters in rat cortex in vivo. *J Pharm Exp Ther* 1995; **273:**.230–240.
31. Enz A, Florsheim P. Cholinesterase inhibitors: an overview of their mechanism of action. In: Giacobini E, Becker R, eds. *Alzheimer's Disease: From Molecular Biology to Therapy*. Boston: Birkhäuser, 1996: 211–215.
32. Mesulam M-M, Geula C. Butyrylcholinesterase reactivity differentiates the amyloid plaques of aging from those of dementia. *Ann Neurol* 1994; **36(5):** 722–727.
33. Giacobini E, Griffini PL, Maggi T *et al.* The effect of MF 8622: a selective BuChE inhibitor. *Soc Neurosci* 1996; **22:** 203 (abstract).
34. Massoulié J, Toutant JP. Vertebrate ChEs: structure and types of interaction. In: Whittaker VP, ed. *Handbook of Experimental Pharmacology*. Berlin: Springer Verlag, 1988: 167–224.
35. Scarsella G, Toschi G, Bareggi SR *et al.* Molecular forms of cholinesterases in cerebrospinal fluid, blood plasma, and brain tissue of the beagle dog. *J Neurosci Res* 1979; **4:** 19–24.
36. Elble R, Giacobini E, Scarsella GF. Cholinesterases in cerebrospinal fluid. A longitudinal study in Alzheimer disease. *Arch Neurol* 1987; **44:** 403–407.
37. Nakano S, Kato T, Nakamura S *et al.* Acetylcholinesterase activity in cerebrospinal fluid

of patients with Alzheimer's disease and senile dementia. *J Neurol Sci* 1987; **75:** 213–223.
38. Kumar V, Giacobini E, Markwell S. CSF choline and acetylcholinesterase in early-onset vs late-onset Alzheimer's disease patients. *Acta Neurol Scand* 1989; **80:** 461–466.
39. Elble R, Giacobini E, Higgins C. Choline levels are increased in cerebrospinal fluid of Alzheimer patients. *Neurobiol Aging* 1989; **10:** 45–50.
40. Giacobini E. Brain acetylcholine — a view from the cerebrospinal fluid (CSF). *Neurobiol Aging* 1986; **7(5):** 392–396.
41. Frölich L, Dirr A, Götz E *et al.* Acetylcholine in human CSF: methodological considerations and levels in dementia of Alzheimer type. *J Neural Transm* 1998; **105:** 961–973.
42. Mattio T, McIlhany M, Giacobini E *et al.* The effects of physostigmine on acetylcholinesterase activity of CSF, plasma and brain. A comparison of intravenous and intraventricular administration in beagle dogs. *Neuropharmacology* 1986; **25(10):** 1167–1177.
43. Nitsch RM, Rossner S, Albrecht C *et al.* Muscarinic acetylcholine receptors activate the acetylcholinesterase gene promoter. *J Physiol* 1998; **92:** 257–264.
44. Kumar V, Giacobini E. Cerebrospinal fluid choline, and acetylcholinesterase activity in familial vs non-familial Alzheimer's disease patients. *Arch Gerontol Geriatr* 1988; **7:** 111–117.
45. Manyam BV, Giacobini E, Ferraro TN *et al.* Cerebrospinal fluid as a reflector of central cholinergic and amino acid neurotransmitter activity in cerebellar ataxia. *Arch Neurol* 1990; **47:** 1194–1199.
46. Manyam BV, Giacobini E, Colliver JA. Cerebrospinal fluid acetylcholinesterase and choline measurements in Huntington's disease. *J Neurol* 1990; **237:** 281–284.
47. Kumar V, Giacobini E. Use of agraphia in subtyping of Alzheimer's disease. *Arch Gerontol Geriatr* 1990; **11:** 155–159.
48. Manyam BV, Giacobini E, Colliver JA. Cerebrospinal fluid choline levels are decreased in Parkinson's disease. *Ann Neurol* 1990; **27(6):** 683–685.
49. Giacobini E. Cholinomimetic therapy of Alzheimer disease: does it slow down deterioration? In: Racagni G, Brunellos N, Langer SZ, eds. *Recent Advances in the Treatment of Neurodegenerative Disorders and Cognitive Dysfunction.* New York: Karger/International Academy of Biomedical Drug Research, 1994: 51–57.
50. Geula C, Mesulam MM. Cholinergic systems and related neuropathological predilection patterns in Alzheimer disease. In: Terry RD, Katzman R, Bick KL, eds. *Alzheimer Disease.* New York: Raven, 1994: 263–291.
51. Tune L, Brandt J, Frost JJ *et al.* Physostigmine in Alzheimer's disease: effects on cognitive functioning, cerebral glucose metabolism analyzed by positron emission tomography and cerebral blood flow analyzed by single photon emission tomography. *Acta Psychiatr Scand* 1991; **366:** 61–65.
52. DeKosky ST, Harbaugh RE, Schmitt FA *et al.* Cortical biopsy in Alzheimer's disease: diagnostic accuracy and neurochemical, neuropathological and cognitive correlations. *Ann Neurol* 1992; **32:** 625–632.
53. Bowen DM. Biochemical assessment of neurotransmitter and metabolic dysfunction and cerebral atrophy in Alzheimer's disease. In: *Branbury Report 15: Biological Aspects of Alzheimer's disease.* Cold Spring Harbor: Cold Spring Harbor Laboratory, 1983: 219–231.
54. Greaney M, Marshall D, During M *et al.* Ultrasensitive measurement of acetylcholine release in the conscious human hippocampus and anesthetized rat striatum using microdialysis. *Soc Neurosci Abstr* 1992; **2:** 137.
55. Giacobini E. Cholinesterase inhibitors. From preclinical studies to clinical efficacy in Alzheimer disease. In: Quinn DM, Balasubramanian AS, Doctor B, Taylor P, eds. *Enzymes of the Cholinesterase Family.* New York: Plenum, 1995: 463–469.
56. Bartus R, Dean R. Developing and utilizing animal models in the search for an effective treatment for age-related memory disturbances. In: Gottfries CG, ed. *Normal Aging, Alzheimer's Disease and Senile Dementia: Aspects on Etiology Pathogenesis, Diagnosis and Treatment.* Brussels: Editions de l'Univer-

site de Bruxelles, 1985: 231–267.
57. Bartus RT, Dean RL, Beer B *et al.* The cholinergic hypothesis of geriatric memory dysfunction. *Science* 1982; **217:** 408–417.
58. Rupniak NM, Field MJ, Samson NA *et al.* Direct comparison of cognitive facilitation by physostigmine and tetrahydroaminoacridine in two primate models. *Neurobiol Aging* 1990; **11:** 609–613.
59. Aigner TG, Mortimer M. The effects of physostigmine and scopolamine on recognition memory in monkeys. *Behav Neural Biol* 1986; **45:** 81–87.
60. O'Neill J, Fitten LJ, Siembeda D *et al.* Reversal of scopolamine effect. *Prog Neuropsychopharmacol Biol Psychiatry* 1998; **22(4):** 665–678.
61. Davis KL, Hollister LE, Overall J *et al.* Physostigmine: effects on cognition and affect in normal subjects. *Psychopharmacology* 1976; **51:** 23–27.
62. Davis KL, Mohs RC, Tinklenberg JR *et al.* Physostigmine: improvement of long-term memory processes in normal humans. *Science* 1978; **201:** 272–274.
63. Drachman DA, Sahakian BJ. Memory and cognitive function in the elderly. A preliminary trial of physostigmine. *Arch Neurol* 1980; **37:** 674–675.
64. Drachman DA, Glosser G, Fleming P. Memory decline in the aged: treatment with lecithin and physostigmine. *Neurology* 1982; **32:** 944–950.
65. Davis KL, Mohs RC, Tinklenberg JR. Enhancement of memory by physostigmine. *N Engl J Med* 1979; **301:** 946–947.
66. Summers WK, Kaufman KR. THA — a review of the literature and its use in treatment of five overdose patients. *Clin Toxicol* 1980; **16(3):** 269–281.
67. Peters BH, Levin HS. Memory enhancement after physostigmine treatment in the amnesic syndrome. *Arch Neurol* 1977; **34:** 215–219.
68. Goldberg E, Gerstman LJ, Mattis S *et al.* Effects of cholinergic treatment on posttraumatic anterograde amnesia. *Arch Neurol* 1982; **39:** 581–589.
69. Catsman-Berrevoets CE, Van Harskamp F, Appelhof A. Beneficial effect of physostigmine on clinical amnesic behaviour and neuropsychological test results in a patient with a post-encephalitic amnesic syndrome. *J Neurol Neurosurg Psychiatry* 1986; **49:** 1088–1090.
70. Rose RP, Moulthrop A. Differential responsivity of verbal and visual recognition memory to physostigmine and ACTH. *Biol Psychiatry* 1986; **21:** 538–542.
71. Levin Y, Elizur A, Korczyn AD. Physostigmine improves ECT-induced memory disturbances. *Neurology* 1987; **37:** 871–875.
72. Kertzman C, Robinson DL, Litvan I. Effects of physostigmine on spatial attention in patients with progressive supranuclear palsy. *Arch Neurol* 1990; **47:** 1346–1350.
73. Albin MS, Bunegin L, Janetta PJ *et al.* Tetrahydroaminoacridine (THA). I. Effect on postanesthetic emergence responses and anesthesia sleep-time after ketamine, phencyclidine and thiamylal in animals. *Excerpta Medica* 1975; **33:** 143–146.
74. Levy ML, Cummings JL, Kahn-Rose R. Neuropsychiatric symptoms and cholinergic therapy for Alzheimer's disease. *Gerontology* 1999; **45(1):** 15–22.
75. Gorman DG, Read S, Cummings JL. Cholinergic therapy of behavioral disturbances in Alzheimer's disease. *Neuropsychiatry Neuropsychol Behav Neurol* 1993; **6:** 229–234.
76. Morris JC, Cyrus PA, Orazem J *et al.* Metrifonate benefits cognitive, behavioral, and global function in patients with Alzheimer's disease. *Neurology* 1998; **50:** 1222–1230.
77. Cummings JL, Cyrus PA, Ruzicka BB *et al.* The efficacy of metrifonate in improving the behavioral disturbances of Alzheimer's disease patients. *Am Acad Neurol* 1998; **524:** 1004.
78. Farlow M, Gracon SI, Hershey LA *et al.* Controlled trial of tacrine in Alzheimer's disease. *JAMA* 1992; **268:** 2523–2529.
79. Knapp MJ, Knopman DS, Solomon PR. A 30 week randomized controlled trial of high-dose tacrine in patients with Alzheimer's disease. *JAMA* 1994; **271:** 985–991.
80. Canal I, Imbimbo BP. Clinical trials and therapeutics: relationship between pharmacodynamic activity and cognitive effects of eptastigmine in patients with Alzheimer's disease. *Clin Pharmacol Ther* 1996; **15:** 49–59.
81. Imbimbo BP. Eptastigmine: a cholinergic

approach to the treatment of Alzheimer's disease. In: Becker R, Giacobini E, eds. *Alzheimer Disease: From Molecular Biology to Therapy*. Boston: Birkhäuser, 1996: 223–230.

82. Rogers SL, Friedhoff T. The efficacy and safety of donepezil in patients with Alzheimer's disease: results of a US multicentre, randomized, double-blind, placebo-controlled trial. *Dementia* 1996; **7**: 293–230.
83. Doody RS. Treatment of Alzheimer's disease. *Neurologist* 1997; **3**: 279–289.
84. Rogers SL, Farlow MR, Doody SR *et al.* A 24 week, double blind placebo controlled trial of donepezil in patients with AD. *Neurology* 1998; **50**: 136–145.
85. Anand R, Hartman RD, Hayes PE. An overview of the development of SDZ ENA 713, a brain selective cholinesterase inhibitor disease. In: Becker R, Giacobini E, eds. *Alzheimer Disease: From Molecular Biology to Therapy*. Boston: Birkhäuser, 1996: 239–243.
86. Rosler M, Anand R, Cicin-Sian A *et al.* Efficacy and safety of rivastigmine in patients with Alzheimer's disease: international randomized controlled trial. *Br Med J* 1999; **318**: 633–658.
87. Becker R, Colliver JA, Markwell SJ *et al.* Double-blind, placebo-controlled study of metrifonate, an acetylcholinesterase inhibitor for Alzheimer disease. *Alzheimer Dis Assoc Disord* 1996; **1**: 124–131.
88. Becker R, Moriearty P, Unni L *et al.* Cholinesterase inhibitors as therapy in Alzheimer's disease: benefit to risk considerations in clinical application. In: Becker R, Giacobini E, eds. *Alzheimer Disease: From Molecular Biology to Therapy*. Boston: Birkhäuser, 1996: 257–266.
89. Morris J, Cyrus P, Orazem J *et al.* Metrifonate: potential therapy for Alzheimer's disease. *American Society of Neurology Meeting, Boston, 1997*. Abstr 155.
90. Keith MC, Dubois B, Collins O *et al.* Efficacy and safety of metrifonate in Alzheimer's disease. *Fifth International Geneva/Springfield Symposium on Advances in Alzheimer Therapy, Geneva, 1998*. Abstr 74.
91. Farlow MR, Cyrus PA, Gulanski B. Metrifonate improves the cognitive deficits of Alzheimer's disease patients in a dose-related manner. *American Geriatric Society Meeting, Seattle,1998*. Abstr A15.
92. Wilkinson D. Galanthamine hydrobromide — results of a group study. *Eighth International Congress Psychogeriatrics, Jerusalem, 1997*. Abstr 70.
93. Gracon S, Goodrich J, Fayad R. A prospective study on a once daily formulation of tacrine in patients with and without an apolipoprotein E4 allele. *Fifth International Geneva/Springfield Symposium on Advances in Alzheimer Therapy, Geneva, 1998*. Abstr 47.
94. Poirier J, Sevigny P. Pharmacogenetic approach to the treatment of Alzheimer's disease: a role for apolipoprotein E polymorphism. *Fifth International Geneva/Springfield Symposium on Advances in Alzheimer Therapy, Geneva, 1998*. Abstr 87.
95. Amberla K, Almqvist O, Jelic V *et al.* Long-term tacrine treatment has positive effects on cognitive functions in mild Alzheimer patients compared to untreated controls. *Fifth International Geneva/Springfield Symposium on Advances in Alzheimer Therapy, Geneva, 1998*. Abstr 136.
96. Farlow MR, Lahiri DK, Poirier J *et al.* Treatment outcome of tacrine therapy depends on apolipoprotein genotype and gender of the subjects with Alzheimer's disease. *Neurology* 1998; **50**: 669–679.
97. Soininen HS, Lehtovirta M, Laakso MP *et al.* ApoE genotype and MRI volumetry: application for therapy in Alzheimer Disease. In: Becker R, Giacobini E, eds. *Alzheimer Disease: From Molecular Biology to Therapy*. Boston: Birkhäuser, 1996: 475–480.
98. Rogers S, Friedhoff L. Long-term efficacy and safety of donepezil in the treatment of Alzheimer's disease. *Euro Neuropsychopharmacol* 1998; **8**: 67–75.
99. Jelic V, Amberla K, Almkvist O *et al.* Long-term tacrine treatment slows the increase of theta power in the EEG of mild Alzheimer patients compared to untreated controls. *Fifth International Geneva/Springfield Symposium on Advances in Alzheimer Therapy, Geneva, 1998*. Abstr 147.

100. Anand R, Hartman R, Messina J *et al.* Long-term treatment with rivastigmine continues to provide benefits for up to one year. *Fifth International Geneva/Springfield Symposium on Advances in Alzheimer Therapy, Geneva, 1998.* Abstr 18.
101. Giacobini E. Cholinesterase inhibitors: from preclinical studies to clinical efficacy in Alzheimer disease. In: Quinn D, Balasubramaniam AS, Doctor BP *et al. Enzymes of the Cholinesterase Family.* New York: Plenum, 1995; 463–469.
102. Mattio T, Mcilhany M, Giacobini E *et al.* The effects of physostigmine on acethylcholinesterase activity of CSF, plasma and brain. *Neuropharmacology* 1986; **25:** 1167–1177.
103. Thal L, Fuld PA, Masur DM *et al.* Oral physostigmine and lecithin improve memory in Alzheimer disease. *Ann Neurol* 1983; **13:** 491–496.
104. Imbimbo BP, Lucchelli PE. A pharmacodynamic strategy to optimize the clinical response to eptastigmine. In: Becker R, Giacobini E, eds. *Alzheimer Disease Therapeutic Strategies.* Boston: Birkhäuser, 1994: 223–230.
105. Becker R, Colliver J, Elbe R. Effects of metrifonate, a long-acting cholinesterase inhibitor. *Drug Dev Res* 1990; **19:** 425–434.
106. Giacobini E, Desarno P, Clark B *et al.* The cholinergic receptor system of the human brain — neurochemical and pharmacological aspects in aging and Alzheimer. In: Nordberg A, Fuxe K, Holmstedt B, eds. *Progress in Brain Research.* Amsterdam: Elsevier, 1989: 335–343.
107. Becker R, Moriearty P, Unni L. The second generation of cholinesterase inhibitors: clinical and pharmacological effects. In: Becker R, Giacobini E, eds. *Cholinergic Basis for Alzheimer Therapy.* Boston: Birkhäuser, 1991: 263–296.
108. Giacobini E. Cholinomimetic therapy of Alzheimer disease: does it slow down deterioration? *Recent Adv Treatment Neurodegen Disord Cogn Dysfunct* 1994; **7:** 51–57.
109. Rupniak NMJ, Field MJ, Samson NA *et al.* Direct comparison of cognitive facilitation by physostigmine and tetrahydroaminoacridine in two primate models. *Neurobiol Aging* 1990; **11:** 609–613.
110. Moss DE, Kobayashi H, Pacheco G *et al.* Methanesulfonyl fluoride: a CNS selective cholinesterase inhibitor. In: Giacobini E, Becker R, eds. *Current Research in Alzheimer Therapy.* New York: Taylor & Francis, 1988: 305–314.
111. Becker R, Giacobini E. Pharmacokinetics and pharmacodynamics of acetylcholinesterase inhibition. *Drug Dev Res* 1988; **14:** 235–246.
112. Gallo MA, Lawryk N. Organic phosphorus pesticides. In: Hayes W, Laws E, eds. *Handbook of Pesticides Toxicology.* San Diego: Academic Press, 1991: 927–929.
113. Von Der Kammer H, Mayhaus M, Albrecht C *et al.* Muscarinic acetylcholine receptors activate expression of the Erg gene family of transcription factors. *J Biol Chem* 1998; **273:** 10–17.
114. Bareggi SR, Giacobini E. Acetylcholinesterase activity in ventricular and cisternal CSF of dogs. *J Neurosci Res* 1978; **3:** 335–339.
115. Scarsella G, Toschi G, Bareggi SR *et al.* Molecular forms of cholinesterase in cerebrospinal fluid, blood plasma and brain tissue of the beagle dog. *J Neurosci Res* 1979; **4:** 19–24.
116. Elble R, Giacobini E, Scarsella GF. Cholinesterase in cerebrospinal fluid. *Neurology* 1987; **44:** 403–407.
117. Nordberg A, Hellstrom-Lindahl E, Almqkvist O *et al.* Acetylcholinesterase activity in CSF of Alzheimer patients following long term tacrine treatment. *Fifth International Geneva/Springfield Symposium, 1998.* Abstr 45, 154.
118. Hinz VC, Kolb J, Schmidt B. Effects of subchronic administration of metrifonate on cholinergic neurotransmission in rats. *Neurochem Res* 1998; **23:** 933–940.
119. Rider JA, Moeller HC, Swader J *et al.* The effect of parathion on human red blood cell and plasma cholinesterase. *AMA Arch Ind Health, Section II* 1958; **18:** 441–445.
120. Genina SA. Dynamics of blood cholinesterase activity in workers exposed to some organophosphorus insecticides during aerochemical application. *Gig Tr Prof Zabol* 1974; **12:** 42–44 [in Russian].

121. Cummings MD, Cyrus PA, Bieber F *et al.* Metrifonate treatment of the cognitive deficits in Alzheimer's disease. *Neurology* 1999; **50:** 1214–1221.
122. Cummings JL, Cyrus PA, Gulanski B. Metrifonate efficacy in the treatment of psychiatric and behavioral disturbances of Alzheimer's disease patients. *American Geriátric Society Meeting, Seattle, 1998*. Abstr A15.
123. Pettigrew LC, Bieber F, Lettieri J *et al.* A study of the pharmacokinetics, pharmacodynamics and safety of metrifonate in Alzheimer's disease patients. *J Clin Pharmacol* 1998; **38:** 236–245.
124. Wilkinson D, Fulton B, Benfield P. Galanthamine. *Drugs Aging* 1996; **1:** 60–66.
125. Lahiri DK, Lewis S, Farlow MR. Tacrine alters the secretion of β-amyloid precursor protein in cell lines. *J Neurosci Res* 1994; **8:** 777–787.
126. Inestrosa N, Alvarez A, Perez CA *et al.* Acetylcholinesterase accelerates assembly of amyloid-β-peptides into Alzheimer's fibrils: possible rôle of the peripheral site of the enzyme. *Neuron* 1996; **16:** 881–891.
127. Saez-Valero J, Sberna G, Small DH. The β-amyloid protein regulates acetylcholinesterase expression, assembly and glycosylation in cell culture, APP transgenic mice and the Alzheimer brain. *The Sixth International Meeting on Cholinesterases, La Jolla, 1998*. Abstr 6.
128. Nordgren IK. Cholinesterase inhibitors — are they all the same? *Intl J Geriatr Psychopharmacol* 1998; **1:** 176–178.
129. Fox NC, Warrington EK, Seiffer AL *et al.* Presymptomatic cognitive deficits in individuals at risk of familial Alzheimer's disease. A longitudinal prospective study. *Brain* 1998; **121:** 1631–1639.
130. Sunderland T. Cholinergic therapy. *Am J Geriatr Psychol* 1998; **6(Suppl 1):** 57–63.
131. Ferris SH, de Leon MJ. MRI and cognitive markers of progression and risk of Alzheimer disease. In: Becker R, Giacobini E, eds, *Alzheimer's Disease: From Molecular Biology to Therapy*. Boston: Birkhäuser, 1997: 463–469.
132. Perry EK, McKeith IG, Perry RH. Dementia with Lewy bodies: a common cause of dementia with therapeutic potential. *Intl J Geriatr Pharmacol* 1998; **1:** 120–125.
133. Shea C, MacKnight C, Rockwood K. Aspects of dementia. Donepezil for treatment of dementia with Lewy bodies: a case series of nine patients. *Intl Psychogeriatr* 1998; **10(3):** 229–238.
134. Levy R, Eagger S, Griffiths M *et al.* Lewy bodies and response to tacrine in Alzheimer's disease. *Lancet* 1994; **343:** 176.
135. Hutchinson M, Fazzini E. Cholinesterase inhibitors in Parkinson's disease. *J Neurol Neurosurg Psychiatry* 1996; **61:** 324–325.
136. Kumar V, Sugaya K, Messina J *et al.* Efficacy and safety of rivastigmine in Alzheimer's disease patients with vascular risk factors. *Neurology* 1999; **52(Suppl 2):** A395.
137. Tang XC, Han YF. Pharmacological profile of huperzine A, a novel acetylcholinesterase inhibitor isolated from Chinese herb. *CNS Drug Rev* 1999; in press.
138. Kaufmann WE, Andreasson KI, Isakson PC *et al.* Cyclooxygenase and the central nervous system. *Prostaglandins* 1997; **54:** 601–624.
139. Eisenach JC. Muscarinic-mediated analgesia. *Life Sci* 1999; **64:** 6–7, 549–554.
140. Kishnai PS, Sullivan JA, Walter KW *et al.* Cholinergic therapy for Down's syndrome. *Lancet* 1999; **353:** 1064–1065.
141. Asthana S, Raffaele KC, Gerig NH *et al.* Neuroendocrine responses to intravenous infusion of physostigmine in patients with Alzheimer disease. *Alzheimer Dis Assoc Disord* 1999; **13:** 102–108.
142. Krupp LB, Elkins LE, Scott SR *et al.* Donepezil for the treatment of memory impairments in multiple sclerosis. *Neurology* 1999; **52(Suppl 2):** A137.
143. Davis KL, Mohs RC, Marin D *et al.* Cholinergic markers in elderly patients with early signs of Alzheimer disease. *JAMA* 1999; **281:** 1401–1406.
144. Moss DE, Berlanga P, Hagan MM *et al.* Methanesulphonyl fluoride (MSF). *Alzheimer Dis Assoc Disord* 1999; **13:** 20–25.
145. Skau KA, Shipley MT. Phenylmethylsulphonyl fluoride inhibitory effects on acetylcholinesterase of brain and muscle. *Neuropharmacology* 1999; **38:** 691–698.
146. Schneider LS, Giacobini E. Metrifonate: a cholinesterase inhibitor for Alzheimer's disease therapy. *CNS Drug Rev* 1999; **5:** 14–27.
147. Doody RS, Pratt RD, Perdomo CA. Clinical

benefits of donepezil: results from a long-term Phase III extension trial. *Neurology* 1999; **52(Suppl 2):** A174.
148. Racchi M, Schmidt B, Koenig G *et al.* Treatment with metrifonate promotes soluble amyloid precursor protein release from SH-SY5Y neuroblastoma cells. *Alzheimer Dis Assoc Disord* 1999; **13:** 679–688.
149. Cummings JL. Changes in neuropsychiatric symptoms as outcome measures in clinical trials with cholinergic therapies for Alzheimer disease. *Alzheimer Dis Assoc Disord* 1997; **11:** S1–S9.
150. Raskind MA, Cyrus PA, Ruzicka BB *et al.* The effects of metrifonate on the cognitive, behavioral, and functional performance of Alzheimer's disease patients. *J Clin Psychol* 1999; **60:** 318–325.
151. Spencer MC, Noble S. Rivastigmine. A review of its use in Alzheimer's disease. *Drugs Aging* 1998; **13(Suppl 5):** 391–411.
152. Wimo A, Winblad B, Grafstrom M. The social consequences for families with Alzheimer disease patients: potential impact of new drug treatment. *Intl J Geriatr Psychol* 1998; **14:** 338–347.
153. Rainer M, Mucke HAM. Long-term cognitive benefit from galanthamine in Alzheimer's disease. *Intl J Geriatr Psychol* 1999; **1:** 197–201.
154. Thal LJ, Ferguson JM, Mintzer J *et al.* Randomized trial of controlled release physostigmine. *Neurology* 1999; **52:** 1146–1152.
155. Mohs R, Doody R, Morris J *et al.* Donepezil preserves functional status in Alzheimer's disease patients. *Eur Neuropsychopharm* 1999; **9(Suppl 5):** S328.
156. Winblad B, Engedal K, Soininen H *et al.* Donepezil enhances global function, cognition and activities of daily living compared with placebo in a one-year double blind trial in patients with mild to moderate Alzheimer's disease. *Ninth Congress of the International Psychogeriatric Association, August 1999*, Vancouver, Canada, Poster 118.

Bibliography

Comprehensive publications in the field of ChEs and ChE inhibitors (from 1948 to 1999, in chronological order):

Augustinsson KB. Cholinesterases. A study in comparative enzymology. *Acta Physiol Scand* 1948; **15(Suppl 52):** 1–182.

Holmstedt B. Pharmacology of organophosphorus cholinesterase inhibitors. *Pharmacol Rev* 1959; **11:** 567–688.

Giacobini E. The distribution and localization of cholinesterases in nerve cells. *Acta Physiol Scand* 1959; **45(Suppl 156):** 1–45.

Holmstedt B. Structure–activity relationship of the organophosphorus anticholinesterase agents. In: Eichler O, Farah A, eds. *Handbuch der experimentellen pharmakologie.* Berlin: Springer-Verlag, 1962: 428–485.

Koelle GB. Cytological distributions and physiological functions of cholinesterases. In: Eichler O, Farah A, eds. *Handbuch der experimentellen pharmakologie.* Berlin: Springer-Verlag, 1962: 187–298.

Karczmar AG, Usdin E, Wills JH. Anticholinesterase agents. In: Karczmar AG, ed. *International Encyclopedia of Pharmacology and Therapeutics*, Vol 1. Oxford: Pergamon, 1970.

Silver A. *The Biology of Cholinesterases.* Amsterdam: North Holland, 1974.

Massoulie J, Bon S. The molecular forms of cholinesterase and acetylcholinesterase in vertebrates. *Ann Rev Neurosci* 1982; **5:** 57–106.

Brzin M, Sketelj J, Klinar B. Cholinesterases. In: Lajtha A, ed. *Handbook of Neurochemistry.* New York: Plenum, 1983: 251–292.

Whittaker M. *Cholinesterase.* New York: Karger, 1986: 86–97.

Soreq H, Zakut H. Cholinesterase genes: multileveled regulation. *Monogr Human Genet* 1990; **13.**

Massoulie J, Bacou F, Chatonnet A, Doctor BP, Quinn DM, eds. *Cholinesterases.* Washington, DC: American Chemical Society, 1991.

Soreq H, Zakut H. *Human Cholinesterases and Anticholinesterases.* San Diego: Academic Press, 1993.

Taylor P, Radic Z. The cholinesterases: from genes to proteins. *Annu Rev Pharmacol Toxicol* 1994; **34:** 281–320.

Quinn D, Balasubramanian AS, Doctor BP, Taylor P, eds. *Enzymes of the Cholinesterase Family.* New York: Plenum, 1995.

Doctor BP, Taylor P, Quinn D, Rotundo R, Gentry M. *Structure and Function of Cholinesterases and Related Proteins.* New York: Plenum, 1998.

13

Cholinesterase inhibitors do more than inhibit cholinesterase

Anne-Lie Svensson, Ezio Giacobini

Introduction

Deficits in the cholinergic system of the brain contribute to loss of cognitive function in Alzheimer's disease (AD). Based upon these deficits, cholinergic strategies have been developed to improve cholinergic function. The most successful strategy has been the use of cholinesterase (ChE) inhibitors, three of which (tacrine, donepezil and rivastigmine) are presently in clinical use.[1] Although the main action of ChE inhibitors is to inhibit degradation of acetylcholine (ACh), other interesting targets may be of importance and contribute to the clinical efficacy seen in AD patients treated with ChE inhibitors.[2] This chapter focuses mainly on the effects of ChE inhibitors other than inhibition of ChEs in the central nervous system (CNS).

Interaction of ChE inhibitors with β-amyloid toxicity, aggregation and APP release

The β-amyloid peptide (Aβ) is one of the major components of the senile plaques detectable in the brain of AD patients.[3,4] It is produced by proteolytic cleavage of a larger amyloid precursor protein (APP).[5] APP is an integral cell membrane glycoprotein which is found in most cells throughout the human body, but most abundantly in the brain, and is metabolized by several alternative pathways. Non-amyloidogenic products of APP are generated by proteolytic cleavage within the Aβ region by a putative α-secretase, whereas intact Aβ is derived from APP by excision of the Aβ region.[6] The endosomal–lysosomal pathway represents a third processing pathway, which produces Aβ bearing amyloidogenic breakdown products. It has been shown that the secreted forms of APP generated by normal processing of APP within the Aβ region are neuroprotective,[7] whereas the abnormal processing of APP to Aβ has neurotoxic actions.[8]

Processing of APP can be modulated by several factors and agents, which may be important for understanding the process of extracellular depositions of Aβ in the brains of AD patients.[9] Several studies suggest a relationship between cholinergic neurotransmission and Aβ, and that Aβ may act as a physiological active neuromodulator.[10] Neurotoxic lesions in the nucleus basalis of the rat forebrain cholinergic system have been shown to decrease cortical ACh levels and ACh release, increase accumulation of APP in cortex and cerebrospinal fluid (CSF), induce APP synthesis, as well as increase cerebrocortical levels of APP mRNA. Low concentrations of soluble Aβ can induce cholinergic hypofunction that is not dependent on concurrent neurotoxicity.[10] For example, Aβ has been found to decrease potassium-stimulated release of ACh from rat hippocampal and cortical slices at picomolar to nanomolar concentrations,[11,12] and to be able to inhibit the high-affinity uptake of choline in rat hippocampal slices. Treatment with Aβ has been shown to reduce the synthesis of ACh in a mouse septal cell line SN56[13] and to impair carbachol-induced muscarinic cholinergic signal transduction in rat cortical neurons,[14] without causing cell death.

Injections of Aβ in rat nucleus basalis have been shown to impair learning and memory, disrupt cortical cholinergic innervation, decrease the activity of acetylcholinesterase (AChE) and choline acetyltransferase (ChAT) in frontal cortex, as well as decrease the number of muscarinic receptors measured by labelled quinuclidinylbenzilate (QNB).[15,16] Continuous infusion of Aβ into rat cerebral ventricle decreases nicotinic-induced stimulation of ACh release in frontal cortex and hippocampus, suggesting that Aβ may impair the function of nicotinic receptors and/or the process of nicotine-induced depolarization.[17]

Cholinergic agonists as well as ChE inhibitors appear to regulate processing and secretion of APP.[18] Cholinergic stimulation of m1 and m3 muscarinic receptor subtypes has been found to increase the secretion of APP in vitro in various cell lines and rat cortical slices.[19–22] Furthermore, treatment of rat phaeochromocytoma cells (PC12) with nicotine increases the release of APP into the conditioned medium without affecting the level of expression of APP mRNA.[23] Long-term inhibition of ChE has been suggested to result in activation of normal APP processing in AD brain, via increased levels of synaptic ACh.[24]

Haroutunian *et al.*[25] reported that treatment for 1 week with the ChE inhibitor phenserine normalized the levels of secreted APP in the CSF of forebrain cholinergic lesioned rats. Lahiri and Farlow[26] found that the levels of soluble APP derivatives normally present in conditioned medium were severely inhibited by treating cells with the ChE inhibitor tacrine, but not with physostigmine (Table 13.1). In addition, to reduce soluble APP levels, tacrine has been found also to reduce levels of total Aβ as well as levels of the two major forms of Aβ (Aβ1-40 and Aβ1-42) in conditioned media of human neuroblastoma cells (SK-N-SH). This suggests that tacrine may reduce levels of neurotoxic Aβ as well as neuroprotective soluble APP, and thereby influence the formation of Aβ aggregates[27] (see Table 13.1). Tacrine and the ChE inhibitors physostigmine, eptastigmine and DDVP (dichlorvos, a metabolite of metrifonate) have been shown to increase the release of soluble APP in superfused cortical slices of the rat[18,28] (see Table 13.1). The level of total APP mRNA and the level of APP-KPI (Kunitz-type) protease inhibitor mRNA in rat cerebral cortex was not found to be significantly changed after treatment with physostigmine and DDVP, whereas eptastigmine did decrease the levels of APP-KPI mRNA.[29] Chong and Suh[30] observed, when measuring APP immunoreactivity by using APP770 as substrate, that tacrine in concentrations lower than 0.5 mM enhanced APP processing, while tacrine concentrations above 0.5 mM significantly blocked APP processing, suggesting that low concentrations of tacrine enhance APP processing and high tacrine concentrations increase APP accumulation (see Table 13.1).

Accumulating evidence indicates that nicotinic receptors are involved in neuroprotection. Nicotine and nicotinic agonists have been shown to attenuate Aβ induced toxicity in cultured neurons,[31,32] an effect that seems to be mediated via the nicotinic receptor. Recently, the ChE inhibitors tacrine and donepezil were found to be able to attenuate Aβ(23–35)-induced toxicity in vitro in rat PC12 cells through a mechanism which, at least for

Drug	Concentration	Preparation	Effects	Ref.
Tacrine	100 μg/ml	Cell lines	Decrease the secretion of sAPP	26
Physostigmine	100 μg/ml	Cell lines	No effect on the secretion of sAPP	26
Tacrine	10 μg/ml	Cell lines	Decrease levels of Aβ, Aβ(1–40), Aβ(1–42)	27
Physostigmine	0.1 μM	Rat cortical slices	Increase sAPP release	28
Eptastigmine	0.1 μM	Rat cortical slices	Increase sAPP release	28
DDVP	0.02 μM	Rat cortical slices	Increase sAPP release	28
Tacrine	0.5 μM	Rat cortical slices	No change in sAPP release	29
Tacrine	0.1 μM	Rat cortical slices	Increase sAPP release	29
Tacrine	<500 μM	APP770	Increase the secretion of sAPP	30
Tacrine	>500 μM	APP770	Increase the APP accumulation	30
Tacrine	0.0001–10 μM	Cell lines	Reduce Aβ-induced toxicity	33
Donepezil	0.01–0.1 μM	Cell lines	Reduce Aβ-induced toxicity	33

Table 13.1
Effect of ChE inhibitors on APP processing in vitro.

tacrine, seems to be mediated via nicotinic receptors[33] (see Table 13.1).

A functional relationship seems also to exist between AChE and APP, as AChE has been found to be co-localized with Aβ deposits in AD brains.[34] Sberna *et al.*[35] have reported that Aβ increases the levels of AChE in cells (P19) by increasing intracellular calcium. Increased levels of AChE have also been found in the brains of transgenic mice expressing human APP,[36] suggesting that Aβ overexpression may disturb calcium homeostasis in the brain of these transgenic mice. Furthermore, it has been demonstrated that recombinant human AChE effectively binds soluble Aβ, forms stable complexes with Aβ, and promotes in vitro aggregation of Aβ into amyloid fibrils.[37,38] The neurotoxicity of stable AChE–Aβ complexes in neuronal cell cultures is higher than the toxicity induced by Aβ alone,[39] indicating that incorporation of AChE into amyloid fibrils changes their neurotoxic properties. Thus, both in vitro and in vivo data suggest that AChE behaves as a potent amyloid-promoting factor and modulates the toxicity of amyloid fibrils.[38] Inhibition of AChE might therefore affect the APP processing and slow down the cognitive deterioration of AD patients. Drugs such as ChE inhibitors used with the purpose of increasing cholinergic neurotransmission may improve or correct the APP metabolism imbalance. This suggests that cholinergic therapy, in addition to its symptomatic effect, might also affect progression of AD.

Effects of ChE inhibitors on cholinergic receptors

A variety of pharmacological actions has been reported for tacrine. Although its main action is to inhibit degradation of ACh, tacrine has been shown to inhibit the uptake and to increase the synthesis and release of several neurotransmitters, and to interact with cholinergic receptors and ion channels.[40] Whether these various pharmacological effects are exerted by other ChE inhibitors has not yet been demonstrated.

In AD, muscarinic receptors have been reported to be preserved or increased in number, although a decrease in the number of receptors has been shown particularly for the m2 muscarinic receptor subtype.[41,42] Tacrine has been reported to interact with muscarinic receptors in human and rodent brain, with similar affinity to m1 and m2 muscarinic receptor subtypes.[43–45] In vitro studies indicate that tacrine can both increase and decrease the release of ACh from rat cortical tissue in a concentration-dependent manner via muscarinic m1 and m2 receptors.[46] Physostigmine has demonstrated less affinity for muscarinic as well as nicotinic receptors than has tacrine.[43]

A consistent loss of nicotinic receptors has been observed in vitro in post-mortem brain tissue obtained from patients suffering from AD[47,48] and in vivo in AD patients by positron emission tomography (PET).[49] An increase in the number of nicotinic receptors has been observed following administration of tacrine to rats.[50] Long-term tacrine treatment in AD patients also shows restoration of nicotinic receptors in brain measured by PET.[51–53] Interestingly, several ChE inhibitors, including tacrine, physostigmine, galantamine and donepezil, have been shown to bind to an allosteric activator site on the nicotinic receptor, which may be of importance for clinical efficacy[54] (Svensson A-L and Nordberg A, unpublished work, 1999). Electrophysiological studies, have demonstrated that physostigmine and galanthamine significantly increase the frequency of opening of nicotinic receptor channels and potentiate agonist-activated currents in cell lines and hippocampal neurons.[55,56] By acting at the allosteric activator site on the nicotinic receptor, ChE inhibitors may protect the channel from desensitization, and thereby enhance cholinergic neurotransmission via functional nicotinic receptors.

The cholinergic agonistic effect of tacrine might be due to a direct action of tacrine on muscarinic and nicotinic receptors[43] and/or to an indirect stimulation of muscarinic and nicotinic receptors via the increased concentration of synaptic ACh caused by AChE inhibition (Fig. 13.1). In fact, it has been shown that measurements of AChE activity are not sufficient to predict extracellular levels of ACh following treatment with ChE inhibitors.[57] This further suggests that additional factors such as affinity for cholinergic receptors may influence the relationship between AChE inhibition and the concentration of extracellular ACh in the brain.[57]

Both muscarinic and nicotinic receptors are present not only on cholinergic nerve terminals but also on monoaminergic nerve terminals (see Fig. 13.1). Warpman *et al.*[58] demonstrated that tacrine enhanced monoaminergic neurotransmission (e.g. dopamine) in rat striatum by interacting with muscarinic and nicotinic receptors on dopaminergic nerve terminals in the striatum. Similarly, other ChE inhibitors investigated (e.g. physostigmine, eptastigmine, metrifonate and donepezil) showed a significant increase in the level of ACh as well as norepinephrine (NE), dopamine (DA) and serotonin (5-HT) levels in rat cortex after systemic administration.[18,59–61] Previous studies have found that tacrine inhibits monoaminergic uptake mechanisms[62,63] and monoamino oxidase (MAO) activities[64] (see Fig. 13.1). These effects of ChE inhibitors on multiple neurotransmitters may be of clinical significance by improving cognitive function in AD.

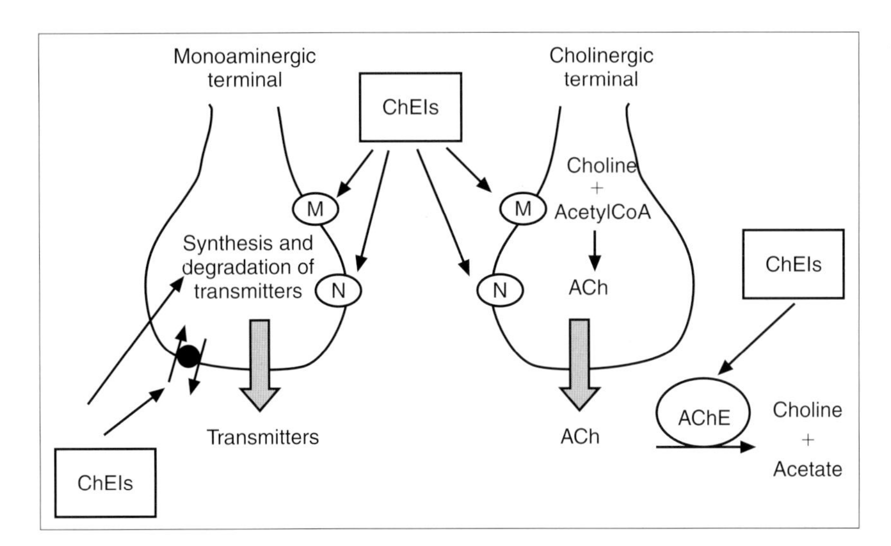

Figure 13.1
Possible site of action of ChE inhibitors on cholinergic and monoaminergic nerve terminals. ChEIs, cholinesterase inhibitors; M, muscarinic receptors; N, nicotinic receptors.

Interaction of ChE inhibitors with oestrogen effects

A number of epidemiological studies suggest that oestrogens may improve cognitive function, and delay the onset and decrease the risk of AD in postmenopausal women.[65,66] Several effects of oestrogens could account for their positive action on cognition. Oestrogens have been shown to stimulate ChAT activity,[67,68] take part in the nerve growth factor signalling pathway,[68] build up and maintain synapses, increase the cerebral blood flow, stimulate the non-amyloidogenic pathway of APP degradation,[69] as well as reduce formation of Aβ in vitro.[70] Oestrogens have also been demonstrated to protect against Aβ-induced toxicity in different cell lines,[69,71,72] but the underlying mechanism(s) for this effect has not been clarified in detail. Recent observations have shown that the neuroprotective effect of certain oestrogens (e.g. 17β-oestradiol) might be mediated by the α_7 nicotinic receptor subtype.[72]

Interestingly from a clinical point of view, women with AD receiving oestrogen replacement therapy have shown a better response to cholinergic inhibitory therapy (e.g. tacrine) than those not receiving oestrogen.[73] In oestrogen treated cells, the neuroprotective effect of the ChE inhibitors tacrine and donepezil was found to be more pronounced than in cells treated with ChE inhibitor alone.[72] It is plausible that the effect observed with co-administration of oestrogen and a ChE inhibitor might be exerted through interactive mechanisms that involve allosteric sites on the nicotinic receptor. In fact, oestrogens have been found to sensitize the nicotinic receptor in vitro.[72]

Presently available data suggest that oestrogen treatment for a number of years after menopause may decrease the risk of developing AD, and that co-administration of oestrogen and a ChE inhibitor could constitute a new therapeutic approach.[65,66,73,74]

Conclusion

In addition to the demonstrated effect of ChE inhibitors on synaptic ACh and improvement of cognition (see Chapters 10 and 11), these drugs may act upon β-amyloid toxicity and aggregation, APP release, increase the release of non-cholinergic neurotransmitters and modulate the effect of oestrogens (see Table 13.1). These 'non-cholinergic' effects may contribute to the long-term clinical efficacy of these compounds in the treatment of AD.

References

1. Nordberg A, Svensson A-L. Cholinesterase inhibitors in the treatment of Alzheimer's disease: a comparison of tolerability and pharmacology. *Drug Safety* 1998; **19:** 465–480.
2. Giacobini E. Cholinesterase inhibitors do more than inhibit cholinesterase. In: Becker R, Giacobini E, eds. *Alzheimer's Disease: From Molecular Biology to Therapy*. Boston: Birkhäuser, 1996: 187–204.
3. Dickson DW. The pathogenesis of senile plaques. *J Neuropathol Exp Neurol* 1997; **56:** 321–339.
4. Terry RD, Hansen LA, DeTeresa R *et al.* Senile dementia of the Alzheimer type without neurocortical neurofibrillary tangles. *J Neuropathol Exp Neurol* 1987; **46:** 262–268.
5. Kang J, Lemaire HG, Unterbeck A *et al.* The precursor of Alzheimer's disease amyloid A4 protein resembles a cell surface receptor. *Nature* 1987; **325:** 733–736.
6. Mills J, Reiner PB. Regulation of amyloid precursor protein cleavage. *J Neurochem* 1999; **72:** 443–460.
7. Mattson MP, Barger SW, Cheng B, Lieberburg I, Smith-Swintosky VL, Rydel RE. β-Amyloid precursor protein metabolites and loss of neuronal Ca^{2+} homeostasis in Alzheimer's disease. *TINS* 1993; **16:** 409–414.
8. Yankner BA, Duffy LK, Kirschner DA. Neurotrophic and neurotoxic effects of amyloid β protein: reversal by tachykinin neuropeptides. *Science* 1990; **250:** 279–282.
9. Checler F. Processing of the β-amyloid precursor protein and its regulation in Alzheimer's disease. *J Neurochem* 1995; **65:** 1431–1444.
10. Auld DS, Kar S, Quirion R. β-amyloid peptides as direct cholinergic neuromodulators: a missing link? *TINS* 1998; **21:** 43–49.
11. Kar S, Seto D, Gaudreau P, Quirion R. β-Amyloid-related peptides inhibit potassium-evoked acetylcholine release from rat hippocampal slices. *J Neurosci* 1996; **16:** 1034–1040.
12. Kar S, Issa AM, Seta D, Auld DS, Collier B, Quirion R. Amyloid β-peptide inhibits high-affinity choline uptake and acetylcholine release in rat hippocampal slices. *J Neurochem* 1998; **70:** 2179–2187.
13. Pedersen WA, Kloczewiak MA, Blusztajn JK. Amyloid β-protein acetylcholine synthesis in a cell line derived from cholinergic neurons of the basal forebrain. *Proc Natl Acad Sci USA* 1996; **93:** 8068–8071.
14. Kelly JF, Furukawa K, Barger SW *et al.* Amyloid β-peptide disrupts carbachol-induced muscarinic cholinergic signal transduction in cortical neurons. *Proc Natl Acad Sci USA* 1996; **93:** 6753–6758.
15. Harkany T, Lengyel Z, Soós K, Penke B, Luiten P, Gulya K. Cholinotoxic effects of β-amyloid (1–42) peptide on cortical projections of the rat nucleus basalis mangocellularis. *Brain Res* 1995; **695:** 71–75.
16. Harkany T, O'Mahony S, Kelly JP *et al.* β-Amyloid (Phe (SO3H)24)25–35 in rat nucleus basalis induces behavioral dysfunctions, impairs learning and memory and disrupts cortical cholinergic innervation. *Beh Brain Res* 1998; **90:** 133–145.
17. Itoh A, Nitta A, Nadai M *et al.* Dysfunction of cholinergic and dopaminergic neuronal systems in β-amyloid protein-infused rats. *J Neurochem* 1996; **66:** 1113–1117.
18. Giacobini E. From molecular structure to Alzheimer therapy. *Jpn J Pharmacol* 1997; **74:** 225–241.
19. Nitsch RM, Slack BE, Wurtman RJ, Growdon JH. Release of Alzheimer amyloid precursor derivatives stimulated by activation of muscarinic acetylcholine receptors. *Science* 1992; **258:** 304–307.
20. Haring R, Gyrwitz D, Barg J *et al.* Amyloid precursor protein secretion via muscarinic receptors: reduced desensitization using the M1-selective agonist AF102B. *Biochem Biophys Res Commun* 1994; **203:** 652–658.
21. Farber SA, Nitsch RM, Schulz G, Wurtman RJ. Regulated secretion of β-amyloid precursor protein in rat brain. *J Neurosci* 1995; **15:** 7442–7451.
22. Pittel Z, Heldman E, Barg J, Haring R, Fisher

A. Muscarinic control of amyloid precursor protein secretion in rat cerebral cortex and cerebellum. *Brain Res* 1996; **742:** 299–304.
23. Kim S-H, Kim Y-K, Jeong S-J, Haass C, Kim Y-H, Suh Y-H. Enhanced release of secreted form of Alzheimer's amyloid precursor protein from PC12 cells by nicotine. *Mol Pharmacol* 1997; **52:** 430–436.
24. Giacobini E. Cholinomimetic therapy of Alzheimer's disease: does it slow down deterioration? In: Racagni G, Brunello N, Langer SZ, eds. *Recent Advances in the Treatment of Neurodegenerative Disorders and Cognitive Dysfunction.* New York: International Academy of Biomedical and Drug Research/Karger, 1994: 51–57.
25. Haroutunian V, Greig N, Pei X-F *et al.* Pharmacological modulation of Alzheimer's β-amyloid precursor protein levels in the CSF of rats with forebrain cholinergic system lesions. *Mol Brain Res* 1997; **46:** 161–168.
26. Lahiri DK, Farlow MR. Differential effect of tacrine and physostigmine on the secretion of the β-amyloid precursor protein in cell lines. *J Mol Neurosci* 1996; **7:** 41–49.
27. Lahiri DK, Farlow MR, Sambamurti K. The secretion of amyloid β-peptides is inhibited in the tacrine-treated human neuroblastoma cells. *Mol Brain Res* 1998; **62:** 131–140.
28. Mori F, Lai C-C, Fusi F, Giacobini E. Cholinesterase inhibitors increase secretion of APPs in rat brain cortex. *NeuroReport* 1995; **6:** 633–636.
29. Giacobini E, Mori F, Buznikov A, Becker R. Cholinesterase inhibitors alter APP secretion and APP mRNA in rat cerebral cortex. *Soc Neurosci Abstr* 1995; **21:** 988.
30. Chong YH, Suh Y-H. Amyloidogenic processing of Alzheimer's amyloid precursor protein in vitro and its modulation by metal ions and tacrine. *Life Sci* 1996; **59:** 545–557.
31. Kihara T, Shimohama S, Sawada H *et al.* Nicotinic receptor stimulation protects neurons against β-amyloid toxicity. *Ann Neurol* 1997; **42:** 159–163.
32. Zamani M, Allen Y, Owen G, Gray JA. Nicotine modulates the neurotoxic effect of β-amyloid protein(25–35) in hippocampal cultures. *NeuroReport* 1997; **8:** 513–517.
33. Svensson A-L, Nordberg A. Tacrine and donepezil attenuate the neurotoxic effect of Aβ(25–35) in rat PC12 cells. *NeuroReport* 1998; **9:** 1519–1522.
34. Moran MA, Mufson EJ, Gómez-Ramos P. Colocalization of cholinesterases with β amyloid protein in aged and Alzheimer's brains. *Acta Neuropathol* 1993; **85:** 362–369.
35. Sberna G, Saez-Valero J, Beyreuther K, Masters CL, Small DH. The amyloid beta-protein of Alzheimer's disease increases acetylcholinesterase expression by increasing intracellular calcium in embryonal carcinoma P19 cells. *J Neurochem* **69:** 1175–1184.
36. Sberna G, Sáez-Valero J, Li Q-X *et al.* Acetylcholinesterase is increased in the brains of transgenic mice expressing the C-terminal fragment (CT100) of the β-amyloid protein precursor of Alzheimer's disease. *J Neurochem* 1998; **71:** 723–731.
37. Alvarez A, Bronfman F, Pérez CA, Vicente M, Garrido J, Inestrosa NC. Acetylcholinesterase, a senile plaque component, affects the fibrillogenesis of amyloid-β-peptides. *Neurosci Lett* 1995; **201:** 49–52.
38. Inestrosa NC, Alvarez A, Pérez CA *et al.* Acetylcholinesterase accelerates assembly of amyloid-β-peptides into Alzheimer's fibrils: possible role of the peripheral site of the enzyme. *Neuron* 1996; **16:** 881–891.
39. Alvarez A, Alarcón R, Opazo C *et al.* Stable complexes involving acetylcholinesterase and amyloid-β-peptide change the biochemical properties of the enzyme and increase the neurotoxicity of Alzheimer's fibrils. *J Neurosci* 1998; **18:** 3213–3223.
40. Wagstaff A, McTavish D. Tacrine: a review of its pharmacodynamic and pharmacokinetic properties and therapeutic potential in Alzheimer's disease. *Drugs Aging* 1994; **4:** 1–31.
41. Giacobini E. Cholinergic receptors in human brain: effects of aging and Alzheimer's disease. *J Neurosci Res* 1990; **27:** 548–560.
42. Nordberg A. Neurochemical changes in Alzheimer's disease. *Cerebral Brain Met Rev* 1992; **4:** 303–328.
43. Nilsson L, Adem A, Hardy J, Winblad B, Nordberg A. Do tetrahydroaminoacridine (THA) and physostigmine restore acetylcholine release in Alzheimer brain via nicotinic recep-

tors? *J Neural Transmitters* 1987; **70:** 357–368.
44. Perry EK, Smith CJ, Court JA, Bonham JR, Rodway M, Atack JR. Interaction of 9-amino-1,2,3,4-tetrahydroaminoacridine (THA) with human cortical nicotinic and muscarinic receptor binding in vitro. *Neurosci Lett* 1988; **91:** 211–216.
45. Flynn DD, Mash DC. Multiple in vitro interactions with and differential in vivo regulation of muscarinic receptor subtypes by tetrahydroaminoacridine. *J Pharmacol Exp Ther* 1989; **250:** 573–581.
46. Svensson A-L, Zhang X, Nordberg A. Biphasic effect of tacrine on acetylcholine release in rat brain via M1 and M2 receptors. *Brain Res* 1996; **726:** 207–212.
47. Nordberg A, Winblad B. Reduced number of ^{3}H-nicotine and ^{3}H-acetylcholine binding sites in the frontal cortex of Alzheimer brains. *Neurosci Lett* 1986; **72:** 115–119.
48. Warpman U, Nordberg A. Epibatidine and ABT-418 reveal selective losses of $\alpha 4\beta 2$ nicotinic receptors in Alzheimer brains. *NeuroReport* 1995; **6:** 2419–2423.
49. Nordberg A, Lundqvist H, Hartvig P, Lilja A, Långström B. Kinetic analysis of regional $(S)(-)^{11}$C-nicotine binding in normal and Alzheimer brains: in vivo assessment using positron emission tomography. *Alzheimer's Disease Assoc Disorders* 1995; **9:** 21–27.
50. Nilsson-Håkansson L, Lai Z, Nordberg A. Tetrahydroaminoacridine induces opposite changes in muscarinic and nicotinic receptors in rat brain. *Eur J Pharmacol* 1990; **186:** 301–305.
51. Nordberg A, Lilja A, Lundqvist H *et al.* Tacrine restores cholinergic nicotinic receptors and glucose metabolism in Alzheimer patients as visualized by positron emission tomography. *Neurobiol Aging* 1992; **13:** 747–758.
52. Nordberg A, Lundqvist H, Hartvig P *et al.* Imaging of nicotinic and muscarinic receptors in Alzheimer's disease: effect of tacrine treatment. *Dement Geriatr Cogn Disorders* 1997; **8:** 78–84.
53. Nordberg A, Amberla K, Shigeta M *et al.* Long-term tacrine treatment in three mild Alzheimer patients: effects on nicotinic receptors, cerebral blood flow, glucose metabolism, EEG and cognitive abilities. *Alz. Disease Assoc Disorders* 1998; **12:** 228–237.
54. Svensson A-L, Nordberg A. Tacrine interacts with an allosteric activator site on $\alpha 4\beta 2$ nAChRs in M10 cells. *NeuroReport* 1996; **7:** 2201–2205.
55. Pereira EFR, Alkondon M, Reinhardt S *et al.* Physostigmine and glanthamine: probes for a novel binding site on the $\alpha 4\beta 2$ subtype of neuronal nicotinic acetylcholine receptors stably expressed in fibroblast cells. *J Pharmacol Exp Ther* 1994; **270:** 768–778.
56. Maelicke A, Coban T, Storch A, Schrattenholz A, Pereira EFR, Albuquerque EX. Allosteric modulation of torpedo nicotinic acetylcholine receptor ion channel activity by noncompetitive agonists. *J Rec Signal Transduc Res* 1997; **17:** 11–28.
57. Messamore E, Warpman U, Ogane N *et al.* Cholinesterase inhibitor effects on extracellular acetylcholine in rat cortex. *Neuropharmacology* 1993; **32:** 745–750.
58. Warpman U, Zhang X, Nordberg A. Effect of tacrine on in vivo release of dopamine and its metabolites in the striatum of freely moving rats. *J Pharmacol Exp Ther* 1996; **277:** 917–922.
59. Cuadra G. Summers K, Giacobini E. Cholinesterase inhibitor effects on neurotransmitters in rat cortex in vivo. *J Pharmacol Exp Ther* 1994; **270:** 277–284.
60. Mori F, Caudra G, Giacobini E. Metrifonate effects on acetylcholine and biogenic amines in rat cortex. *Neurochem Res* 1995; **20:** 1081–1088.
61. Giacobini E, Zhu X-D, Williams E, Sherman KA. The effect of the selective reversible acetylcholinesterase inhibitor E2020 on extracellular acetylcholine and biogenic amine levels in rat cortex. *Neuropharmacology* 1996; **35:** 205–211.
62. Druckarch B, Leysen JE, Soof JC. Further analysis of the neuropathological profile of 9-amino-1,2,3,4-tetrahydroacridine (THA), an alleged drug for the treatment of Alzheimer's disease. *Life Sci* 1988; **42:** 1011–1017.
63. Jossan SS, Adem A, Winblad B *et al.* Characterization of dopamine and serotonin uptake inhibitory effects of tetrahydroaminoacridine in rat brain. *Pharmacol Toxicol* 1992; **71:** 213–215.

64. Adem A, Sing-Jossan S, Oreland L. Tetrahydroaminoacridine inhibits human and rat brain monoamine oxidase. *Neurosci Lett* 1989; **14:** 243–248.
65. Henderson VW. Estrogen replacement therapy for the prevention and treatment of Alzheimer's disease. *CNS Drugs* 1997; **8:** 343–351.
66. Yaffe K, Sawaya G, Lieberburg I, Grady D. Estrogen therapy in postmenopausal women: effects on cognitive function and dementia. *JAMA* 1998; **279:** 688–695.
67. Luine VN, Park D, Joh T, Reis D, McEwen BS. Immunochemical demonstration of increased choline acetyltransferase concentration in rat preoptic area after estradiol administration. *Brain Res* 1980; **191:** 273–277.
68. Singh M, Meyer EM, Huang FS, Millard WJ, Simkins JW. Ovariectomy reduces ChAT activity and NGF mRNA levels in the frontal cortex and hippocampus of the female Sprague Dawley rat. *Abstr Soc Neurosci* 1993; **19:** 1254.
69. Green PS, Gridley KE, Simpkins JW. Estradiol protects against β-amyloid(25–35)-induced toxicity in SK-H-SH human neuroblastoma cells. *Neurosci Lett* 1996; **218:** 165–168.
70. Xu H, Gouras GK, Greenfield JP *et al.* Estrogen reduces neuronal generation of Alzheimer β-amyloid peptides. *Nature Med* 1998; **4:** 447–451.
71. Behl C, Skutella T, Lezoualc'h F *et al.* Neuroprotection against oxidative stress by estrogens: structure–activity relationship. *Mol Pharmacol* 1997; **51:** 535–541.
72. Svensson A-L, Nordberg A. β-estradiol attenuate amyloid-β-peptide toxicity via nicotinic receptors and promotes neuroprotection of cholinesterase inhibitors. *NeuroReport* 1999; in press.
73. Schneider LS, Farlow MR, Henderson VW, Pogoda JM. Effects of estrogen replacement therapy on response to tacrine in patients with Alzheimer's disease. *Neurology* 1996; **46:** 1580–1584.
74. Giacobini E. Aging, Alzheimer's disease, and estrogen therapy. *Exp Gerontol* 1998; **33:** 865–869.

14

The effect of cholinesterase inhibitors studied with brain imaging

Agneta Nordberg

Introduction

Alzheimer's disease (AD) is one of the most devastating brain disorders of elderly humans. The last decade has witnessed a steadily increasing effort directed at the discovery of the aetiology and the neuropathological and neurochemical mechanisms involved in the disease, but still there is no cure.[1-4] Extensive research activities have stimulated the development of new treatment strategies in AD, and so far the cholinesterase (ChE) inhibitors, representing symptomatic transmitter therapy, have reached clinical use. Although arguments have been raised that the transmitter dysfunction might be secondary to the course of AD, the treatment strategy seems to play a significant role in tackling both the symptoms[5] and the progression of the disease.[6] These effects might be most significant if the drug is initiated early in the course of the disease, when temporary restoration of communicative processes between transmitter pathways is possible and neuroprotective mechanisms and disease-modifying mechanisms are still susceptible to interaction.

Symptomatic treatment, mainly focusing on cholinergic therapy, has been clinically evaluated in randomized, double-blind, placebo-controlled, parallel-group studies measuring performance-based tests of cognitive function, activity of daily living and behaviour. An important question concerning clinical trials of antidementia drugs is how to evaluate the efficacy of the drug. Different outcome measures have been used, including scales such the AD Assessment Scale (ADAS), the Clinical Interview-based Impression (CIBI) scale, the Global Deterioration Scale (GDS), the Caregiver-Related Clinical Gobal Impression of Changes, the Gottfries-Bråne Scale (GBS), and the Mini Mental State Examination.

Significant progress has been made in recent years in the development and application of functional brain imaging techniques, allowing early diagnosis of dementia and evaluation of treatment efficacy. The availability of symptomatic treatment in AD increases the pressure for imaging measures which are sensitive and specific early in the course of the disease. Positron emission tomography (PET) has been found to be a suitable method for functional studies of pathological changes in brain. As a clinical instrument PET both reveals dysfunctional changes early in the course of disease and provides deep insight into the functional mechanisms of new potential drug treatment strategies. The advantage of PET is that, besides its capacity to measure changes in glucose metabolism and cerebral blood flow, it can also be used to obtain knowledge about cell communicative processes (transmitter–receptor interactions) and pharmacokinetic events. Studies of the functional effect in brain of drugs evaluated by neuropsychological measures, electroencephalography and imaging techniques will be valuable complements to the results of conventional clinical trials. The application of the PET technique to the evaluation of cholinergic drug therapy in AD is the focus of this chapter.

Functional studies of cholinergic activity in AD

Several brain imaging techniques are presently used in the assessment of patients with memory disorders. Structural imaging techniques such as computed tomography (CT) and magnetic resonance imaging (MRI) are important tools in morphological investigations of the brain which can be used for different diagnostic purposes (e.g. AD versus vascular dementia).[7,8] Functional imaging by PET and single photon emission tomography (SPECT) will increase the understanding of the functional correlates of structural and biological changes in diseased brain. These techniques have so far been extensively used in studies of changes in cerebral blood flow and glucose metabolism disturbances in dementia.[9–11] Although dynamic changes in cerebral blood flow and glucose metabolism can be used as rough indicators of neurotransmitter activity, the established cholinergic hypothesis gives rise to the need to search for in vivo markers reflecting cholinergic activity in brain. A limiting factor in this regard is the availability of radiolabelled compounds suitable for studies with SPECT and PET (Table 14.1).

AChE activity

Several PET ligands have been developed for visualizing acetylcholinesterase (AChE) activity in brain, including [^{11}C]physostigmine, [^{11}C]*N*-methyl-3-piperidyl acetate ([^{11}C]MP3A), [^{11}C]*N*-methyl-4-piperidyl acetate ([^{11}C]MP4A) and *N*-[^{11}C]-methylpiperidin-4-yl proprionate ([^{11}C]PMP).[12–15] Studies were initially performed ex vivo in rats and in PET studies in baboons.[13,16]

[^{11}C]Physostigmine, [^{11}C]MP4A and [^{11}C]PMP

Cholinergic parameter	*Radioligand*	*Imaging technique*	*Ref.*
AChE	[^{11}C]MP4A	PET	14
	[^{11}C]PMP	PET	18
Cholinergic terminals	[^{123}I]IBVM	SPECT	22
Nicotinic receptors	[^{11}C]Nicotine	PET	33, 34, 37
Muscarinic receptors	[^{11}C]Benztropine	PET	23
	[^{11}C]Scopolamine	PET	24
	[^{11}C]Tropanylbenzilate	PET	25
	[^{11}C]NMPB	PET	26
	[^{123}I]QNB	SPECT	27, 28
	[^{123}I]4-Iododexetimide	SPECT	29
	[^{123}I]4-Iodolevetimide	SPECT	29

[^{11}C]NMPB, [^{11}C]*N*-methyl-4-piperidyl benzilate.

Table 14.1
PET and SPECT ligands for the visualization of cholinergic activity in human brain in vivo.

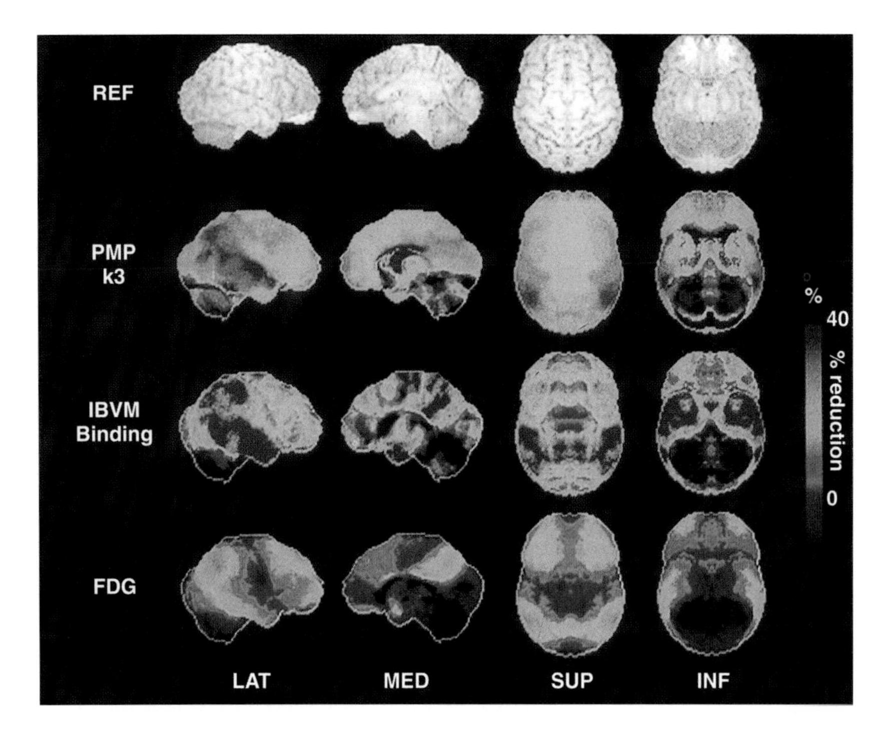

Figure 14.1
Percentage reduction in brain AChE activity in AD patients compared to normal subjects as measured using the PET ligand [^{11}C]PMP. The data are compared to corresponding data for cholinergic terminal loss ([^{123}I]IBVM) and glucose hypometabolism ([^{18}F]FDG). (From Kuhl et al.[18] with permission from Lippincott Williams & Wilkins.)

have been used in PET studies in healthy volunteers.[16–18] For the radioligands [^{11}C]MP4A and [^{11}C]PMP a quantitative measurement was achieved by using a three-compartment kinetic model.[17] The rate constant k_3 expresses the hydrolysis of the radiotracer by AChE. The selectivity of [^{11}C]MP4A and [^{11}C]PMP for AChE in human brain is estimated to be more than 90% and the k_3 value is proportional to the AChE activity in brain.[17,18] The regional k_3 values measured in human brain in vivo by PET agree fairly well with earlier values obtained in autopsy brain tissue.[17,18] No significant decrease in AChE activity was found in human cerebral cortex with normal ageing.[17,18]

A decrease in cortical AChE activity has been measured by PET in AD patients.[14,18] The decrease in AChE found in mildly to moderately demented AD patients was less pronounced than in autopsy material and did not strictly correlate with the cerebral glucose impairment[18] (Fig. 14.1). A crucial question is whether cholinergic enzyme activity is an early marker in AD. Recently Davis *et al.*[19] have attempted to correlate choline acetyltransferase activity in autopsy brain tissue with cognitive testing performed in the AD patients prior death. In addition, Traykov *et al.*[20] have studied the distribution and kinetics of [^{11}C]methyltetrahydroaminoacridine in human brain by PET, and found that the uptake of this compound did not parallel the distribution of AChE in brain. This finding emphasizes the fact that not all ChE inhibitors are suitable PET ligands for the measurement of AChE activity in human brain.

Cholinergic terminal density

To quantify the presynaptic vesicular ACh transporter (VAChT) and use it as an in vivo marker of presynaptic cholinergic density, [^{123}I]iodobenzovesamicol ([^{123}I]IBVM) has been used as radioligand in SPECT studies.[21–22] Greater reduction in [^{123}I]IBVM binding was observed throughout the cerebral cortex in patients with early-onset compared to late-onset of AD.[22]

Muscarinic receptors

Imaging of muscarinic receptors in human brain has been performed by PET using several different radioligands (see Table 14.1). Using [^{11}C]benztropine and [^{11}C]tropanylbenzilate (TRP) as radioligands, an age-related decrease in muscarinic receptors was observed in cortical brain regions of healthy volunteers.[25,30] Studies using [^{123}I]quinuclidinylbenzilate ([^{123}I] QNB) and SPECT have shown both a decreased and a relatively unchanged number of muscarinic receptors in brain of AD patients.[27,28,31]

Nicotinic receptors

The use of [^{11}C]nicotine in PET studies has recently been reviewed by Mazière and Delforge,[32] and some of the problems encountered by using nicotine as a PET tracer were identified. Deficits in nicotinic receptors have been observed in AD brains with [^{11}C]nicotine and PET.[33–35] The distribution in brain of [^{11}C]nicotine was initially analysed using a simple two-compartment model.[33,36] Muzic *et al.*[37] studied the uptake of [^{11}C]nicotine in brain of normal volunteers using the distribution volume as an index of specific binding. A dual tracer method using [^{11}C]nicotine and [^{15}O]water has recently been developed.[35] The dual tracer model has been evaluated and verified in a rhesus monkey model.[38] The rate constant k_2* for [^{11}C]nicotine is assumed to provide a quantitative measure of [^{11}C]nicotine binding in human brain; the rate constant is inversely related to the tissue binding of nicotine.[35] A significantly lower binding of [^{11}C]nicotine (i.e. a significantly higher k_2* value) was found in cortical brain regions of AD patients compared to age-matched controls. The k_2* value in the temporal cortex is significantly correlated with the cognitive function of the AD patient,[35] suggesting that losses of cortical nicotinic receptor binding sites might be a very early phenomenon in the progression of the AD disease.

Attempts have recently been made to develop PET and SPECT ligands which allow the visualization of subtypes of nicotinic receptors. Azetidine analogues such as [^{11}C]MPA and 5-[^{76}Br]bromo-3-[2(S)-acetidinyl]methoxy]pyridine ([^{76}Br]BAP) might be promising nicotinic receptor ligands for further human studies.[39,40]

Effect of acute and chronic treatment with ChE inhibitors in AD patients

Imaging techniques such as PET and SPECT offer a unique opportunity to study functional effects in the brain induced by drug treatment. Relatively few drug treatments in AD patients have so far been evaluated by these techniques. The ChE inhibitors are reviewed below (Table 14.2).

Acute treatment with ChE inhibitors

Imaging studies in humans have revealed that ChE inhibitors can influence the cerebral blood flow and glucose metabolism in brain. SPECT measurements have shown that acute treatment with physostigmine,[41,46–48,52,53] vel-

Imaging technique and drug	*Ligand*	*Effect exerted upon*	*Ref.*
PET			
Physostigmine	[^{18}F]FDG	Metabolism	41, 42
	[^{15}O]H_2O	Blood flow	42
Tacrine	[^{18}F]FDG	Metabolism	34
	[^{15}O]H_2O	Blood flow	43
	[^{11}C]NIC	Nicotinic receptors	34, 43, 44
	[^{11}C]Benz	Muscarinic receptors	44
Donepezil	[^{18}F]FDG	Metabolism	45
SPECT			
Physostigmine	[^{99m}Tc]HMPAO	Blood flow	41, 46, 47
Velnacrine	[^{99m}Tc]HMPAO	Blood flow	48
Tacrine	[^{99m}Tc]HMPAO	Blood flow	49–52

[^{11}C]NIC, [^{11}C]nicotine.
[^{11}C]Benz, [^{11}C]benztropine.
[^{18}F]FDG, [^{18}F]fluorodeoxyglucose.
[^{99m}Tc]HMPAO, [^{99m}Tc]hexamethylpropylene amine oxide.

Table 14.2
Cholinergic drug therapies in AD as evaluated by PET and SPECT.

nacrine[48] and tacrine[52] increases cortical blood flow in AD patients. Opposite changes in cerebral blood flow and glucose metabolism were found by Blin *et al.*[42] in a study involving intravenous infusion of physostigmine to AD patients. The study was double-blind and each subject was investigated twice, under placebo and physostigmine; the maximum tolerated dose of physostigmine was determined for each subject before the PET investigations were performed.[42] The infusion of physostigmine caused a 5–10% increase in the cerebral blood flow in AD patients, while the glucose metabolism was decreased (by 16–20%).[42] The cerebral blood flow response was suggested to consist of two components, the larger one (increase in cerebral blood flow) representing a vascular response induced by ACh.[42] The uncoupling between blood flow and glucose metabolism effects after physostigmine infusion might therefore be caused by the vascular component of ACh, which increases the cerebral blood flow.[42,54] It has been reported that tacrine can overcompensate for the decreased blood flow seen following experimental basal forebrain lesions in the rat.[55]

Infusion of 1.5 mg physostigmine salicylate to healthy subjects was found to cause a 50% decrease in AChE activity was measured using [^{11}C]PMP and PET.[18] The decline in enzyme activity following physostigmine infusion was of the same order of magnitude as expected in AD patients clinically treated with ChE

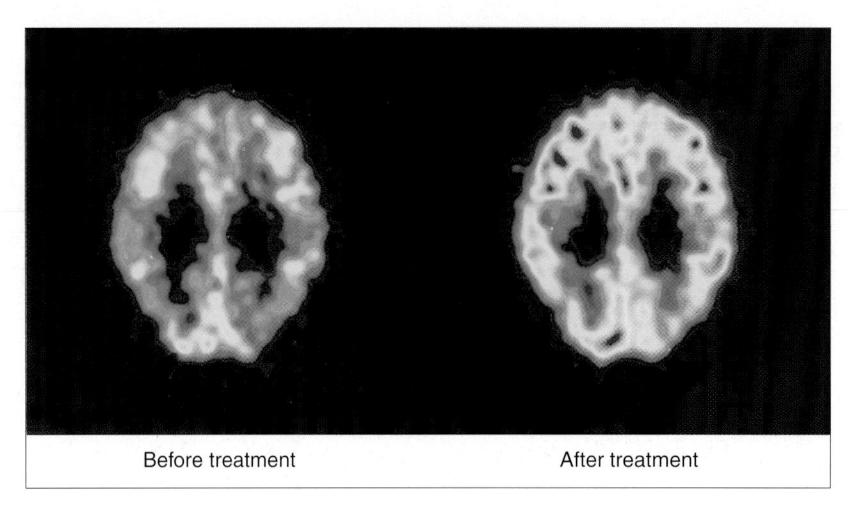

Figure 14.2
The effect on cerebral blood flow in an AD patient of 3 months of treatment with rivastigmine. PET sections through the basal ganglia of a patient receiving an intravenous tracer dose of [^{15}O]H_2O. The colour scale indicates regional cerebral blood flow (ml/min per 100 g). The patient underwent PET studies prior to (left) and after 3 months treatment with rivastigmine 6 mg/day (right). Red indicates high, yellow medium and blue low glucose metabolism. (Photo: Uppsala University PET centre, Uppsala, Sweden.)

inhibitors. PET studies may be used to study the clinical efficacy of ChE inhibitors in AD patients.

The effect of physostigmine on working-memory performance and cereberal blood flow was studied in healthy subjects.[56] The magnitude of the physostigmine-induced reduction in reaction time was correlated to a decrease in cerebral blood flow in the mid-frontal cortex.[56]

Subchronic and long-term treatment with ChE inhibitors

Cerebral blood flow

An increase in cerebral blood flow has been measured, using ^{133}Xe inhalation[50–51] and PET,[43] in AD patients responding to long-term tacrine treatment. Cohen *et al.*[49] treated AD patients with tacrine and lecithin for some weeks, but no change in cerebral blood flow was observed. A possible explanation for the difference in the effects observed might be the length of tacrine treatment. The positive effect on cerebral blood flow after 3 months of treatment with rivastigmine is shown in Fig. 14.2.

Glucose metabolism

Long-term treatment of AD patients with tacrine improves the cerebral glucose metabolism; the improvement seems to be dependent on both the length of treatment and the dose of tacrine.[34,43] We have observed a similar general increase in glucose metabolism in brain of AD patients treated with donepezil (Fig. 14.3). Tune *et al.*[45] recently reported a double-blind, parallel treatment study in 28 AD patients where the glucose metabolism was measured prior to and after 24 weeks of treatment with donepezil 10 mg/day. The donepezil-treated AD patients maintained or slightly increased

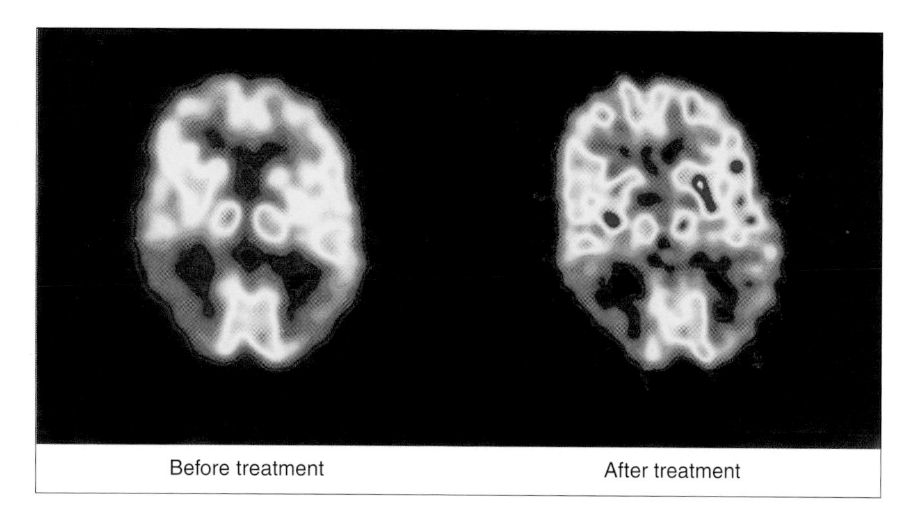

Figure 14.3
The effect on glucose metabolism in an AD patient of 3 months of treatment with donepezil. PET sections through the basal ganglia of a patient receiving an intravenous tracer dose of [^{18}F]fluorodeoxyglucose. The colour scale indicates regional glucose metabolism (μmol/min per 100 g). The patient underwent PET studies prior to (left) and after 3 months treatment with donepezil 10 mg/day (right). Red indicates high, yellow medium and blue low glucose metabolism. (Photo: Uppsala University PET centre, Uppsala, Sweden.)

their global glucose metabolism, while the placebo-treated patients showed a decreased global glucose metabolism.[45]

Nicotinic receptors

An improved binding of [^{11}C]nicotine has been observed in the temporal cortex of AD patients treated with tacrine 80 mg/day for 3 months.[34,43] Treatment studies with other ChE inhibitors such as NXX-066 have shown similar increases in [^{11}C]nicotine binding (reduced k_2* values) as found for tacrine (Fig. 14.4). The improvement in neuropsychological tests generally parallels the improvement in nicotine binding in AD patients during long-term treatment.[43] The effect of ChE inhibitor treatment seems to be more regionally specific with regard to the nicotinic receptors than does the effect on the glucose metabolism. The restoration of cortical nicotinic receptors following tacrine treatment might be due to a stimulatory effect of tacrine on the nicotinic receptors,[57] but it is also plausible that the effect might be due to an indirect stimulation of the nicotinic receptors via the increased amount of endogenous ACh in the synaptic cleft in the presence of the AChE inhibitor. Interestingly, recent data indicate that ChE inhibitors may activate nicotinic receptors via an allosteric site separately located from the ACh binding sites on the nicotinic receptors.[58,59]

Muscarinic receptors

While long-term treatment with tacrine causes an up-regulation of the nicotinic receptors in brain, a down-regulation of the muscarinic receptors is expected, in accordance with treatment studies using ChE inhibitors in

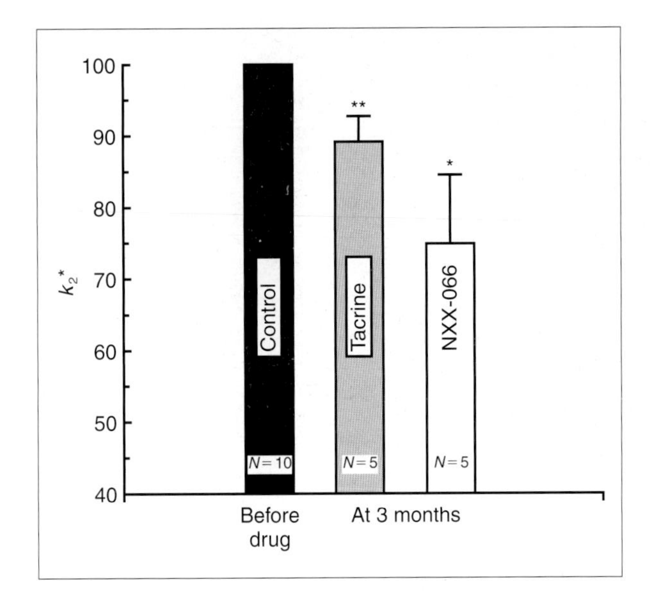

Figure 14.4
The effect of the ChE inhibitors tacrine and NXX-066 on [^{11}C]nicotine binding in the temporal cortex of AD patients. Five AD patients were treated with tacrine 80 mg/day for 3 months, and another five patients were treated with NXX-066 ?? mg/day for a similar length of time. The patients received a tracer dose of (S)(−)[^{11}C]nicotine intravenously and the uptake and distribution in brain were measured by PET. The [^{11}C]nicotine binding is expressed as k_2 where an increase in [^{11}C]binding corresponds to a decrease in the k_2* value. The data are expressed as a percentage of the control value. *$p < 0.05$; **$p < 0.01$; N, number of patients.*

rodents.[60] When [^{11}C]benztropine was injected intravenously an increased uptake of ^{11}C radioactivity was observed in the brain of AD patients after 3 months of tacrine treatment, compared to prior to treatment, reflecting a drug-induced increase in blood flow.[44] A decreased [^{11}C]benztropine binding was observed in the temporal cortex following 3 months of tacrine treatment, reflecting a transient down-regulation of the muscarinic receptors, which was normalized after 10 months of tacrine treatment.[44]

Acknowledgements

This study was supported by grants from the Swedish Medical Research Council (project No. 05817), Loo and Hans Ostermans's foundation, and KI foundations.

References

1. Braak H, Braak E. Evolution of neuronal changes in the course of Alzheimer's disease. *J Neural Transm* 1998; **53(Suppl):** 127–140.
2. Hardy J. Amyloid, the presenilins and Alzheimer's disease. *Trends Neurosci* 1997; **20:** 154–159.
3. Hardy J, Duff K, Hardy KG, Perez-Tyr J, Hutton M. Genetic dissection of Alzheimer's disease and related demetias: amyloid and this relationship to tau. *Nature Neurosci* 1998; **1:** 355–358.
4. Master CL, Beyreuther K. Alzheimer disease. *Br Med J* 1998; **316:** 446–448.
5. Cummings JL, Kaufer D. Neuropsychiatric aspects of Alzheimer's disease: the cholinergic hypothesis revisited. *Neurology* 1996; **47:** 876–883.
6. Knopman DS, Schneider L, Davis K *et al.* Long-term tacrine (Cognex) treatment: effect on nursing home placement and mortality. *Neurology* 1996; **47:** 166–177.
7. Fontaine S, Nordberg A. Brain imaging. In: Gauthier S, ed. *Clinical Diagnosis and Management of Alzheimer's Disease*. London: Martin Dunitz, 1996: 83–105.
8. Fox N. Magnetic resonance imaging in Alzheimer's disease: from diagnosis to measuring therapeutic effect. *Alzheimer Rep* 1999; **2:** 5–12.
9. Rapoport S. Positron emission tomography in

Alzheimer's disease in relation to disease pathogenesis: a critical review. *Cerebrovasc Brain Metab Rev* 1991; **3:** 297–335.
10. Nordberg A. Clinical studies in Alzheimer patients with positron emission tomography. *Behav Brain Res* 1993; **57:** 215–224.
11. Waldemar G. Functional brain imaging with SPECT in normal aging and dementia — methodological, pathophysiological, and diagnostic aspects. *Cerebrovasc Brain Metab Rev* 1995; **7:** 89–130.
12. Tavitian B, Pappata S, Planas A M *et al.* In vivo visualization of acetylcholinestesterase with positron emission tomography. *NeuroReport* 1993; **4:** 535–538.
13. Namba H, Irie T, Fukushi K, Iyo M. In vivo measurement of acetylcholinesterase activity in the brain with a radioactive acetylcholine analog. *Brain Res* 1994; **667:** 278–282.
14. Iyo M, Namba H, Fukushi K *et al.* Measurement of acetylcholinesterase by positron emission tomography in the brain of healthy controls and patients with Alzheimer's disease. *Lancet* 1997; **349:** 1805–1809.
15. Planas AM, Crouzel C, Hinnen F *et al.* Rat brain acetylcholinesterase visualized with [^{11}C]physostigmine. *Neuroimage* 1994; **1:** 173–180.
16. Pappata S, Tavitian B, Traykov L *et al.* In vivo imaging of human cerebral acetylcholinesterase. *J Neurochem* 1996; **67:** 876–879.
17. Namba H, Masaoimi I, Fukushi K *et al.* Human cerebral acetylcholinesterase activity measured with positron emission tomography procedure, normal values and effect of age. *Eur J Nucl Med* 1999; **25:** 135–143.
18. Kuhl DE, Koeppe RA, Minoshima S *et al.* In vivo mapping of cerebral acetylcholinesterase activity in aging and Alzheimer's disease. *Neurology* 1999; **52:** 691–699.
19. Davis P. Challenging the cholinergic hypothesis in Alzheimer's disease. *JAMA* 1999; **281:** 1433–1441.
20. Traykov L, Tavitian B, Jobert A *et al.* In vivo PET studies of cerebral [^{11}C]methyltetrahydroaminoacridine distribution and kinetics in healthy subjects. *Eur J Neurology* 1999; **3:** 273–278.
21. Van Dort ME, Jung YW, Gidersleve DL *et al.* Synthesis of the ^{123}I- and ^{125}I-labeled cholinergic nerve marker (−)-5-iodobenzovesamicol. *Nucl Med Biol* 1993; **20:** 929–937.
22. Kuhl DE, Minoshima S, Fessler JA *et al.* In vivo mapping of cholinergic terminals in normal aging, Alzheimer's disease, and Parkinson's disease. *Ann Neurol* 1996; **40:** 399–410.
23. Dewey SL, MacGregor RR, Brodie JD *et al.* Mapping muscarinic receptors in human and baboon brain using [*N*-^{11}C-methyl]-benztropine. *Synapse* 1990; **5:** 213–223.
24. Frey KA, Koeppe RA, Mulholland GK *et al.* In vivo muscarinic cholinergic receptor imaging in human brain with [^{11}C]scopolamine and positron emission tomography. *J Cereb Blood Flow Metab* 1992; **12:** 147–154.
25. Lee KS, Frey KA, Koeppe RA *et al.* In vivo quantification of cerebral muscarinic receptors in normal human aging using positron emission tomography and [^{11}C]tropanylbenzilate. *J Cereb Blood Flow Metab* 1996; **16:** 303–310.
26. Zubieta J-K, Koeppe RA, Mulholland GK *et al.* Quantification of muscarinic cholinergic receptors with [^{11}C]NMPB and positron emission tomography: method development and differentiation of tracer delivery from receptor binding. *J Cereb Blood Flow Metab* 1998; **18:** 619–631.
27. Weinberger DR, Gibson RE, Coppola R *et al.* The distribution of cerebral muscarinic receptors in-vivo in patients with dementia. *Arch Neurol* 1991; **48:** 169–176.
28. Wyper DJ, Brown D, Patterson J *et al.* Deficits in iodine-labelled 3-quinuclidinyl benzilate binding in relation to cerebral blood flow in patients with Alzheimer's disease. *Eur J Nucl Med* 1993; **20:** 379–386.
29. Müeller-Gärtner H-W, Wilson AA, Dannals RF, Wagner HN, Frost JJ. Imaging muscarinic cholinergic receptors in human brain in vivo with SPECT. [^{123}I]4-Iododexetimide and [^{123}I]4-iodolevetimide. *J Cereb Blood Flow Metab* 1992; **12:** 562–570.
30. Dewey SL, Volkow ND, Logan J *et al.* Age-related decrease in muscarinic cholinergic receptor binding in human brain measured with positron emission tomography (PET). *J Neurosci Res* 1990; **27:** 569–575.
31. Weinberger DR, Jones D, Reba RC *et al.* A comparison of FDG PET and IQNB SPECT in normal subjects and in patients with dementia.

J Neuropsychiatry Clin Neurosci 1992; **4:** 239–248.
32. Mazière M, Delforge J. PET imaging of [^{11}C]nicotine: historical aspects. In: Domino E, ed. *Brain Imaging of Nicotine and Tobacco Smoking*. Ann Arbor, MI: NPP, 1995: 13–28.
33. Nordberg A, Hartvig P, Lilja A *et al.* Decreased uptake and binding of ^{11}C-nicotine in brain of Alzheimer patients as visualized by positron emission tomography. *J Neural Transm* 1990; **2:** 215–224.
34. Nordberg A, Lilja A, Lundqvist H *et al.* Tacrine restores cholinergic nicotinic receptors and glucose metabolism in Alzheimer patients as visualized by positron emission tomography. *Neurobiol Aging* 1992; **13:** 747–758.
35. Nordberg A, Lundqvist H, Hartvig P, Lilja A, Långström B. Kinetic analysis of regional (*S*)(−)^{11}C-nicotine binding in normal and Alzheimer brains — in vivo assessment using positron emission tomography. *Alzheimer Dis Assoc Disord* 1995; **1:** 21–27.
36. Yokoi F, Komiyama T, Ito T, Hayashi T, Iio M, Hara T. Application of carbon-11 labelled nicotine in the measurement of human cerebral blood flow and other physiological parameters. *Eur J Nucl Med* 1993; **20:** 4652.
37. Muzic Jr RF, Berridge MS, Friedland RP, Zhu N, Nelson AD. PET quantification of specific binding of carbon-11-nicotine in human brain. *J Nucl Med* 1998; **39:** 2048–2054.
38. Lundqvist H, Nordberg A, Hartvig P, Långström B. (*S*)(−)[^{11}C]nicotine binding assessed by PET: a dual tracer model evaluated in the rhesus monkey brain. *Alzheimer Dis Assoc Disord* 1998; **12:** 238–246.
39. Sihver W, Fasth KJ, Ögren M *et al.* In vivo positron emission tomography studies on the novel nicotinic receptor agonist [^{11}C]MPA compared with [^{11}C]ABT and (*S*)(−)[^{11}C]nicotine in Rhesus monkeys. *Nucl Med Biol* 1999; in press.
40. Sihver W, Fasth KJ, Horti A *et al.* The synthesis and characterization of a novel nicotinic acetylcholine receptor ligand, 5-[^{76}Br]bromo-3-(2(*S*)-azetidinyl)methoxy)pyridine, in the rat brain. *J Neurochem* 1999; in press.
41. Tune L, Brandt J, Frost J *et al.* Physostigmine in Alzheimer's disease: effect on cognitive functioning cerebral glucose metabolism analysed by positron emission tomography and cerebral blood flow analysed by single photon emission tomography. *Acta Psychiatry Scand* 1991; **366(Suppl):** 61–65.
42. Blin J, Ivanoiu A, Coppens A *et al.* Cholinergic neurotransmission has different effects on cerebral glucose consumption and blood flow in young normals, aged normals, and Alzheimer's disease patients. *Neuroimage* 1997; **6:** 335–343.
43. Nordberg A, Amberla K, Shigeta M *et al.* Long-term tacrine treatment in three mild Alzheimer patients: effects on nicotine receptors, cerebral blood flow, glucose metabolism, EEG and cognitive abilities. *Alzheimer Dis Assoc Disord* 1998; **12:** 228–237.
44. Nordberg A, Lundqvist H, Hartvig P *et al.* Imaging of nicotinic and muscarinic receptors in Alzheimer's disease: effect of tacrine treatment. *Dementia Geriatr Cogn Disord* 1997; **8:** 78–84.
45. Tune LE, Hoffman JM, Tiseo PJ *et al.* Donepezil HCL maintains global glucose activity in patients with Alzheimer's disease: results of a 24 weeks study. *J Cereb Blood Flow Metab* 1999; **19(Suppl 1):** S838.
46. Geany DP, Soper N, Shepstone BJ, Cowen PJ. Effect of central cholinergic stimulation on regional cerebral blood flow in Alzheimer disease. *Lancet* 1990; **335:** 1484–1487.
47. Hunter R, Wyper DJ, Patterson J, Hansen MT, Goodwin GM. Cerebral pharmacodynamics of physostigmine in Alzheimer's disease investigated using ^{99m}Tc-HMPAO SPECT imaging. *Br J Psychiatry* 1991; **158:** 351–357.
48. Ebmeier KP, Hunter R, Curran SM *et al.* Effect of single dose of the acetylcholinesterase inhibitor velnacrine on recognition memory and regional cerebral blood flow in Alzheimer's disease. *Psychopharmacology* 1992; **108:** 103–109.
49. Cohen MB, Fitten J, Lake RR. SPECT brain imaging in Alzheimer's disease during treatment with oral tetrahydroaminoacridine and lecithin. *Clin Nucl Med* 1992; **17:** 312–315.
50. Minthon L, Gustafson L, Dalfelt G *et al.* Oral tetrahydroaminoacridine treatment of Alzheimer's disease evaluated clinically and by regional cerebral blood flow and EEG. *Dementia* 1993; **4:** 32–42.
51. Minthon L, Nilsson K, Edvinson L, Wendt PE,

Gustafson L. Long-term effects of tacrine on regional cerebral blood flow changes in Alzheimer's disease. *Dementia* 1995; **6:** 245–251.
52. Riekkinen P, Kuikka J, Soininen H, Helkala EL, Hallikainen M, Riekkinen P. Tetrahydroaminoacridine modulates technetium-99m labelled ethylene dicysteinate retention in Alzheimer's disease measured with single photon emission computed tomography imaging. *Neurosci Lett* 1995; **195:** 53–56.
53. Gustafson L, Edvinsson L, Dahlgren N *et al.* Intravenous physostigmine treatment of Alzheimer's disease evaluated by psychometric testing, regional cerebral blood flow (rCBF) measurement, and EEG. *Psychopharmacology* 1987; **93:** 31–35.
54. Scremin OU, Allen K, Torres C, Scremin E. Physostigmine enhances blood flow metabolism ratio in neocortex. *Neuropsychopharmacology* 1988; **1:** 297–303.
55. Peruzzi P, Birredib J, Seylaz J, Lacombe P. Tacrine overcompensates for the decreased blood flow induced by basal forebrain lesion in the rat. *NeuroReport* 1996; **8:** 103–108.
56. Furey ML, Pietrini P, Haxby JV *et al.* Cholinergic stimulation alters performance and task-specific regional cerebral blood flow during working memory. *Proc Natl Acad Sci USA* 1997; **94:** 6512–6516.
57. Nilsson L, Adem A, Hardy J, Winblad B, Nordberg A. Do tetrahydroaminoacridine (THA) and physostigmine restore acetylcholine release in Alzheimer brains via nicotinic receptors? *J Neural Transm* 1987; **70:** 357–368.
58. Maelicke A, Schrattenholz A, Schröder H. Modulatory control by non-competitive agonists of nicotinic cholinergic neurotransmission in the central nervous system. *Semin Neurosci* 1995; **7:** 103–114.
59. Svensson AL, Nordberg A. Tacrine interacts with an allosteric activator site on $\alpha 4\beta 2$ AChRs in m10 cells. *NeuroReport* 1996; **7:** 2201–2205.
60. Nilsson-Håkansson L, Lai Z, Nordberg A. Tetrahydroaminoacridine induces opposite changes in muscarinic and nicotinic receptors in rat brain. *Eur J Pharmacol* 1990; **186:** 301–305.

15

Use of cholinesterase inhibitors in the therapy of myasthenia gravis

Morris A Fisher

Introduction

Myasthenia gravis (MG) is an acquired autoimmune disorder affecting neuromuscular transmission due to acetylcholine (ACh) receptor deficiency at the neuromuscular junction (NMJ). The illness is characterized by weakness aggravated by exertion and relieved by rest and anticholinesterase (AChE) medication.

In 1672, Thomas Willis was the first to describe an illness with fluctuating weakness. However, the first detailed description of MG did not appear until the latter half of the 19th century. In 1879 Erb[1] described the classic signs and recognized that the weakness with its bulbar findings and absence of anatomic lesions was different from that seen in other diseases. In 1895 Jolly[2] first used the term MG (myasthenia gravis pseudoparalytica) and demonstrated that stimulation with alternating ('faradic') current could produce the weakness, even though the muscles would still then respond to direct ('galvanic') current. Jolly also suggested that neostigmine might be used in the treatment of this condition. The association of MG and abnormalities in the thymus was noted by Buzzard in 1905,[3] and was well described by Castleman and Norris in 1949.[4] An autoimmune etiology for MG was initially postulated by Simpson in 1960.[5] His arguments included the association of MG with other putative autoimmune disorders. The demonstration of a fluctuating and neostigmine responsive weakness in rabbits immunized with ACh receptor protein,[6] the subsequent finding of antibodies to ACh receptor in patients with MG,[7] and the localization of immune complexes to postsynaptic muscle membranes[8] provided the experimental basis for our current understanding of MG as an immunologically mediated illness affecting the postsynaptic membrane at the neuromuscular junction.

In 1941, Harvey and Masland[9] described the decrementing muscle electrical response to low-frequency stimulation which is now the most commonly used electrodiagnostic test for the diagnosis of MG. Starting in the mid-1960s, single fiber electromyography (SFEMG) was developed as a sensitive clinical physiological test for evaluating NMJ disorders including MG.[10–12]

Anticholinesterases (anti-ChEs) were the first effective treatment for MG. The first reported use of anti-ChE treatment (neostigmine) for MG was by Remen in 1932.[13] These drugs, however, attained common use only after Mary Walker's description of their value in 1934 (physostigmine)[14] and 1935 (neostigmine).[15] Walker was a somewhat reclusive general physician. When the prominent neurologist Derek Denny-Brown was visiting her hospital, she showed him a patient with MG. When Denny-Brown explained that the symptoms were similar to curare poisoning, she asked for permission to treat the patient with subcutaneous physostigmine. Although not encouraged to do so by Denny-Brown, she nevertheless proceeded and thereby helped establish both a treatment and a diagnostic test for MG.[16] Walker's findings also pointed to the NMJ as the region of abnormality.

The therapeutic value of thymectomy in MG gained general currency until Blalock's reports starting in 1939.[17,18] With the increasing recognition of the immunological basis for MG, immunomodulating therapies have become standard. Because of the initial unfavorable experience with steroids in the 1950s, these drugs fell into disfavor until their therapeutic effectiveness was recognized in the 1970s.[19] Treatment with other immunosuppressive drugs was also introduced.[20] Since then other immunomodulating therapies, particularly plasmapheresis[21] and, later, intravenous immunoglobulin (IVIG),[22] have become commonplace in treating MG.

Clinical aspects

The incidence of MG is estimated to be 2–10 per 100 000 and the prevalence at least 4 per 100 000.[23,24] MG may present at any age, but has a peak incidence in the second and third decades, predominantly in females, and a second peak in the seventh and eighth decades, mainly in males. There are different human leucocyte antigen (HLA) associations in these two patient groups,[25] which is consistent with different hereditary influences.

Since MG is not common, sensitivity to the clinical features is important if the condition is to be recognized and managed. The clinical features have been well described in standard texts. The characteristic complaint is muscle fatiguability with use and relief by rest. This may be associated with a baseline muscle weakness. The onset is usually insidious but may be precipitous. The initial symptoms are usually focal and involve particularly muscles innervated by the cranial nerves (bulbar muscles). In only about 10% are initial complaints due to weakness of limb muscles, while in about 60% the initial symptom is related to weakness of eye muscles, namely dropping of eyelids (ptosis) and double vision (diplopia).[26]

The natural history of MG has not been fully defined, partially because of the longstanding use of therapeutic thymectomies. Based on data obtained prior to the use of steroids,[27] it is known that spontaneous improvement can occur, especially during the first several years, and the illness can remit for varying periods. Although nearly all patients develop ocular symptoms, in only about 10% does the illness remain strictly confined to the ocular muscles. Generally, progression beyond the ocular stage occurs within the first year of the illness. The 'active' progressive stage of the disease tends to occur during the first several years, this being followed by a prolonged (15–20 years) period of fluctuating weakness. Subsequently, there is usually an 'inactive' phase in which the muscle weakness may become fixed. Many factors can worsen MG, often both suddenly and severely. These include emotional upset, the menses, viral respiratory infections, thyroid dysfunction, and drugs with neuromuscular blocking effects. The latter include commonly used antiarrthymics (e.g. quinine, quinidine, and procainmide), local anesthetics (e.g. lidocaine), and aminoglycoside antibiotics (e.g. streptomycin, gentamycin, and kanamycin).[28] Drugs that depress respiration may also aggravate the respiratory problems in MG.

MG is a serious illness with mortality largely related to respiratory failure. Between 1940 and 1960 a mortality of 33% was reported, but this had declined to 12% by 1960–1980 mainly due to improved respiratory care.[27] A recent study reported no deaths in 100 consecutive patients with MG referred between 1985 and 1989 and subsequently followed. In these patients, who were followed for a mean of 9.6 years, 88% had either improved (45%) or were in remission (43%).[29] These encouraging statistics reflect the effect of better medical therapy.

Examination usually reveals preserved reflexes and intact sensation but fatiguable weakness, usually focal and asymmetrical. Ptosis and diplopia are common, as is difficulty with swallowing and hoarseness. MG does not affect all the limb muscles equally. Neck muscles as well as the deltoids and wrist/finger extensors tend to be particularly involved.

The association of MG with autoimmune disorders was one of the reasons why Simpson[5] postulated an autoimmune mechanism in MG. About 4% of patients with MG have an associated illness such as rheumatoid arthritis, lupus erythematosis, or pernicious anemia.[30] Furthermore, 13% of patients with MG have thyroid disorders,[31] many of which are also autoimmune. Relatives of patients with MG are also more likely to have autoimmune disorders and, although familial MG is rare, relatives of patients with MG are much more likely than otherwise to have the disease.[32] Transient neonatal MG occurs in about 12% of infants born to mothers with MG. This is due to transfer of ACh receptor antibodies from the mother to the fetus.[33] These features of MG emphasize the immunological nature of the illness. This is also consistent with the occasional collection of lymphocytes in the muscles of patients with MG.

Although ACh receptor antibodies have been reported in up to 95% of patients with MG,[34] a more accurate figure is probably of the order of 75%. This figure is even lower (about 56%) in those with purely ocular MG.[26] In general, patients with antibody-negative MG have milder disease. Although, with rare exceptions, the presence of ACh receptor antibodies is specific for MG, the clinical utility of their presence is limited by the frequent absence of equivocal clinical cases where the study might be most helpful. Furthermore, absolute levels of the antibody do not necessarily correlate with the severity of the disease. MG patients without ACh receptor antibodies respond satisfactorily to current treatments including AChE medication.

About 50% of patients with MG have thymic abnormalities[4] and about 15% have thymic tumors (thymomas).[35] The highest levels of ACh receptor antibodies are found in patients with thyomas, especially where the tumors are invasive. All patients with MG should undergo radiographic studies to evaluate for thymomas. The high incidence of thymic abnormalities implies a role of the thymus in the pathogenesis of MG, but this remains incompletely defined.

Pathogenesis

Immunological aspects

The ACh receptor antibody is polyclonal, immunoglobulin G (IgG).[36] ACh receptor antibody can bind to a number of sites on the receptor, but in general binds to a circumscribed portion known as the main immunogenic region.[37] This site is distinct from the cholinergic binding site. There is evidence that ACh receptor antibody is present at the NMJ in patients with MG.[38,39]

The ACh receptor antibody is not known to have a direct effect on ion channels; nor does it act primarily as a blocking antibody.[40,41] Complement-mediated attack of the NMJ has, however, been observed consistently in

MG.[38,42–44] The antibody also accelerates destruction of the NMJ by enhancing cross-linking of ACh receptor.[45–47] Furthermore, processes related to these destructive effects inhibit NMJ repair.

Pathology

The primary pathology in MG follows from the immunologically mediated postsynaptic injury. The postsynaptic membrane is distorted and simplified with sparse, shallow folds.[48,49] There has been ultrastructural localization of both IgG[50] and lytic C9 complement[51] at the NMJ in MG. At the same time, the presynaptic nerve terminals contain normal numbers of ACh vesicles of unremarkable size.

Physiology

As with the pathology, the physiological abnormalities in MG are consistent with the immunological attack of the postsynaptic NMJ. Due to loss of postsynaptic receptor sites, there is a decrease in the 'safety factor' for transmission across the NMJ. Normally, motor fibers can be stimulated at rates of 40–50/s without fatigue or decrease in size (decrement) of the postsynaptic evoked motor response recorded from muscle. In normal subjects, as well as in myasthenics, the amount of ACh released declines after the first few impulses.[52,53] In patients with MG, this decline in ACh release combined with the decreased number of postsynaptic receptors results in a characteristic decrementing response of greater than 10%. This occurs at low rates of stimulation (i.e. 3/s), and the largest decrement is seen after the third or fourth response. These studies have been reported positive in up to 95% of patients with MG,[54] but the results obtained depend on the number of muscles studied as well as which muscles are examined. MG is not a uniform illness. Although hand muscles are most commonly studied for repetitive stimulation, evaluating proximal muscles such as the trapezius is more sensitive.

In MG, the failure of postsynaptic activation of individual muscle fibers can be observed during needle electromyography (EMG). Using this technique, the electrical activity from muscle discharge in motor units can be observed. In MG, the configuration of these potentials may vary due to variable discharge of muscle fibers in a motor unit. It is indeed this failure of muscle fiber activation which accounts for the weakness in MG.

Single-fiber EMG (SFEMG)[12] is a sensitive electrophysiological method for evaluating NMJ abnormalities. Because of particular characteristics of the recording electrodes as well as the recording techniques, the discharges from single muscle fibers are recorded and with this it is possible to monitor transmission across NMJs. Characteristically, in MG there is a prolonged 'jitter' that reflects a delay in transmission and, in others with more severe involvement, intermittent 'blocking' due to failure of neuromuscular transmission.

ChE inhibitor therapy

History

As discussed above, anti-ChE therapy in MG followed Walker's report[15] of its effectiveness. Neostigmine was recognized to be preferable to physostigmine. Physostigmine readily crosses the blood–brain barrier with resultant possible central nervous system (CNS) effects; this is not true for neostigmine, since it contains a quaternary ammonium ion. A number of other reversible anti-ChE agents with quaternary ammonium ions were tested in the 1950s. The hope was to find drugs with longer action and fewer side-effects in comparison to neostigmine. As a result of such testing, pyridostigmine bromide (Mestinon)[55,56] and ambenonium chloride (Mytelase)[57] became available. Longer acting quaternary compounds were found unsuitable for general use because of problems with drug accumulation and overdosage. These problems were even more pronounced with a number of irreversible organophosphorus anti-ChE inhibitors. Such tested drugs included isopropylmethylphosphorofluoridate (Sarin).[58] About the same time, edrophonium chloride (Tensilon) was introduced[59] as a short-acting anti-ChE useful in the diagnostic testing of MG. Interest in anti-ChE therapy in MG subsequently waned due to the emphasis on immunomodulating therapies starting in the 1960s.

Current usage

ChE inhibitors increase ACh by retarding the enzymatic hydrolysis of ACh. This in turn prolongs the effect of ACh at the NMJ, allowing for persistent depolarization of the motor end-plate and thereby increasing the 'safety factor' for neuromuscular transmission. Anti-ChE drugs also increase the radius of the postsynaptic ACh receptor reached by ACh and allow single ACh molecules to bind more than once to ACh receptors.[24] At the same time, these drugs have been shown experimentally to decrease the availability of ACh receptors,[60,61] as well as to reduce the amplitudes of miniature end-plate potentials.[62,63] These 'curare-like' effects have a slower time-course and therefore may outlast the effect of the inhibition of AChE.[28] Neostigmine given to rats for up to 149 days resulted in degeneration of postsynaptic folds, most noticeably in red muscle fibers.[63] There are, however, no established long-term adverse effects of anti-ChE therapy in patients with MG.

ChE inhibitors produce symptomatic improvement in the majority of patients, but these drugs by themselves rarely produce normal strength. The appropriate clinical use of these drugs follows from the pathophysiology of MG. There are no 'fixed' recommended doses. The drugs should be used at doses needed to produce relief of symptoms without producing meaningful side-effects. The amount of medication required may, therefore, vary from patient to patient, as well as within the same patient even during a day. The medication may have to be increased with activity and decreased at times of relative inactivity. With education about the nature of their illness and the medication, patients can usually monitor their own medication needs most effectively.

Side-effects can be common and disabling. The potential side-effects of anti-ChE medication are those of excess ACh. The most common side-effects are gastrointestinal and include nausea, vomiting, abdominal cramping, flatulence, and diarrhea. Other muscarinic effects include miosis, salivation, sweating, lacrimation, urinary frequency, bradycardia, and hypotension. Because of the cardiac and hypotensive effects, it has been recommended that anti-ChE medication be used cautiously in patients with heart problems, including those with myocardial ischemia and disease of the cardiac conduction system.[64] Nicotinic actions include fasciculations as well as fatigability and weakness — fatigability and weakness of course being similar to the findings in MG itself. Because ChE inhibitors can increase oral and bronchial secretions, they can aggravate the swallowing and respiratory problems in patients with MG. Anti-ChEs should be used with circumspection in patients with bronchial asthma. There may be prolonged neuromuscular blockade if the anti-ChEs are used with succinylcholine.[65] This seems particularly true in patients with pseudo-ChE deficiency or who are receiving succinylcholine infusions or

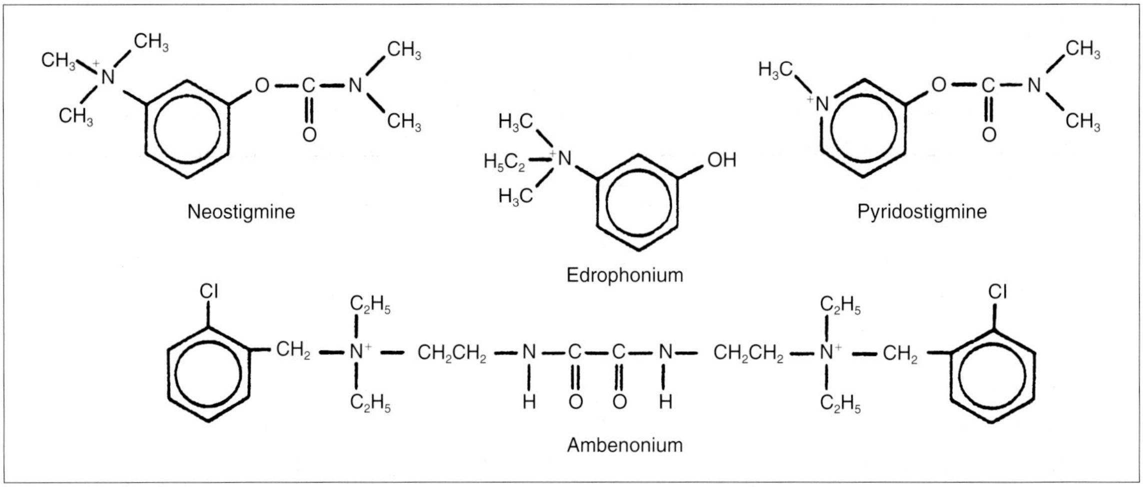

Figure 15.1
ChE inhibitors used clinically in treating MG.

frequent boluses.[66,67] They are contraindicated in those with mechanical urinary or intestinal obstructions, as well as those with known hypersensitivity to the drugs. Allergic reactions, anaphylaxis, and skin rashes have been cited. The Lambert–Eaton syndrome is an immunologically mediated presynaptic neuromuscular junction disorder with clinical features that at times are similar to those of MG. Because of this clinical similarity, these patients have at times been treated with anti-ChE medication. It has been our experience that some of these patients will then develop prominent side-effects on relatively low doses of anti-ChE medication, presumably due to the associated autonomic neuropathy and resultant denervation hypersensitivity.

All of the currently used anti-ChEs used for the treatment of MG are structurally related (Fig. 15.1) and contain permanently charged quarternary ammonium groups.[28] As such, they are highly water soluble but do not cross cell membranes easily. They are not well absorbed from the gastrointestinal tract, and this at times accounts for their seemingly erratic therapeutic effects. Absorption is most rapid and regular on an empty stomach.[28] The oral bioavailabilities of these drugs is about 10% or less.[68] They also do not enter the CNS. but this limitation by the blood–brain barrier may be lowered with stress.[69] Plasma clearances of the reversible quaternary ChE inhibitors is 0.5–1.0 l/h per kg and the plasma elimination half-lives are about 30–60 min.[68] Although all the ChE inhibitors currently used in MG have similar basic pharmacological actions, patients at times may prefer one of the drugs to another.

As mentioned previously, neostigmine (Prostigmine) was the first anti-ChE to be widely used for the treatment of MG. Neostigmine is metabolized in the liver, but 50% is excreted unchanged by the kidney.[70] Although there are no formal recommendations to decrease the drug in patients with renal failure, prudence would deem monitoring patients on neostigmine under these circumstances. Neostigmine bromide (Prostigmine) is available for oral use, and neostigmine methylsulfate (Prostigmine) for parenteral use. The standard oral dose of neostigmine is 15 mg. The time to onset of action following an oral dose is 10–20 min and the duration of effect is about 3–4 h. The dose varies from about 7.5 mg every 4 h to 60 mg every 2–3 h,[25] with an average effective dose in MG of 150 mg/day in divided doses. Neostigmine 0.5 mg given parenterally (subcutaneously or intramuscularly) is equivalent to 15 mg p.o. The pediatric dose is 0.4 mg/kg p.o. or 0.04 mg/kg i.m. every 4–6 h. The elimination half-life of neostigmine in infants and children is less than that in adults; and in infants and children the dose of neostigmine required to reverse neuromuscular blockade is about half that required in adults.[71] Neostigmine can be given intravenously, but this use is largely restricted for testing for MG (see below). Successful use of intranasal neostigmine has been reported.[72] The therapeutic effect occurs within 5–15 min. Intranasal delivery could be helpful in emergencies or in those with severe bulbar dysfunction.

Due to its somewhat longer duration of action (3.0–6.0 h) and fewer adverse effects, pyridostigmine bromide (Mestinon) is generally considered the drug of choice for oral administration. The standard oral dose of pyridostigmine is 60 mg and the daily dose usually ranges from 600 to 1500 mg in divided doses. The time to onset of action is 10–30 min.[28] Pyridostigmine is excreted unchanged primarily from the kidney,[73,74] and lower doses may therefore be required in patients with renal disease. As always, the dosage required is based on titration of the patient's needs. The peak concentration after an oral dose has been reported at about 2 h, but may be delayed up to 2 h by eating.[75] Methylcellulose may prevent absorption completely.[76] Pyridostigmine is absorbed from the gastrointestinal tract better than is neostigmine. The peak concentration of pyridostigmine following a 60 mg dose is 40–60 µg/l, while that following a 30 mg oral dose of neostigmine is 1–5 µg/l.[68] Pyridostigmine has been available in a chloride form (50 mg pyridostigmine chloride is equivalent to 60 mg pyridostigmine bromide) for those who cannot tolerate the bromide. The main side-effect of the bromides is skin rashes. One case of bromide intoxication with psychosis from the use of pyridostigmine has been reported.[77] Pyridostigmine is available in a liquid form (12 mg/ml) as well as a 180 mg 'time-release' tablet. The latter has a duration of action about 2.5 times that of the standard preparation. The extended release tablets should be given no more than once every 6 h. Due to variable patterns of absorption, these tablets should be largely restricted to a night-time dose for those who have difficulty sleeping due to the MG. There is a parenteral form of pyridostigmine. The standard parenteral dose is 2 mg, namely about one-thirtieth of the oral dose. The duration of action when given parenterally is about 2–3 h. In one patient, the half-life of pyridostigmine was about 2.5 times greater when given orally than when given intramuscularly.[78] A nebulized form of pyridostigmine has been used successfully in a patient with primary bulbar MG.[79] The usual dose of pyridostigmine for treating neonatal MG is 5 mg every 4–6 h, and the parenteral dose is about four times that for neostigmine. Pyridostigmine is considered compatible with breast-feeding, since negligible amounts are transmitted to the infant.[80]

Although the intraindividual variation in MG plasma pyridostigmine is small, there is a wide interpatient variability.[76,81] For this reason, and because anti-ChE dosage is based on patient need, plasma pyridostigmine levels have not been found useful in the routine clinical management of patients with MG. Since erythrocyte-bound AChE determinations are technically easier to perform and appear to correlate with plasma pyridostigmine levels, blood AChE has been suggested as the preferable method for monitoring plasma pyridostigmine.[82]

Ambenonium chloride (Mytelase) is available as a capsule. The usual adult oral dose is 5–25 mg 3 or 4 times daily up to a maximum of 200 mg/day. The time to onset of action is 20–30 min with a duration of action (4–5 h) somewhat longer than that for pyridostigmine. The drug is rarely used now because of the relatively narrow margin between the appearance of mild in comparison to severe side-effects.

Edrophonium chloride (Tensilon) is available for parenteral use (10 mg/ml ampules). The drug is generally given intravenously, but may be administered intramuscularly or subcutaneously. The time to onset of action is 20–30 s and the duration of action is less than 10 min when given intravenously. Intramuscularly, the time to onset of action is 2–10 min and the duration of action is 5–30 min. The short action of the edrophonium may be due to its rapid uptake by the liver and kidneys.[83] Because of its characteristics, edrophonium is used primarily for diagnosing MG. The technique has changed little since the early descriptions.[84] Edrophonium (1 ml; 10 mg) is drawn into a tuberculin syringe; 0.2 ml (2 mg) is injected intravenously and the needle left in situ. A clinical end-point such as ptosis, vital capacity, or muscle strength is monitored. After 45 s, if no improvement has occurred and the patient tolerates the medication, the remaining 0.8 ml (8 mg) is injected. Saline (1 ml) may be used as a test injection. Lingual fasciculations are commonly observed in non-myasthenic patients. Side-effects are those of cholinergic excess. Atropine sulfate (0.6–1.2 mg) should be available if needed for intravenous administration to treat side-effects. For children weighing less than 34 kg the intravenous dose is 1 mg; this may be repeated every 30–45 s to a maximum dose of 5 mg. For children weighing more than 34 kg the initial dose is 2 mg, with incremental 1 mg doses to a maximum of 10 mg. Alternatively, children weighing less than 34 kg may receive an intramuscular dose of 2 mg, and heavier children a dose of 5 mg. Infants can be given 0.5–1.0 mg by the intramuscular or subcutaneous route.[85] The elimination half-life of edrophonium in anephric patients is increased;[86] the dose of edrophonium may have to be decreased in those with renal failure.

A positive Tensilon test is not diagnostic for MG. The test can also be positive in other NMJ disorders, as well as some neurogenic atrophies due to reduced quantal content of ACh at reinnervated NMJs.[25] To diagnose MG, therefore, other clinical features of the illness must be considered as outlined above.

Neostigmine can be given intravenously *slowly* in a dose of 0.05 mg/kg, with atropine available if needed. The onset of action is within 1–20 min and the duration of action is 1–2 h. As such, neostigmine has also been used in the past to test for MG. Neostigmine, however, is clearly less desirable than edrophonium, given its longer time to onset of action and longer duration of action.

Using methods similar to that for diagnosing MG, short-acting anti-ChEs have been used to test patients being treated with long-term anti-ChE therapy to determine whether they have too much medication or too little. This has been done in the context of patients faring poorly with regard to whether their poor condition relates to too much ACh ('cholinergic crisis') or too little medication ('myasthenic crisis'). This type of testing is no longer indicated and may be dangerous. With a better understanding of the pathogenesis of MG, treatment modalities other than anti-ChEs are available and the need for high anti-ChE doses diminished. In an era of good respiratory care units, if there is any question as to a patient's need for medication and there is concern about the patient's respiration, the preferred management is simply to stop the anti-ChE medication and then to restart the medication at a dose determined by the patient's clinical status. Finally, as has been mentioned, MG is a non-uniform illness. With anti-ChE testing, apparent improvement in one area such as ptosis or diplopia may give misleading information, since other muscles subserving, for example, respiration may be overmedicated.

Although anti-ChE medication need not be stopped routinely prior to electrodiagnostic testing, this may be helpful in patients with mild NMJ abnormalities or equivocal electrodiagnostic abnormalities.[87]

Role with other therapies

Despite other therapies, ChE inhibitors retain an important role in the management of MG. They provide symptomatic relief and can decrease the need for more toxic medications. If used appropriately, the risks of these medications are low. However, the effective and safe use of anti-ChEs requires familiarity with the anti-ChEs, including an understanding of their mode of action as well as an understanding of the pathogenesis of MG and the various types of treatment available.

A critical factor in the use of anti-ChEs is patient education. With some guidance, the patients will recognize when the medications take effect, the duration of the effect, and the dose required. However, the dose needed may fluctuate suddenly and will vary depending on the stage of the illness. The ChE inhibitors are rarely appropriate as the sole therapy, but this can occur such as in a patient who has had a good response to thymectomy. The use of ChE inhibitors may also decrease the required doses of more harmful medications, such as steroids or other immunosuppressive agents. They may improve the status and thereby aid in the management of patients being treated acutely with plasmapheresis or IVIG. In myasthenic crises, pyridostigmine alone may be as effective as pyridostigmine and steroids or plasmapheresis.[88]

ChE inhibitors in non-immunological myasthenias

There are myasthenias the pathogenesis of which is not due to immunological abnormalities. Some of these will respond positively to edrophonium and benefit from anti-ChE therapy. These myasthenias are primarily genetic in origin, and those of genetic origin will not benefit from the immunomodulating treatments currently used for MG.

A congenital myasthenic syndrome caused by a presynaptic defect in ACh resynthesis or mobilization has been described.[89] The inheritance is probably autosomal recessive. Hypotonia, bouts of respiratory insufficiency, and feeding problems are present, particularly during the first 2 years of life.[90] The illness then tends to improve, but in some weakness and easy fatigability continues throughout childhood. Bouts of weakness and respiratory distress may occur into adult life. When present, the muscle weakness responds well to ChE inhibitors, and these may be life-saving.

Another congenital myasthenic syndrome responsive to anti-ChE medication is due to a paucity of postsynaptic secondary clefts.[91] The condition may present at birth with contractures and arthrogryposis, respiratory and feeding problems exacerbated by febrile illness, and delayed motor development.[92,93] A milder adult form has been described.[94]

A girl has been described who had poor feeding since birth as well as respiratory difficulties, facial diplegia, and ophthalmoparesis responsive to pyridostigmine. Later she developed facial abnormalities, including a high arched palate, malocclusion, and an elongated face. Analyses revealed a short channel open time and a decreased number of ACh receptors per NMJ.[95]

An autosomal recessive illness with proximal muscle weakness responsive to anti-ChE medication presenting in childhood or the early teens has been described.[96] Ocular and other cranial muscles are not involved. A phenotypically distinct congenital myasthenic syndrome with autosomal inheritance has been reported in inbred Iraqi and Iranian Jews.[97] All patients have facial abnormalities, including malocclusion, a high arched palate, and an elongated face which is associated with ptosis as well as weakness of facial and masticatory muscles. The course is stable and benign, and the weakness is responsive to pyridostigmine.

D-Penicillamine is used in the treatment of rheumatoid arthritis, Wilson's disease, and cystinuria. After taking the drug for several months, patients may develop antibody-positive MG.[98,99] The condition is usually mild and usually resolves within a year after D-penicillamine is stopped.[100] Treatment with anti-ChEs alone is usually adequate, although immunomodulating therapies are effective. D-Penicillamine is thought to stimulate or enhance immunological reactions against the NMJ.

References

1. Erb W. Über einen neuen, wahrscheinlich bulbären Symptomemcomplex. *Arch Psychiatr* 1879; **9:** 336–350.
2. Jolly F. Über Myasthenia gravis pseudoparalytica. *Berl Klin Wschr* 1895; **32:** 1–7.
3. Buzzard EF. The clinical history and post-mortem examination of five cases of myasthenia gravis. *Brain* 1905; **28:** 438–483.
4. Castleman B, Norris EH. Pathology of the thymus in myasthenia gravis. *Medicine* 1949; **28:** 27–58.
5. Simpson JA. Myasthenia: a new hypothesis. *Scot Med J* 1960; **5:** 419–436.
6. Patrick J, Lindstrom J. Autoimmune response to acetylcholine receptor. *Science* 1973; **180:** 871–872.
7. Lindstrom JM, Seybold ME, Lennon VA, Whittingham S, Duane D. Antibody to acetylcholine receptor in myasthenia gravis. *Neurology* 1976; **26:** 1054–1059.
8. Engel AM, Lambert EH, Howard F. Immune complexes (IgG and C3) at the motor endplate in myasthenia gravis: ultrastructural and light microscopic localization and electrophysiological correlation. *Mayo Clin Proc* 1977; **52:** 267–280.
9. Harvey AM, Masland RL. The electromyogram in myasthenia gravis. *Bull Johns Hopkins Hosp* 1941; **69:** 1–13.
10. Ekstedt J. Human single muscle fiber action potentials. *Act Physiol Scand* 1964; **61 (Suppl 226):** 1–96.
11. Stälberg E, Trontelj JV, Schwartz EM. Single-muscle fiber recording of the jitter phenomenon in patients with myasthenia gravis and in members of their families. *Ann NY Acad Sci* 1976; **274:** 189–202.
12. Stälberg E, Trontelj J. *Single Fiber Electromyography*. Old Woking: Miravalle Press, 1994.
13. Remen L. Pathogenese and Therapie der Myasthenia Gravis Pseudoparalytica. *Dtsch Z Nervenheilk* 1932; **128:** 66–78.
14. Walker MB. Treatment of myasthenia gravis with physostigmine. *Lancet* 1934; **i:** 1200–1201.
15. Walker MB. Case showing effect of prostigmine on myasthenia gravis. *Proc R Soc Med* 1935; **28:** 759–761.

16. Keesey JC. Contemporary opinions about Mary Walker: a shy pioneer of therapeutic neurology. *Neurology* 1998; **51:** 1433–1439.
17. Blalock A, Mason MF, Morgan HJ, Riven SS, Myasthenia gravis and tumors of the thymus region. *Ann Surg* 1939; **110:** 544–559.
18. Blalock A, Harvey AM, Ford FF, Llienthal JJ Jr. The treatment of myasthenia gravis by removal of the thymus gland. *J Am Med Assoc* 1941; **117:** 1529–1533.
19. Warmolts JR, Engel WK, Whitaker JN. Alternate-day prednisone in myasthenia gravis. *Lancet* 1970; **ii:** 1198–1199.
20. Mertens HG, Balzereit F, Leipert M. The treatment of severe myasthenia gravis with immunosuppressive agents. *Eur Neurol* 1969; **2:** 324–339.
21. Pinching AJ, Peters DK, Newton-Davis J. Remission of myasthenia gravis following plasma exchange. *Lancet* 1976; **ii:** 1373–1376.
22. Arsura EL, Bick A, Brunner NG, Grob D. Effects of repeated doses of intravenous immunoglobulin in myasthenia gravis. *Am J Med Sci* 1988; **295:** 438–443.
23. Kurtzke J. Epidemiology of myasthenia gravis. *Adv Neurol* 1978; **19:** 545–564.
24. Grob D, Brunner NG, Namba T. The natural course of myasthenia gravis and effect of therapeutic measures. *Ann NY Acad Sci* 1981; **377:** 652–669.
25. Engel AG. Disturbances of neuromuscular transmission. In: Engel AG, Franzini-Armstrong C, eds. *Myology*, 2nd edn. New York: McGraw-Hill, 1994: 1769–1797.
26. Sanders DB, Howard JF Jr. Disorders of neuromuscular transmission. In: Bradley WG, Daroff RB, Fenichel GM, Marsden CD, eds. *Neurology in Clinical Practice*. Boston: Butterworth-Heinemann, 1991: 1819–1842.
27. Grob D, Arsura EL, Brunner NG, Namba T. The natural course of myasthenia gravis and therapies affecting outcome. *Ann NY Acad Sci* 1987; **505:** 472–499.
28. Flacke W. Treatment of myasthenia gravis. *N Engl J Med* 1973; **288:** 27–31.
29. Beekman R, Kuks JB, Ooserhuis HJ. Myasthenia gravis: diagnosis and follow-up of 100 consecutive patients. *J Neurol* 1997; **244:** 112–118.
30. Sorensen TT, Holm E-B. Myasthenia gravis in the county of Viborg, Debmark. *Eur Neurol* 1989; **27:** 177–179.
31. Osserman KE, Tsairis P, Weiner LB. Myasthenia gravis and thyroid disease. *J Mount Sinai NY* 1967; **34:** 469–483.
32. Pirskanen R. Genetic aspects in myasthenia gravis: a family study of 264 Finnish patients. *Acta Neurol Scand* 1977; **56:** 365–388.
33. Keesey J, Lindstrom, J, Cokeley H. Anti-acetylcholine receptor antibody in neonatal myasthenia gravis. *N Engl J Med* 1977; **296:** 55.
34. Toyka KV, Henninger K. Acetylcholine-receptor antibodies in the diagnosis of myasthenia gravis. *Dsch Med Wochenshr* 1986; **111:** 1435–1439.
35. Viets HR, Schwab RS. *Thymectomy for Myasthenia Gravis.* Springfield, IL: CC Thomas, 1960.
36. Vincent A, Newsome-Davis J. Acetylcholine receptor antibody characteristics in myasthenia gravis. 1. Patients with generalized disease or disease related to ocular muscles. *Clin Exp Immunol* 1982; **49:** 257–265.
37. Tzartos SJ, Kokla A, Walgrove S, Conti-Tronconi BM. Localization of the main immunogenic region of human acetylcholine receptor to residues 67–76 of the α-subunit. *Proc Natl Acad Sci* 1988; **85:** 2899–2903.
38. Engel AG, Lambert EH, Howard FM. Immune complexes (IgC and C3) at the motor end-plate in myasthenia gravis: ultrastructural and light microscopic localization and electrophysiologic correlations. *Mayo Clin Proc* 1977; **52:** 267–280.
39. Lindstrom JM, Lambert EH. Content of acetylcholine receptor and antibodies bound to receptor in myasthenia gravis, experimental autoimmune myasthenia gravis, and Eaton–Lambert syndrome. *Neurology* 1978; **28:** 130–138.
40. Green DPL, Miledi R, Vincent A. Neuromuscular transmission after immunization against acetylcholine receptors. *Proc R Soc London, Ser B* 1975; **189:** 57–68.
41. Albuquerque EX, Lebeda FJ, Appel SH *et al.* Effects of normal and myasthenic serum factors on innervated and chronically denervated mammalian muscles. *Ann NY Acad Sci* 1976; **274:** 475–492.

42. Shahashi K, Engel AK, Lambert EH, Howard FM. Ultrastructural localization of the terminal and lytic ninth complement component (C9) at the motor end plate in myasthenia gravis. *J Neuropathol Exp Neurol* 1980; **39:** 162–172.
43. Engel AG, Arahata K. The membrane attack complex of complement at the end-plate in myasthenia gravis. *Am NY Acad Sci* 1987; **505:** 333–345.
44. Nakano S, Engel AC. Myasthenia gravis: quantitative immunocytochemical analysis of inflammatory cells and detection of complement attack complex at the end-plate in 30 patients. *Neurology* 1993; **43:** 1167–1172.
45. Heinemann S, Merlie J, Lindstrom J. Modulation of acetylcholine receptor in rat diaphragm by anti-receptor sera. *Nature* 1978; **274:** 65–68.
46. Stanley EF, Drachman DB. Effects of myasthenic immunoglobulin in acetylcholine receptors of intact mammalian NMJs. *Science* 1978; **200:** 1285–1287.
47. Fumagalli G, Engel AG, Lindstrom J. Ultrastructural aspects of acetylcholine receptor turnover at the normal end-plate and in autoimmune myasthenia gravis. *J Neuropathol Exp Neurol* 1982; **41:** 567–579.
48. Woolf AL. Morphology of the myasthenic neuromuscular junction. *Ann NY Acad Sci* 1966; **135:** 35–58.
49. Engel AG, Santa T. Histometric analysis of the ultrastucture of the neuromuscular junction in myasthenia gravis and in the myasthenic syndrome. *Ann NY Acad Sci* 1971; **135:** 46–63.
50. Engel AG, Lambert EH, Howard FM. Immune complexes (IgG and C3) at the motor end plate in myasthenia gravis. Ultrastructural and light microscopic localization and electrophysiological correlations. *Mayo Clin Proc* 1977; **52:** 267–280.
51. Sahashi K, Engel AK, Lambert EH, Howard FM. Ultrastructural localization of the terminal and lytic ninth complement component (C9) at the motor end plate in myasthenia gravis. *J Neuropathol Exp Neurol* 1980; **39:** 162–172.
52. Thies RE. Neuromuscular depression and the apparent depletion of transmitter in mammalian muscle. *J Neurophysiol* 1965; **28:** 427–442.
53. Barrett EF, Magleby KL. Physiology of cholinergic transmission. In: Goldman AM, Hanin I, eds. *Biology of Cholinergic Function.* New York: Raven Press, 1976: 29–100.
54. Özdemir C, Young RR. The results to be expected from electrical testing in the diagnosis of myasthenia gravis. *Ann NY Acad Sci* 1976; **274:** 203–222.
55. Osserman KE. Progress report on mestinon bromide (pyridostigmine bromide). *Am J Med* 1955; **19:** 737–739.
56. Tether JE. Treatment of myasthenia gravis with mestinon bromide. *JAMA* 1956; **160:** 156–158.
57. Schwab RS. WIN 8077 in the treatment of sixty myasthenia gravis. *Am J Med* 1955; **19:** 734–736.
58. Grob D. Myasthenia gravis: a review of pathogenesis and treatment. *Arch Intern Med* 1961; **108:** 171–194.
59. Osserman KE, Kaplam LI. Rapid diagnostic test for myasthenia gravis: increased muscle strength without fasciculations after intravenous administration of edrophonium (Tensilon) chloride. *JAMA* 1952; **150:** 265–268.
60. Fambrough DM, Drachman DB, Satyamurti S. Neuromuscular junction in myasthenia gravis: decreased acetylcholine receptor. *Science* 1973; **182:** 293–295.
61. Chang CC, Chen TF, Chuang S-T. Influence of chronic neostigmine treatment on the number of acetylcholine receptors and the release of acetylcholine from the rat diaphragm. *J Physiol* 1973; **230:** 613–618.
62. Roberts DV, Thesleff S. Acetylcholine release from motor-nerve endings in rats treated with neostigmine. *Eur J Pharmacol* 1969; **6:** 281–285.
63. Engel AG, Lambert EH, Santa T. Study of long-term anticholinesterase therapy: effects on neuromuscular transmission and on motor end-plate fine structure. *Neurology* 1973; **23:** 1273–1281.
64. Asura EL, Brunner NG, Namba T, Grob T. Adverse cardiovascular effects of anticholinesterase inhibitors. *Am J Med Sci* 1987; **193:** 18–23.

65. Baraka A. Suxamethonium-neostigmine interaction in patients with normal or atypical cholinesterase. *Br J Anaesth* 1977; **49:** 479–484.
66. Kopman AF. Prolonged response to succinylcholine following physostigmine. *Anesthesiology* 1978; **49:** 142–143.
67. Sunew KY, Hicks RG. Effects of neostigmine and pyridostigmine on duration of succinylcholine action and pseudocholinesterase activity. *Anesthesiology* 1978; **49:** 188–191.
68. Aquilonius SM, Hartvig P. Clinical pharmacokinetics of cholinesterase inhibitors. *Clin Pharmacokinet* 1986; **11:** 236–249.
69. Friedman A, Kaufer D, Shemer J, Hendler I, Soreq H, Tur-Kaspa I. Pyridostigmine brain penetration under stress enhances neuronal excitability and induces immediate transcriptional response. *Nature Med* 1966; **2:** 1382–1385.
70. Cronnelly R, Stanski DR, Miller RD, Sheiner LB, Sohn YJ. Renal functions and the pharmacokinetics of neostigmine in anesthetized man. *Anesthesiology* 1976; **55:** 222–226.
71. Fisher DM, Cronnelly R, Miller RD, Sharma M. The clinical pharmacology of neostigmine in infants and children. *Anesthesiology* 1983; **59:** 220–225.
72. Sghirlanzoni A, Pareyson D, Benvenuti C *et al.* Efficacy of intranasal administration of neostigmine in myasthenic patients. *J Neurol* 1992; **239:** 165–169.
73. Connelly R, Stanski DR, Miller RD, Sheiner LB. Pyridostigmine kinetics with and without renal function. *Clin Pharmacol Ther* 1980; **28:** 87–81.
74. Breyer-Pfaff U, Maier U, Brinkmann AM, Schumm F. Pyridostigmine kinetics in healthy subjects and patients with myasthenia gravis. *Clin Pharmacol Ther* 1985; **5:** 495–501.
75. Aquilonius S-M, Eckaräns S-Å, Hartvig P, Lindström B, Osterman PO. Pharmacokinetics and oral bioavailability of pyridostigmine in man. *Eur J Clin Pharmacol* 1980; **48:** 423–428.
76. White MC, DeSilva P, Havard CWH. Plasma pyridostigmine levels in myasthenia gravis. *Neurology* 1981; **31:** 145–150.
77. Rothenberg DM, Berns AS, Barkin R, Glantz RH. Bromide intoxication secondary to bromide therapy. *JAMA* 1990; **263:** 1121–1122.
78. Galvey TN, Chan K. Plasma pyridostigmine levels in patients with myasthenia gravis. *Clin Pharmacol Ther* 1977; **21:** 187–193.
79. Dooley JM, Goulden KJ, Gatien JG, Gibson EJ, Brown BS. Topical therapy for oropharyngeal symptoms of myasthenia gravis. *Ann Neurol* 1986; **19:** 192–194.
80. Briggs GG, Freeman RK, Yaffe SJ. *Drugs in Pregnancy and Lactation.* Baltimore: Williams and Wilkins, 1994.
81. Sorenson PS, Flachs H, Friis ML, Hvidberg EF, Paulson OB. Steady state kinetics of pyridostigmine in myasthenia gravis. *Neurology* 1984; **34:** 1020–1024.
82. Henze T, Nenner M, Michaelis HC. Determination of erythrocyte-bound acetylcholinesterase activity for monitoring pyridostigmine therapy in myasthenia gravis. *J Neurol* 199?; **238:** 225–229.
83. Calvey TN, Williams NE, Muir KT, Barber HE. Plasma concentration of edrophonium in man. *Clin Pharmacol Ther* 1976; **19:** 813–820.
84. Osserman KE, Teng P. Studies in myasthenia gravis — a rapid diagnostic test: further progress with edrophonium (Tensilon) chloride. *JAMA* 1956; **160:** 153–155.
85. *AHFS: American Hospital Formulary Service Drug Information 94*. Bethesda, MD: American Society of Hospital Pharmacists, 1994.
86. Morris RB, Cronnelly R, Miller RD, Stanski DR, Fahey MR. Pharmacokinetics of edrophonium in anephric and renal transplant patients. *Br J Anaesth* 1981; **53:** 1311–1314.
87. Massey JM, Sanders DB, Howard JF Jr. The effect of cholinesterase inhibitors on SFEMG in myasthenia gravis. *Muscle Nerve* 1989; **12:** 154–155.
88. Berrouschot J, Baumann I, Kalischewski P, Sterker M, Schneider D. Therapy of myasthenic crisis. *Crit Care Med* 1997; **25:** 1228–1235.
89. Mora M, Lambert EH, Engel AG. Synaptic vesicle abnormality in familial myasthenia. *Neurology* 1987; **37:** 206–214.
90. Robertson WC, Chun RW, Kornguth SE. Familial infantile myasthenia gravis. *Arch Neurol* 1980; **37:** 117–119.
91. Smit LME, Jennekens FGI, Veldman H, Barth

PG. Paucity of secondary clefts in a case of congenital myasthenia with multiple contractures: ultrastuctural morphology of a developmental disorder. *J Neurol Neurosurg Psychiatry* 1984; **47:** 1091–1097.
92. Smit LME, Barth PG. Arthrogryposis multiplex congenita due to congenital myasthenia. *Dev Med Child Neurol* 1980; **22:** 371–374.
93. Smit LME, Hageman G, Voldman H. Molenaar PC, Oen BS, Jenneken FG. A myasthenic syndrome with congenital paucity of secondary synaptic clefts: CPSC syndrome. *Muscle Nerve* 1988; **11:** 337–348.
94. Wokke JHJ, Jennekens FGI, Molenaar PC, van den Oord CJ, Oen BS, Bsuch HF. Congenital paucity of secondary synaptic clefts (CPSC) syndrome in 2 adult sibs. *Neurology* 1989; **39:** 648–654.
95. Engel AG, Nagel A, Walls TJ, Harper CM, Waisburg HA. Congenital myasthenic syndromes. I. Deficiency and short open-time of the acetylcholine receptor. *Muscle Nerve* 1993; **16:** 1284–1292.
96. McQuillen MP. Familial limb-girdle myasthenia. *Brain* 1996; **89:** 121–132.
97. Goldhammer Y, Blatt I, Sadeh M, Goodman RM. Congenital myasthenia associated with facial malformations in Iraqi and Iranian Jews: a new genetic syndrome. *Brain* 1990; **113:** 1291–1306.
98. Bucknall RC, Dixon AJ, Gluck EN, Woodland, Zutshi DW. Myasthenia gravis associated with penicillamine treatment for rheumatoid arthritis. *Br Med J* 1975; **1:** 600–602.
99. Vincent A, Newsome-Davis J, Marvin V. Anti-acetylcholine receptor antibodies in D-penicillamine associated myasthenia gravis. *Lancet* 1978; **i:** 1254.
100. Albers JW, Hodach BJ, Kimmel DW, Treacy WI. Penicillamine-induced myasthenia gravis. *Neurology* 1980; **30:** 1246–1250.

INDEX

acetylcholine 182
acetylcholinesterase (AChE) 65
  activity in brain 184–6
    Alzheimer's disease 185, 186, 239
    neuroimaging studies 238–9
  alternative mRNA processing 72, 73
  Alzheimer's disease 94–5, 134, 185, 186, 239
    cerebrospinal fluid 190–2
    loss in cholinergic pathways 129–31
    plaques, tangles and amyloid angiopathy 130, 131–3
  Caenorhabditis elegans 65, 68
  catalytic domain 81-2
  catalytic mechanism 112–13
  cerebrospinal fluid 189–90
    Alzheimer's disease 190–2
  conjugates
    'aged' 16–17
    with ENA713 17
    with MF268 17–18
    with organophosphate nerve agents 16–17
  differences in glycosylation, maturation and catalysis 93–4
  Drosophila 63, 65, 66, 73
  Electrophorus 65, 66
    structure 13–14
  electrostatic characteristics
    'back'door' controversy 15
    in relation to non-cholinergic functions 15–16
    role in catalysis 14–15
  genes 47, 48
    alternative mRNA processing 72–4
    amplification 50
    expression 68–74, 191, 206
    regions of alternative mRNA splicing 68
    relationships between gene structures 68–74
    sequencing 63
    stabilization of mRNA associated with muscle differentiation 71–2
    structure 66
    transcription control and sites of initiation 68–71
  hydrolysis of substrates 103–4
  inhibition
    carbamates 28, 29
    inhibitor specificity 65
    physostigmine analogues 35
    relationship between inhibition levels and efficacy of anticholinesterases 203–5, 207
  inhibitors *see* cholinesterase inhibitors
  interactions with reversible ligands and covalent agents 16–21
  measurement of activity 139–44
  mechanism of hydrolysis 28, 29, 30
  molecular forms 68, 69, 81–101, 121
    central nervous system/brain 84–5, 92–3, 186–7
    physiological assembly and significance of heteromeric forms 89–92
    selective inhibitors for 186–7
  mouse 65, 66
  neuroglia 128–9
  neurons
    Alzheimer's disease 129–31
    cholinergic 122–3, 134
    cholinoceptive 123–7, 134
  non-catalytic roles 50
  non-cholinergic functions, electrostatic characteristics and 15–16
  overproduction
    cortical neuropathology and 54–5
    effect of antisense therapy 55–7
    multi-organ consequences 53
  phosphorylated, oxime reactivators 107, 114
  readthrough (AChE-R) 51
    stress-induced accumulation 52–4
    transgenics 54
  snake 65, 73
  splicing variants 50–1
  stress responses 52–5
  structure 9–25
    α/β-hydrolase fold 9, 63, 65
    disulfide bonds 64
    quaternary 12–14
    site-directed mutagenesis and 12
    three-dimensional 9, 10, 63
  synaptic (AChE-S) 51
    inhibition 52
    transgenics 54
  Torpedo californica (*Tc*AChE)
    'aged' conjugates 16–17
    E2020–*Tc*AChE complex 21, 36, 38, 82
    gene structure 65, 66–7
    HupA-*Tc*AChE complex 17–18, 39
    structure 9, 10–12, 13, 37, 63
    TMFA-*Tc*AChE complex 31, 33
    transcription 68–70
acylation 103, 104
addiction, cholinesterase inhibitors and 166
adversive syndrome 159–60
ageing animals, cognitive impairment, effects of cholinesterase inhibitors 151–2

aggression, effects of cholinesterase inhibitors 164–5
α/β-hydrolase fold 9, 10–12, 63–4, 65
Alzheimer's disease (AD)
brain
AChE forms 94–5
changes in ChE activity 183–9
review of cholinergic hypothesis 217–18
brain imaging studies 237, 238–40
effect of cholinesterase inhibitors 237, 240–4
cerebrospinal fluid, changes in ChE activity 190–2
cholinergic denervation 192–3
cholinesterase inhibitors 145, 197–206, 237
brain imaging studies 237, 240–4
clinical trials 199, 200, 202, 237
cognitive effects 150–2, 200–1, 204
combination therapy 216–17
cost-benefit 218–19
development/history 181–3
effect on behaviour 197–9
effect on cerebral blood flow 240–1, 242
effect on glucose metabolism 240–1, 242–3
efficacy 201–3, 205, 216
long-term effects 202–3, 204, 209–10, 218
mechanism of action 183
organophosphorus 36
physostigmine analogues 35–8
preventive 212–14
side-effects 200, 201, 202, 205–6, 210–12
structure—activity relationships 36–8
synthesis and activity 34–6
tolerance to 206–8
see also names of specific cholinesterase inhibitors
cholinesterases in 129–34
loss of AChE in cholinergic pathways 129–31
plaques, tangles and amyloid angiopathy 130, 131–3
familial, preventive treatment 212
ambenonium 28, 29, 30, 110, 111, 112
myasthenia gravis 254, 256
structure 110, 254
amiridine 38
ammonium compounds, quaternary 39
Amphioxus cholinesterases 67
amyloid angiopathy, cholinesterases and 131–3
β-amyloid peptide (Aβ) toxicity 227–8
effect of cholinesterase inhibitors 228–9
amyloid precursor protein (APP) 227–8
interaction with cholinesterase inhibitors 228–9
analgesia, cholinergic 165–6
anticholinesterases *see* cholinesterase inhibitors
antidotes to cholinesterase inhibitors 169–70
atropine 6, 169–70
reactivators, development 7–8
antinociception, cholinesterase inhibitors 165–6
antioxidants, combined with cholesterase inhibitors 216
antisense oligodeoxynucleotides, suppression of AChE production 55–7
apathy in Alzheimer's disease 198
effect of cholinesterase inhibitors 198, 199
Aricept *see* donepezil
atropine as antidote to anticholinesterases 6, 169
side effects 170
audition, effects of cholinesterase inhibitors 166

bambuterol 108, 112, 188
behaviour
cholinergic alert non-mobile (CANMB) 167
compulsive, anticholinesterase-induced 160
effects of cholinesterase inhibitors 162–6
Alzheimer's disease patients 197–9
bioassay, measurement of ChE activity 140, 141
bispyridinium compounds 8
Bladan 4
blood–brain barrier, penetration by cholinesterase inhibitors 37
effect of stress 53
brain
ACh levels after acute and chronic ChE inhibitor administration 146–7
AChE in 133–4
Alzheimer's disease 94–5, 129–33, 134
cholinergic neurons 123
cholinoceptive neurons 123–7
forms of 92–3, 94–5
neuroglia 128–9
Alzheimer's disease
AChE 94–5, 129–33, 134
BuChe 131–3
changes in ChE activity 183–9
BuChE 133, 134
Alzheimer's disease 131–3
neuroglia 128–9
neurons 127–8
cholinergic system, adaptation to prolonged ChE inhibition/increased ACh levels 148
corkscrew pathology 54
effect of excess AChE 54–5
rhythms, effects of cholinesterase inhibitors 162–3
see also forebrain
brain imaging, Alzheimer's disease 237–8
functional studies of cholinergic activity 238–40
butyrylcholinesterase (BuChE) 65, 82
activity in human brain 184–5
changes in Alzheimer's disease 185, 186
atypical 48–50
cerebrospinal fluid 190
Alzheimer's disease 190
function for 187–9
gene 47, 48, 63
amplification 50
gene structure 66
hydrolysis of substrates 103–4
inhibition
inhibitor specificity 65
physostigmine analogues 35

measurement of activity 139–44
molecular forms 81
mutations 47, 49
neuroglia 128–9
neuromuscular junction 92
neurons 127–8
non-catalytic roles 50
selective inhibitors 187–8
structure, disulfide bonds 64
substrate activation 104
BW284C51 110, 112

C-terminal domains
alternative splicing and multiplicity 82–5
H subunits 83, 84, 85, 86
addition of GPI anchor 85
tissue-specific expression in vertebrates 88–9
R subunits 83, 84, 85
S subunits 83, 84
T subunits 83, 84, 85–7, 95
associations with structural subunits 87–8
tissue-specific expression in vertebrates 88–9
Caenorhabditis elegans cholinesterases 65, 67, 68
Calabar bean 1, 27
calcineurin 72
calcitonin gene related peptide (CGRP), effect on AChE expression 72
carbamates
AChE inhibition 28, 29
Alzheimer's disease 199
effect on respiration 160
functional effects 161
historical development 1–3
as insecticides 2
mechanism of action 108–9
structure—activity relationships 114
toxicity 108
comparison with OP agents 169
8-carbaphysostigmine analogues 31, 36
Carbaryl (Sevin) 2, 108
carbofuran 108
cardiovascular system, effects of cholinesterase inhibitors 161–2
Cartesian diver 140, 142
catalepsy 159
catalytic domain 81–2
central nervous system (CNS)
AChE forms 92–3
effect of AChE accumulation 53
functions, effects of cholinesterase inhibitors 159–62
cerebral blood flow, effect of cholinesterase inhibitors 240–1, 242
cerebrospinal fluid (CSF)
AChE activity, effect of cholinesterase inhibitors 206–8
cholinesterases 189–90
Alzheimer's disease 190–2
chlorpyrifos 34, 148
cholinergic alert non-mobile behavior (CANMB) 167
cholinergic receptors, effects of cholinesterase inhibitors 157–9, 229–31, 243–4
cholinergic system
in Alzheimer's disease 192–3, 217–18
ChE inhibition
adaptation to 148
changes in other neurotransmitter systems brought about by increased cholinergic activity 148–9
cholinesterase inhibitors 1–8
AChE accumulation and 52–3, 54
acylating 104, 105–9
enzymes hydrolysing 114–16
Alzheimer's disease *see* Alzheimer's disease, cholinesterase inhibitors
antidotes 169–70
atropine 6, 169–70
development of reactivators 7–8
β-amyloid toxicity and 228–9
behavioural effects 162–6, 197–9
binding sites 81–2
brain ACh levels after acute and chronic administration 146–7
cerebrospinal fluid 189–90
Alzheimer's disease 190
cholinergic effects, differences in 203–6
classification 81–2
clinical trials 199, 200, 202
cognitive effects 200–1, 204
combination therapy 212–14
and consciousness of self 166–7
differential inhibition 187
dose—behaviour relationships 205–6
drug interactions 211–12, 213, 214
effect on APP processing 228, 229
effect on cholinergic receptors 229–31
effect on neurotransmitter systems 148–9
fast equilibrating 11, 109
functional effects 159–62
hypothalamically-evoked 160–2
ideal 39, 218
interaction with oestrogen effects 231
long-term effects 55, 57, 209–10
mechanism of action 103–19, 183
structure–activity relationships 111–14
myasthenia gravis 249–50, 252–8
non-Alzheimer's dementias 214–16
non-cholinergic effects 227–35
organophosphorus *see* organophosphates
pharmacokinetic differences 208–9, 212
pharmacology 145–55
rational design 27–46
reversible 38–9, 105, 109, 111
side-effects 201, 202, 205–6, 210–12
myasthenia gravis 253–4
slow-release administration 205

cholinesterase inhibitors *continued*
  synaptic effects 157–9
  tight binding 111
  tolerance to 206–8
  toxic actions 2, 167–72, 205–6
    for detailed entry see toxicity
cholinesterases
  Alzheimer's disease 129–33, 134
  C-terminal domains 82–5
  catalytic domain 81–2
  evolutionary relationships 67
  genes 47, 48, 63
    alternative mRNA processing 72–4
    amplification 50
    organization in relation to protein structure 65–7
    regulation of expression 68, 71
    relationship between gene and structures 68–74
    sequences 63–4
    structure 63–5, 65–7
  measurement of activity 139–44
  molecular forms 81–101
    in human brain 186–7
  neuroanatomy 121–37
    vizualization methods 122
  non-catalytic properties 122
  non-catalytic roles 50
  phosphorylated, ageing 107
  structure
    α,β fold 63–4, 65
    disulfide bonds 64–5
  see also acetylcholinesterase; butyrylcholinesterase
circling behavior 159–60
cognition/cognitive deficits, effects of cholinesterase inhibitors 149–50
  ageing animals 151–2
  Alzheimer's disease 150–2, 193, 200–1
  animal models 150–2
  level of AChE inhibition and 204–5, 207
  young humans 194–5
colorimetry, measurement of ChE activity 140, 141
computed tomography (CT) 238
consciousness of self, cholinesterase inhibitors and 166–7
coumarin 110
CP-118,954 200
cyclooxygenase inhibitors (COX-2 I), combined with cholinesterase inhibitors 217
cyclosarin, toxic actions 169

dark adaptation, effects of cholinesterase inhibitors 166
DDVP (2,2–dichlorvinyldimethylphosphate; dichlorvos) 36, 105, 208–9
  effect on APP processing 228, 229
  effects on brain cholinesterases 188, 189
  phosphorylation by 113
  structure 106
deacylation 103, 104–5
decamethonium 28, 30, 110
  mechanism of action 109
  structure 110
delayed cognitive toxicity (DCT) 168
delusions in Alzheimer's disease 198
  effect of cholinesterase inhibitors 198, 199
demecarium 28
dementias
  dementia with Lewy bodies (DLB) 214–15
  non Alzheimer's, anticholinesterase therapy 214–16
  see also Alzheimer's disease
depression in Alzheimer's disease 198
  effect of cholinesterase inhibitors 198, 199
DFP (diisopropylphosphorofluoridate) 4, 32, 34
  effect on brain ACh levels 146
  effect on brain cholinergic system 148
  functional effects 161
  structure 106
DFPase 114
diazinon 34
2,2–dichlorvinyldimethylphosphate *see* DDVP
dichlorvos *see* DDVP
diethylamidoethoxyphosphoryl cyanide 3
diisopropylphosphorofluoridate *see* DFP
dimetan 2
dimpylate 34
distigmine 28, 31
donepezil (Aricept;E2020) 19–21, 110, 112
  Alzheimer's disease 199, 204, 207
    brain imaging studies 241
    clinical trials 200, 202
    effect on behavioural symptoms 198, 199
    effect on glucose metabolism 242–3
    efficacy 201, 203
  analogues 38–9
  complex with *Tc*AChE 21, 38, 82
  dementia with Lewy bodies 215
  drug interactions 214
  effect on
    APP processing 228, 229
    β-amyloid toxicity 228–9
    brain cholinesterases 188, 189
    cholinergic receptors 230
    cognition/cognitive defects 149, 151
    neurotransmitter systems 149
  long-term effect 202–3, 204, 209
  mechanism of action 109
  pharmacokinetic properties 208, 209, 212
  structure 20, 21, 110
dopamine levels, effect of cholinesterase inhibitors 148–9
Down's syndrome, early anticholinesterase therapy 213
Drosophila cholinesterases 63, 65, 66, 67, 73
drug receptors, concept of 7

E600 *see* paraoxon
E605 *see* parathion

E2020 *see* donepezil
echothiophate 32, 34
edrophonium 29, 30, 31, 110
  mechanism of action 109
  myasthenia gravis 253, 254, 256
  preferential inhibition of AChE molecular forms 187
  side-effects 256
  structure 110254
EEG
  arousal 163, 167
  effects of cholinesterase inhibitors 162–3
effect on neurotransmitter systems 149
electrometry, measurement of ChE activity 140, 141
Electrophorus AChE 65, 66
electrotactins 16, 82
emesis, anticholinesterase-induced 160
ENA713 *see* rivastigmine
entactin 64
enzyme histochemistry 122
eptastigmine (heptylphysostigmine) 35, 36–7
  Alzheimer's disease 35, 204, 207
    clinical trials 200, 202
  drug interactions 214
  effect on
    APP processing 228, 229
    brain cholinesterases 187, 188
    cholinergic receptors 230
    cognitive deficits, animal models of Alzheimer's disease 150
    neurotransmitter systems 148–9
  long-term effect 209
  preferential inhibition of AChE molecular forms 187
  side-effects 210
  toxic effects 35, 199
eserine *see* physostigmine
ESTER 63, 64
ethopropazine 112
ethyl-*N*,*N*-dimethylphosphoramidocyanidate *see* Tabun
Exelon *see* revastigmine

fasciculins 12, 82, 111
fasculin 2 (Fas2) 109, 110, 111
fenthion 34
fluorometry, measurement of ChE activity 141, 143
forebrain, cognitive deficits, effect of cholinesterase inhibitors 151

galantamine 39
  Alzheimer's disease 182, 199, 207
    clinical trials 200, 202
    efficacy 203
  effect on
    brain cholinesterases 188, 189
    cholinergic receptors 230
    cognitive deficits in animal models 151
  long-term effect 202–3, 204, 209
  pharmacokinetic properties 208, 209, 212
gasometry, measurement of ChE activity 140, 141, 142
GEN 2819 200
'ginger-Jake' paralysis 168
gliotactin 15, 64
glucose metabolism, effect of cholinesterase inhibitors 240–1, 242–3
glutactin 15, 64
glycophosphatidynositol (GPI) anchoring of AChE 85
Gulf War syndrome 171–2

hallucinations in Alzheimer's disease 198
head injury, closed, AChE overproduction and 54, 57
hearing, effects of cholinesterase inhibitors 166
heptastigmine
  effect on brain ACh levels 147
  effect on cognitive deficits, ageing animals 152
heptylphysostigmine *see* eptastigmine
hetopropazine 188
histochemical techniques, in situ localization of ChE 142
histochemistry, enzyme 122
hunger control, effects of cholinesterase inhibitors 162
huperazine A (HupA) 18, 39, 110, 111
  Alzheimer's disease, clinical trials 200
  effect on brain cholinesterases 189
  effect on cognitive deficits in animal models 151
  HupA-*Tc*AChE complex 18–19, 39
  mechanism of action 109
  pharmacokinetic properties 212
  structure 18, 110
hybridization, in situ 122
hydroxamic acids 8
hydroxylamine 7–8
hypothalamus, effects of cholinesterase inhibitors 161–2

immunocytochemistry 122
indifference *see* apathy
insecticides, anticholinesterase 170
  carbamates 2
  organophosphate 4, 31
    structure–activity relationships 31–4
  toxic actions 170–1
isopropylmethylphosphonofluoridate *see* sarin

KA-672 200

Lambert–Eaton syndrome 254

magnetic resonance imaging (MRI) 238
malaoxon 106
malathion 32, 34, 105
  toxicity 171

mating behaviour, effects of cholinesterase inhibitors 161
MDL 73,745 200
measurement of cholinesterase activity 139–44
memory, effect of cholinesterase inhibitors
  aged normal subjects 195–6
  animal models of Alzheimer's disease 150, 151, 152
  non-demented neurological patients 196–7
  non-human primates 193–4
  normal animals 149
  young adults 194–5
methanesulphonyl fluoride (MSF) 200, 202, 207, 209
7–methoxytacrine 38
N-methylacridinium 109, 110
methylphosphoryl dichloride, synthesized 3
metrifonate 36, 105
  Alzheimer's disease 36, 182, 199, 204, 207
    clinical trials 200, 202
    effect on behavioural symptoms 198, 199
    efficacy 201, 203, 216
  effect on
    brain ACh levels 146, 147
    brain cholinergic system 148
    brain cholinesterases 188, 189
    cholinergic receptors 230
    cognition/cognitive deficits 149, 150–2
    neurotransmitter systems 149
  long-term effect 202–3, 204, 209
  pharmacokinetic properties 208–9, 212
  preferential inhibition of AChE molecular forms 187
  side-effects 211
MF268 37
  Alzheimer's disease 35–6
  interaction with AChE 17–18
MF8622, effects on brain cholinesterases 188, 189
Michaelis constant 103
miotine 28, 29, 31, 38
motor agitation in Alzheimer's disease 198
motor effects of cholinesterase inhibitors 159–60, 167
MSF (methanesulphonyl fluoride) 200, 202, 207, 209
muscarinic receptors
  cholinergic inhibitors and 158–9, 243–4
  neuroimaging 240
muscle
  differentiation, stabilization of associated AChE mRNA 71–2
  effect of AChE accumulation 53
  physiological assembly and significance of heteromeric AChE forms 89–92
myasthenia gravis (MG) 249–50
  cholinesterase inhibitors 249–50, 252–8
    adverse effects 253–4
    current usage 253–7
    dosage 253
    history 252–3
    with other therapies 257
myasthenias, non immunological, cholinesterase inhibitors 257–8

neostigmine 2, 28, 31, 108
  effect on respiration 160
  myasthenia gravis 249–50, 252, 253, 254–5, 256
  structure 108, 254
nerve gases 3, 5, 34
neurexin 64
neuroanatomy of cholinesterases 121–37
  Alzheimer's disease 129–33
  vizualization methods 122
neurofibrillary tangles, cholinesterases in 130, 131–3
neuroglia, cholinesterases in 128–9
neuroimaging *see* brain imaging
neuroligin 64
neuromuscular junction
  BuChE 92
  localization of AChE forms 90, 91
neurons
  BuChE in 127–8
  cholinergic 122–3
    AChE 123, 134
    loss in Alzheimer's disease 192–3
  cholinoceptive 122
    AChE 123–7, 134
  sites of action of cholinesterase inhibitors 157–8
neuropathies, anticholinesterase-induced 168
neuropsychiatric symptoms in Alzheimer's disease 198
  effect of cholinesterase inhibitors 198, 199
neurotactin 15, 64
neurotoxicity, organophosphorus ester induced delayed (OPIDN) 168, 170
neurotropin receptors, AChE expression 72
nicotinic receptors
  cholinergic inhibitors and 158–9, 243
  neuroimaging 240
NIK 247 200
nociception, blocking by cholinesterase inhibitors 165
non-steroidal anti-inflammatory drugs (NSAIDs), combined with cholesterase inhibitors 216, 217
noradrenaline levels, effect of cholinesterase inhibitors 148–9
NX-066 199, 200

octamethyl pyrophosphortetramide (OMPA) 4
oestrogen interactions with cholinesterase inhibitors 231
  combination therapy 216–17, 231
oligodeoxynucleotides, antisense, suppression of AChE production 55–7
OMPA (octamethyl pyrophosphortetramide) 4
organophosphates
  Alzheimer's disease treatment 36
  antidotes 7–8, 169–70
  detoxification enzymes 114–16
  development 3–6
  effect on brain ACh levels 146
  effect on respiration 160
  insecticides 4, 31, 170
    structure–activity relationships 31–4

toxicity 170–1
mechanism of action 105–8
structure—activity relationships 112–13, 114
nerve agents, conjugates with AChE 16–17
pharmacology 5
toxicity 5, 107–8, 167, 168, 170–2
comparison with carbamates 169
neuropathies 168
treatment/antidotes 169–70
war agents 3, 4–5, 171–2
organophosphorus ester induced delayed neurotoxicity (OPIDN) 168, 170
oximes
quaternary, treatment of anticholinesterase poisoning 170
reactivation of phosphorylated AChE 107, 114

PAM (pyridine-2–aldoxime methiodide) 8, 170
paralysis, 'ginger-Jake' 168
paraoxon (E600) 4, 32, 34
effect on brain ACh levels 146
effect on brain cholinergic system 148
structure 106
paraoxonases 114–16
parathion (E600) 4, 32, 34, 105
Parkinson's disease dementia 215
peripheral nervous system, toxic effects of cholinesterase inhibitors 167–8
perlecan 91, 94
Persian Gulf War syndrome 171–2
pesticides, anticholinesterase 170
toxic actions 170–1
PET *see* positron emission tomography
phenserine 36, 188
phosphoric triester hydrolases 114–16
phosphorylation 107–8, 113
photometry, measurement of ChE activity 140, 141
Physostigma venenosum 1, 27
physostigmine 1, 27, 108, 182
adverse effects 35
Alzheimer's disease 34–5, 181–2, 204, 207
brain imaging studies 240–2
clinical trials 200, 202
effect on behavioural symptoms 197–8, 199
anticholinesterase activity 27–8
atropine as antidote 6
dose–behaviour relationships 205–6
drug interactions 214
effect on APP processing 228, 229
effect on brain ACh levels 146
effect on brain cholinesterases 188, 189
effect on cholinergic receptors 230
effect on cognition/cognitive deficits 149–50, 193
animal models of Alzheimer's disease 150–1
effect on memory
aged normal subjects 195–6
non-demented neurological patients 196–7
non-human primates 194
normal young adults 194–5
effect on respiration 160
preferential inhibition of AChE molecular forms 187
structure 1–2, 27, 108
toxicity 28
physostigmine analogues 27, 28–30
Alzheimer's disease 35–8
interaction with AChE 17–18
structure–activity relationship 30–1, 36–8
toxicity 28
physostol *see* physostigmine
physovenine 29, 31
pinacolylmethylphosphonofluoridate *see* soman
plaques, senile, cholinesterases 94, 130, 131–3
polarography, measurement of ChE activity 141, 143
positron emission tomography (PET) 237–8
Alzheimer's disease
anticholinesterase therapy 240–3
functional studies of cholinergic activity 238–40
pralidoxime 170
propidium 110
propylmethylphosphorofluoridate *see* sarin
psychotic symptoms in Alzheimer's disease 198
effect of cholinesterase inhibitors 198, 199
pyridine-2–aldoxime methiodide *see* PAM
pyridostigmine 28, 31, 161
myasthenia gravis 254, 255–6
prophylactic use, anticholinesterase poisoning 170
side-effects 170
structure 254

radiometric approach to measuring of ChE activity 141, 143
receptor theory of drug action 7
red blood cells, AChE activity, effect of cholinesterase inhibitors 207–8
reflexes, effects of cholinesterase inhibitors 159
respiration, effects of cholinesterase inhibitors 160, 167, 169
rivastigmine (ENA713;Exelon) 17, 36, 38
Alzheimer's disease 199, 207
clinical trials 200, 202
effect on cerebral blood flow 242
efficacy 201, 203
effects on brain cholinesterases 188, 189
interaction with AChE 17
long-term effect 202–3, 204, 209–10, 211
non-Alzheimer's dementias 215, 216
pharmacokinetic properties 209, 212
preferential inhibition of AChE molecular forms 187
RNA, messenger, AChE genes
alternative processing 72–4
stabilization of AChE mRNA associated with muscle differentiation 71–2
Ro-02–0683 108

sarin 3, 4–5, 32, 34
  myasthenia gravis 253
  poisoning/toxic actions 169
    Tokyo incidents 171
  structure 106
schizophrenia, effect of cholinesterase inhibitors 167
scopolamine-induced cognitive deficits, effects of cholinesterase inhibitors 150–1
seizures, anticholinesterase-induced 162
self, consciousness of, cholinesterase inhibitors and 166–7
Sevin (Carbaryl) 2, 108
sexual actions, effects of cholinesterase inhibitors 161
single photon emission tomography (SPECT), Alzheimer's disease
  anticholinesterase therapy 240–1
  functional studies of cholinergic activity 238, 240
sleep phases 163–4
  effects of cholinesterase inhibitors 164
SM10888 38
soman 5, 32, 34
  structure 106
  toxic actions 169
SPECT *see* single photon emission tomography
spectrophotometry, measurement of ChE activity 141, 142–3
stem cells, pleuripotent, effect of AChE accumulation 53
stress, AChE accumulation 52–5
succinylcholine 48, 68
suronacrine 200

tabun 3, 4, 5–6, 32, 34
  structure 106
  toxic actions 169
tacrine 34, 38, 109, 110
  Alzheimer's disease 182, 199, 204, 207
    brain imaging studies 241
    clinical trials 200, 202
    effect on behavioural symptoms 197–8, 199
    effect on cerebral blood flow 242
    effect of cholinesterase inhibitors 243–4
    effect on glucose metabolism 242
    efficacy 201, 203
  dementia with Lewy bodies 215
  dose–behaviour relationships 205–6
  drug interactions 214
  effect on
    acute amnesic syndromes and delirium after drug intoxication/overdose 196–7
    APP processing 228, 229
    ß-amyloid toxicity 228–9
    brain ACh levels 147
    brain cholinesterases 188, 189
    cholinergic receptors 229–30
    cognition/cognitive deficits 149, 151
    memory, non-human primates 194
  long-term effect 202–3, 204, 209
  mechanism of action 109
  pharmacokinetic properties 208, 212
  preferential inhibition of AChE molecular forms 187
  structure 110
  tricyclic antidepressant overdose 197
tacrine analogues 38
tactins 15, 64
Tak147 compound 39, 200
Tensilon test, myasthenia gravis 256
tetraethylmonothionopyrophosphate 4
tetraethylpyrophosphate (TEPP) 3
  effect on brain ACh levels 146
  synthesis 3, 4
tetramethylammonium salt 39
TFK 110, 111
thermoregulation
  effect of AChE accumulation 53
  effect of cholinesterase inhibitors 162
thiocholine in measurement of ChE activity 143
thirst control, effects of cholinesterase inhibitors 162
titrimetry, measurement of ChE activity 140, 141
TMFA, complexed with *Tc*AChE 31, 33
tolerance to cholinesterase inhibitors 206–8
Torpedo californica acetylcholinesterase *see* acetylcholinesterase, *Torpedo californica*
toxicity of cholinesterase inhibitors 2, 167–72, 205–6
  comparison of carbamates and OP agents 169
  delayed cognitive toxicity 168
  fatal consequences 168–9
  insecticides/pesticides 170–1
  strategic sites of 167–9
  treatment/antidotes 169–70
  war gases 171–2
trifluoroacetophenones (TFK) 110, 111

velnacrine 38, 200
  Alzheimer's disease, brain imaging studies 240
  side-effects 210
VX
  organophosphorus 171–2
  structure 106
  toxic actions 169

war gases
  effect on respiration 160
  toxic actions 169, 171–2

xanomeline 94 203